WASTE WATER TREATMENT

Third Edition

WASTE WATER TREATMENT

Rational Methods of Design and Industrial Practices

M. Narayana Rao

Emeritus Professor

Jawaharlal Nehru Technological University, Hyderabad

(Formerly, Principal, Technical Teachers. Training Institute, Chennai)

AND

Amal K. Datta

Former Professor

Bengal Engineering College, Howrah

Oxford & IBH Publishing Co. Pvt. Ltd.

New Delhi

(A Unit of CBS Publishers & Distributors Pvt Ltd **)**

CBS

Dedicated to Education

CBS Publishers & Distributors Pvt Ltd

New Delhi • Bengaluru • Chennai • Kochi • Kolkata • Mumbai

Hyderabad • Jharkhand • Nagpur • Patna • Pune • Uttarakhand

Waste Water Treatment

Third Edition

ISBN-13: 978-81-204-1712-0
ISBN-10: 81-204-1712-0

OXFORD & IBH

New Delhi
(A Unit of CBS Publishers & Distributors Pvt Ltd)

Published by Satish Kumar Jain and Produced by Varun Jain for
CBS Publishers & Distributors Pvt Ltd
4819/XI Prahlad Street, 24 Ansari Road, Daryaganj, New Delhi 110 002, India.

Ph: 23289259, 23266861, 23266867   Fax: 011-23243014   Website: www.cbspd.com
                                                               e-mail: delhi@cbspd.com;
                                                               cbspubs@airtelmail.in.

Corporate Office: 204 FIE, Industrial Area, Patparganj, Delhi 110 092, India
Ph: 4934 4934   Fax: 4934 4935   e-mail: publishing@cbspd.com;
                                                               publicity@cbspd.com

Branches

- **Bengaluru:** Seema House 2975, 17th Cross, K.R. Road, Banasankari 2nd Stage, Bengaluru 560 070, Karnatakc
  Ph: +91-80-26771678/79   Fax: +91-80-26771680   e-mail: bangalore@cbspd.com
- **Chennai:** 7, Subbaraya Street, Shenoy Nagar, Chennai 600 030, Tamil Nadu
  Ph: +91-44-26680620, 26681266   Fax: +91-44-42032115   e-mail: chennai@cbspd.com
- **Kochi:** Ashana House, 39/1904, AM Thomas Road, Valanjambalam, Ernakulam 682 016, Kochi, Kerala
  Ph: +91-484-4059061-65,67   Fax: +91-484-4059065   e-mail: kochi@cbspd.com
- **Kolkata:** 6/B, Ground Floor, Rameswar Shaw Road, Kolkata-700014 (West Bengal), India
  Ph: +91-33-2289-1126, 2289-1127, 2289-1128   e-mail: kolkata@cbspd.com
- **Mumbai:** 83-C, Dr E Moses Road, Worli, Mumbai-400018, Maharashtra
  Ph: +91-22-24902340/41   Fax: +91-22-24902342   e-mail: mumbai@cbspd.com

Representatives

- **Hyderabad** 0-9885175004
- **Jharkhand** 0-9811541605
- **Nagpur** 0-9021734563
- **Patna** 0-9334159340
- **Pune** 0-9623451994
- **Uttarakhand** 0-9716462459

Printed at Chaman Enterprises, Daryaganj, New Delhi, India

Dr. B. SEN
*Director of Technical
Education, West Bengal*

GOVERNMENT OF WEST BENGAL
Directorate of Technical Education
New Secretariate Buildings (12th Floor)
Kolkatta-700 001

Dated, the 5th September, 1979

Foreword

It is a matter of great pleasure to note that Sri A.K. Datta, one of my ex-students, is bringing out a book on "Waste Water Treatment", along with his ex-senior colleague, Dr. M.N. Rao. Dr. Rao is now a well-known figure in the field of Environmental Engineering, with a large number of valuable technical papers in journals of repute, to his credit. Sri Datta has made his entry into the field of Environmental Engineering literature comparatively recently, but with a glorious mark. He was awarded the Nawab Zain Yar Jung Bahadur Memorial Prize along with Dr. Rao for their joint authorship of the best paper published in the Environmental Engineering Divisional Journal of the Institution of Engineers (India), in 1973-74. It is not quite unlikely that a joint venture of the workers like Dr. Rao and Sri Datta will produce something which may be considered as a valuable and timely addition to the field of Environmental Engineering literature.

The book, as pointed out by the authors is mostly a compilation work. However, the materials have been presented keeping in view Indian condition and contains the Indian data. And for this reason, it is expected that this book will have an edge over other books, particularly on "Industrial Waste Treatment", by the foreign authors.

I wish the authors all success in this, as well as in their subsequent efforts in bringing out the updated editions of this book and many more such "compilation work", as mentioned by the authors, to make the Indian literature on Environmental Engineering rich and self-sufficient.

B. Sen
*Director of Technical Education
West Bengal*

Preface to The Third Edition

The publication of the third edition of Waste Water Treatment marks the 27th year in print of this title. In the recent past, many environmental questions, once of interest mainly to Scientists and Engineers, have become serious issues of public policy and have sustained a steadily growing public awareness. Concern about the environment and waste water treatment is now worldwide.

It is with these and other changes in mind that this book is revised thoroughly. Removal of odour and colour from effluents is discussed in this revised edition. Industrial waste audit-which is an important tool for effective environmental management is included. In the 21st century, there will be a shift from clean-up technologies to cleaner technologies in waste water treatment. This calls for a change of attitude by the people in the industry. They should know the effects of their industrial operations on the environment and resort to waste minimization and cleaner production techniques. These concepts are highlighted at appropriate places in this book.

It is hoped that this edition meets the needs of the students, teachers and people engaged in the area of pollution control and continues to enjoy the patronage it has been receiving over the years.

M. Narayana Rao

Preface to The First Edition

This book is meant to meet the requirements of Civil Engineering and Chemical Engineering students opting for a specialised course in Environmental Engineering. An attempt has been made to elucidate the latest developments made in Environmental Engineering and their application in India.

It is also hoped that, this book will be useful to Engineering personnel who are new to the Industry in coordinating their theoretical knowledge with the actual Engineering practice and that, it will help them in understanding special treatises on the subject better. More experienced engineers may use it to renew their theoretical knowledge of the subject.

As the title indicates, this book contains two parts. In the first part, consisting of first eight chapters, the rational methods of design of waste water treatment units are presented. The current and suggested methods of industrial waste treatment are dealt with in the second part.

As the biological method of treatment is the most effective and economical method of treatment for organic and certain inorganic waste in India, more emphasis is given to this area. The newer methods of design, particularly those of biological treatment units, are now widely used by the leading consulting engineers and laboratories, but are not taught in the undergraduate level, probably due to the lack of textbooks covering the above. Though some of the design offices have already attained a rational outlook towards the design of facilities for both domestic and industrial waste water treatment, others and the educational institutions still follow the older thumb rule methods. This book is expected to help the teachers and students of this subject and also the design engineers to overcome this deficiency.

The industrial processes and the sources of their wastes are discussed briefly in the chapters on industrial wastes, and explained with simple line diagrams. Treatment and proper disposal of industrial waste water is a neglected subject in our country. An attempt has been made in this book to highlight the ill effects of wastes disposed off into the streams by different Indian industries. Suitable methods of treating these wastes under Indian conditions are discussed at length.

The major portion of the material presented in this book has been derived from the works of others. Authors may only be given credit for the compilation of all the available information in one volume. The authors are grateful to all the scientists and engineers, whose contribution has made this book complete and meaningful. While every attempt has been made to acknowledge all the sources of information, we apologise in advance for any inadvertant omission in this connection. The authors are also thankful to the Indian Standards Institution and M/s Hindustan Dorr-Oliver Ltd., for allowing them to use their data and photographs of equipments respectively in the text.

The authors also gratefully acknowledge the help rendered by many others who offered various suggestions in the preparation of the book.

Great appreciation should also be given to Mrs. Datta and Mrs. Rao, for their patience in bearing with the situation with the authors were more than fully occupied during the preparation, of this book.

Durgapur   M. Narayana Rao
August 31, 1978.   Amal K. Datta

Contents

1 | Characterisation and Degree of Treatment of Waste Water

1.0 INTRODUCTION

The ultimate disposal of waster water can only be onto the land or into the water. But whenever the water courses are used, for the ultimate disposal, the waste water is given a treatment to prevent any injury to the aquatic life in the receiving water. Normally the treatment consists of the removal of suspended and dissolved solids through different units of the treatment plant. Engineers can design a treatment plant to accomplish any degree of treatment. But a complete treatment or 100% removal of the pollution load is uneconomical and never aimed at in any waste treatment plant. The water courses can assimilate certain portion of the pollution load without affecting seriously the water quality and the environment. In fact, the treatment plant accomplishes a major part of the cleaning work, and then the nature, after the disposal of the effluent into the water courses, completes the work and thus rids the water of pollution.

Therefore, like any other Civil Engineering design, before proceeding with the design of the treatment plant itself, it is essential to determine:

(*i*) the characteristics of the raw waste water,

and (*ii*) the required characteristics of the treatment plant effluent, which will not pollute the receiving water course beyound certain acceptable limits. In most of the cases, however, the required characteristics of the effluents are established by the law of the local authority, instead of that by vigorous engineering analysis.

1.1 CHARACTERISTICS OF WASTE WATER

As mentioned above, the characterization of the raw waste is essential in the planning for effective and economical methods of water pollution control. Due to the varying nature of the industrial wastes, many of the recent installations have designed their treatment units with due consideration to the raw waste characteristics, and the effluent characteristics, as established by the Indian Standards Institution (ISI), State Pollution Control Boards, or by the local administrative authorities. But the characterization of the municipal waste water prior to a treatment plant design have not received the attention it deserves, probably because of its lower pollution potential compared to that of Industrial Wastes.

The characteristics of the municipal waste water vary from place to place, and depend on various factors like ecoomic status and food habits of the community, water supply position, and the weather condition of the locality. The characteristics of the waste from an Indian ciry may not be similar to that from a city in the USA.

Suspended Solids (SS) and 5 day 20°C BOD (BOD₅) are the usual parameters of pollution, particularly in domestic waste waters. As the pollution load varies according to the dilution offered by the water used, it is a convenient and common practice to express the SS and BOD_5 in terms of per capita contribution. However, with a greater amount of water use, more amount of solids per capita are expected to join the waste water flow, but it remains fairly constant beyond a waste flow rate of about 450 litres/capita/day.

The per capita contributions of SS, Total Solids, and BOD_5 in different Indian cities are shown in table 1.1. The composition of domestic waste water in some Indian cities are given in Table 1.2. The composition of different industrial wastes will be presented afterwards in the chapters dealing with the types of wastes under question.

In the absence of other information per capita SS and BOD_5 may be taken as 90 to 95 gms and 40 to 45 gms/day respectively in Indian cities. BOD associated with the suspended solids, in normal domestic waste water, is usually at a rate of 0.25 kg of BOD per kg of SS. The 1st order BOD removal rate constant may be much higher than the usually assumed value of 0.23 per day at 20°C.

TABLE 1.1. Per Capita Contributions to the Pollution in Different Indian Cities

Parameters / Cities	BOD$_5$ gms/day	Total Solids gms/day	SS gms/day	BOD removal rate constant, per day	Flow, l/day
Housing Estate near Kolcutta	38-44	246-270	88-100	0.232-0.251	327
Bombay	32-64	—	—	0.281	122
Kanpur (i)	70	—	64	—	580
(ii)	32-39	—	—	—	—
(iii)	58	—	198	0.345	490
Madras	27	—	—	—	—
Nagpur	18	—	—	—	—
Jamshedpur	27	—	—	—	—
Ahmedabad (i)	27-45	—	—	—	—
(ii)	63	—	70	—	340
Durgapur	45	—	—	—	—
Jaipur	35	—	35	—	81
UK & Germany average	54	—	—	—	—
USA average	54	—	90	—	300

TABLE 1.2. Composition of Domestic Waste Water in Different Indian Cities

Parameters	Kolkatta	Delhi	Madras	Kanpur	Nagpur	Ahmedabad	USA Average
Temperature,°C	24	—	—	27	30	30	—
pH	7.0	7.4	7.4	7.0	7.0	—	—
Total solids, mg/1	785	—	—	—	—	—	—
Suspended solids, mg/1	287	354	530	560	292	206	295
Chlorides, mg/1	—	147	259	114	40	352	—
BOD$_5$ mg/1	125	203	352	255	334	186	180
COD, mg/1	290	377	—	532	—	—	—

TABLE 1.3. ISI Tolerance Limits for the Sewage and Industrial Effluents and that of Inland Surface Water.

Characteristics	Tolerance limits for sewage effluents discharged into inland surface water. IS : 4764-1973	Tolerance limits for industrial effluents discharged into		Tolerance limits for Inland surface water, when used as raw water for public water supplies and bathing ghats IS : 2296-1974
		Inland surface water IS: 2490-1974	Public sewers IS: 3306-1974	
1	2	3	4	5
BOD (5 day 20°C), mg/l	20	30	500	3
COD, mg/l	—	250	—	—
pH	—	5.5-9.0	5.5-9.0	6.0-9.0
Total suspended solids, mg/l	30	100	600	—
Temperature, °C	—	40	45	—
Oil and Grease, mg/l	—	10	100	0.1
Phenolic compounds, mg/l	—	1.0	5	0.005
Cyanides (as CN), mg/l	—	0.2	2.0	0.01
Sulphides (as S), mg/l	—	2.0	—	—
Fluorides (as F), mg/l	—	2.0	—	1.5

(*Contd.*)

TABLE 1.3. (*Contd.*)

1	2	3	4	5
Total residual cholorine, mg/1	—	1.0	—	—
Insecticides, mg/1	—	zero	—	zero
Arsenic (as AS), mg/1	—	0.2	—	0.2
Cadmium (as Cd), mg/1	—	2.0	—	—
Chromium, hexavalent (as Cr), mg/1	—	0.1	2.0	0.05 (Total chromium)
Copper, mg/1	—	3.0	3.0	—
Lead, mg/1	—	0.1	1.0	0.1
Mercury, mg/1	—	0.01	—	—
Nickel, mg/1	—	3.0	2	—
Selenium, mg/1	—	0.05	—	0.05
Zinc, mg/1	—	5.0	15.0	—
Chloride (as Cl), mg/1	—	—	600	600
Sulphates, mg/1	—	—	—	1000
% Sodium	—	—	60	—
Ammoniacal Nitrogen, mg/1	—	50	50	—

(*Contd.*)

TABLE 1.3. (*Contd.*)

1	2	3	4	5
Nitrates (as No$_3$), mg/l	—	—	—	50
Radioactive materials:				
α-emitters, μc/ml—	10^{-7}	—	10^{-9}	
β-emitters, μc/ml—	10^{-6}	—	10^{-8}	
Dissolved Oxygen, mg/l	—	—	—	40% of the saturation value, or 3 mg/l, which ever is higher.
Coliform organism— (monthly average)— MPN per 100 ml	—	—	—	Should not exceed 5000. (should not exceed 20000 with less than 5% samples, and 5000 with less than 20% samples).

Note: Tables 1.3 and 16.1 of this book have been reproduced with the permission of ISI, from Indian Standard nos. 2296-1974, 2490 (pt 1)-1974, 3306-1974, 4764-1973, and 4903-1968, to which reference is invited for further details. These standards are available for sale from Indian Standards Institution, New Delhi and its Regional and Branch Offices at Ahmedabad, Bangalore, Bhubaneswar, Bombay, Kolkatta, Chandigarh, Hyderabad, Jaipur, Kanpur, Madras, Patna, and Trivandrum.

1.2 CHARACTERISTICS OF THE TREATMENT PLANT EFLUENTS

The required quality of the treatment plant effluents is solely dictated by the quality requirements of the receiving water. As stated earlier the quality of the receiving water may be established either by law or by a vigorous engineering analysis giving due consideration to the natural purification or self-purification that occurs in the receiving water. Effluent standards as well as river water qualities, as recommended by I.S.I. are given in Table 1.3. Once the quality required to be maintained in the receiving water at a particular cross-section is established (either at the point of discharge, or at a point downstream to the point of discharge), the quality required of the effluent being discharged can be computed.

The natural purification is a slow process, and depends on various factors. In the following sections, some of the significant forces of natural purification will be discussed; it will be then demonstrated how the knowledge of the same can be utilized in establishing the required quality of the effluent, and thus the degree of treatment of the treatment plants.

1.3 PATTERN OF POLLUTION AND SELF-PURIFICATION IN A STREAM

When pollutants are discharged into a stream, a succession of changes in water quality takes place, in the down stream side of the point of pollution. The resulting pattern of change along the stream establishes a well defined profile of pollution and self purification, which again changes with seasons and hydrography.

Whenever a single, heavy charge of putrescible organic matter is added into a clean stream, depending on the hydrography of the stream, the suspended matter is either settled at the bed near the point of discharge, or is carried along with the water to the downstream side. If the wetted surface of the river bed is sufficiently large, a major portion of the organic load is also removed from the main stream by adsorption. At the same time, the aerobic micro-organisms, which utilize the organic pollutants as the source of their food and energy, grow till the food supply is adequate for them, and thus the organic matter is stabilized under aerobic condition. The removal of organics are accomplished by (*i*) settling and adsorption, and by (*ii*) microbiological activities.

The intensity of the life activities of the micro-organisms is reflected by the biochemical oxygen-demand (BOD). Due to the microbial activities, the oxygen resources of the water are heavily drawn upon; in an overloaded stream, the dissolved oxygen (DO) may be completely exhausted due to these activities.

In course of time and flow, the food supply gets exhausted. The life activities of the microbial population come to an end; and as such the BOD is decreased. The rate of reaeration, or the absorption of oxygen from the atmosphere, which at first has lagged behind the rate of oxygen consumption by the micro-organisms, assumes a momentum and very soon takes the lead. The water becomes clear and the stream returns to its original condition. The self purification is thus complete.

As stated earlier, the self-purification is a very slow process. A heavily polluted stream may have to traverse quite a long distance for many days for the attainment of a significant degree of purification. It may also be noted that, beside the factor noted earlier, many other natural forces play an important role in the natural purification either in favour or against the process. These will be discussed briefly in the subsequent sections.

It should be recognized that, the natural forces of purification enter in one way or the other into the artificial treatment processes, where they are purposely intensified in order to accomplish the desired degree of treatment in a verty short time and small space.

1.3.1 Parameters of Pollution and Self-purification

The extent or degree of pollution and that of self-purification can be measured in various ways. The method of measurement depends on the nature of the pollution and the subsequent uses of the receiving water. When the nuisance potential of the discharged effluent is of great concern, the BOD and DO of the receiving water, taken together, are measured to trace the profile of pollution and self-purification in the stream. The BOD represents the intensity of biodegradable matter remaining in the stream at any time, and the DO shows the ability of the stream to purify itself through the biochemical processes. Other parameters of pollution include pH, Suspended solids, and Toxic substances, and will be dealt with, briefly afterwards. It may however be noted that the BOD and DO taken together, and the other parameters mentioned above, provide no information on the state of eutrophication and the possibility of algal bloom. Conditions sometimes demand the removal of inorganic nutrients in addtion to organic materials. As such, the nitrogenous nutrients may also be considered as the parameter of pollution.

1.4 DISSOLVED OXYGEN

Organic waste normally undergoes aerobic decomposition in the stream. Only when the rate of supply of oxygen (mainly from atmosphere) cannot keep pace with the rate of oxygen demand, the condition within the stream becomes anaerobic. The anaerobic condition is however not desirable, as while the aerobic receiving water look reasonably clean and are free from odour, the anaerobic condition makes it black, unsightly and malodourous. The biochemical reactions within the stream exert BOD, resulting in the deoxygenation of the stream.

Apart from the above, a certain portion of the biodegradable organics get deposited at the bed of the stream. They undergo anaerobic and benthic decomposition. The products of such decomposition are organic acids and reduced gases. These are further stabilized by the aerobic microorganisms in the upper layer, thereby increasing the BOD of the stream. A small amount of oxygen is also utilized by the higher animals for their respiratios.

But, as stated earlier, the simultaneous replenishment of oxygen in the stream occurs due to the absorption of oxygen from the atmosphere and also due to the release of oxygen by the green plants during photosynthesis. This is known as reaeration or reoxygenation of the stream. The interplay between the deoxygenation and reaeration produces a well defined profile of the dissolved oxygen in the stream as shown in Fig. 1.1.

The simplest way of estimating the required effluent quality, however, merely takes into account the available DO of the stream and ignore the amount of oxygen available through reaeration. The

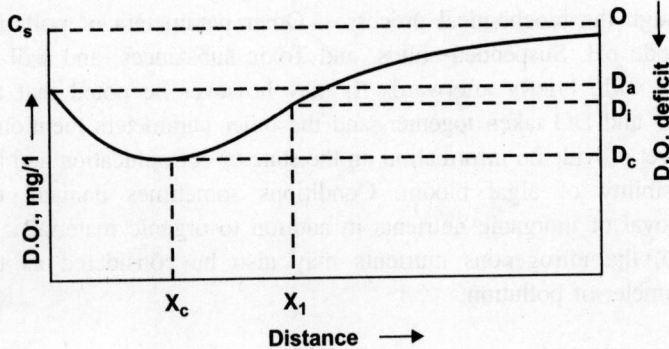

Figure 1.1: Oxygen Sag Curve.

permissible effluent BOD is calculated using the following simple relationship.

$$Q. \, DO_r - L.q = DO_m \, (Q + q) \qquad \text{... (1.1)}$$

in which Q = rate of flow of the stream,

q = rate of waste flow,

DO_r = DO of the stream, upstream to the point of discharge,

L = Permissible BOD_L of the waste water,

and DO_m = Permissible minimum DO of the stream, downstream to the point of discharge.

The more accurate and rational method of estimating the maximum permissible BOD load requires the knowledge of the DO profile. The forecasting of a DO profile downstream to a point of discharge again requires thorough knowledge of different natural forces in deoxygenation and reaeration.

1.4.1 Deoxygenation Through Exertion of BOD

The instantaneous rate of deoxygenation at any time t is given by the following 1st order equation.

$$\frac{dC}{dt} = K_2 \, (C_3 - C) = K_2 D \qquad \text{... (1.2)}$$

in which C = concentration of DO at any time t, mg/1,

           L = ultimate BOD of the polluted water, remaining after time t, mg/1,

     and K_1 = deoxygenation constant, per day.

The Eqn. 1.2 is identical to the 1st order BOD removal rate equation

$$\frac{dL}{dt} = - K_1 L \qquad \qquad \text{... (1.3)}$$

which is valid when the BOD removal is mediated only through the microbiological activities.

When the BOD removal is also accomplished by physical forces like sedimentation and adsorption, the Eqn. 1.3 is modified to the following form:

$$\frac{dL}{dt} = - K_1 L - K_3 L \qquad \qquad \text{... (1.4)}$$

in which K_3 = rate constant for BOD removal by sedimentation and adsorption, per day.

The deoxygenation constant, K_1, varies with temperature as per the following relationship

$$K_1 = (K_1)_{20} \cdot \left(\theta_{K_1}\right)^{T-20} \qquad \qquad \text{... (1.5)}$$

in which $(K_1)_{20}$ = deoxygenation constant at a temperature of 20°C,

         K_1 = deoxygenation constant of a temperture of T°C,

   and θ_{K_1} = a temperature coefficient = usually 1.047 for K_1 in the temperature range of 15° to 30°C.

The value of K_1 may vary from 0.12 per day for treated waste effluent to 0.39 per day for a strong waste.

The ultimate BOD, L, was previously assumed to vary with temperature, and equations similar to the Eqn. 1.5 were also derived for the temperature variation of the ultimate BOD of the wastes.

Later work suggests that no such variation should occur with temperature.

A small fraction of the BOD is removed by settling and adsorption; if the condtions favour, the settled particulate wastes under-go anaerobic digestion. As stated earlier, the products of the anaerobic digestion, *viz.*, organic acids, gases etc. are utilised as the source of food and energy by the aerobic microorganisms. So the products of digestion of settled particles become part of the BOD of the stream. Apart from this, certain amount of BOD is also added to the stream by the local run off. The rate at which the BOD is added to the stream may be given by the following relationship:

$$\frac{d\text{L}}{dt} = \text{L}_b \qquad \qquad ... \ (1.6)$$

in which L_b = rate of addition of BOD by local run off and by the products of anaerobic digestion of bottom sludge deposits.

1.4.2 Reaeration of the Polluted Streams

The rate at which water absorbs oxygen from the atmosphere can be given by the following equation:

$$\frac{d\text{C}}{dt} = \text{K}_2 \, (\text{C}_3 - \text{C}) = \text{K}_2\text{D} \qquad \qquad ... \ (1.7)$$

in which   D = DO deficit after time t, mg/1,
   K_2 = reaeration coefficient, per day,
   and C_3 = DO saturation concentration.

The magnitude of K_2 is dependent on various factors like the surface exposure, volume of water flow and the turbulance of the stream. Isaac and Gaudy have given the following dimensionally homogeneus equation for the theoretical value of K_2:

$$\text{K}_2 = 2.3 \; k_2 = 2.3 \; c \; \frac{\text{D}_m^{\frac{1}{2}}}{\text{V}^{1/6} g^{1/6}} \cdot \frac{v}{\text{H}^{3/2}} \qquad \qquad ... \ (1.8)$$

in which K_2 = reaeration coefficient (expressed in $\log_e$),

        k_2 = reaeration coefficient (expressed in $\log_{10}$),

        c = proportionality constant (dimensonless),

     D_m = molecular diffusion coefficient for oxygen in water equal to 2.037×10^{-9} m^2/sec. at 20°C (it varies with temperature and the temperature coefficient is usually taken as 1.037),

        v = kinematic viscosity of water,

        g = gravitational constant,

       V = mean velocity of flow,

and H = average depth of flow.

In a simulated stream the value of c was found to be equal to 0.06339. Natural stream data in a particular case however indicates that c may be taken as 0.07762, which takes the irregular channel geometry into account.

By substituting the known values of D_m, v and g, and the assumed value of c (equal to 0.07762) in Eqn. 1.8, the expression for K_2 at 20°C may be given by the following expression:

$$(K_2)_{20} = 4.75 \ V/(H)^{\frac{3}{2}} \qquad \ldots (1.9)$$

in which $(K_2)_{20}$ = reaeration coefficient at 20°C, per day,

         V = mean velocity in m/sec.,

and H = average depth of the stream, m.

The Eqn. 1.8 or 1.9 should only be used in the absence of any actual test result. The value of K_2 not only varies from stream to stream, but also varies along the length of a particular stream. The variation of K_2 with temperature can be given by the following relationship:

$$K_2 = (K_2)_{20} \ \theta \ K_2^{(T-20)} \qquad \ldots (1.10)$$

in which K_2 = reaeration coefficient at T°C,

   $(K_2)_{20}$ = reaeration coefficient at 20°C,

and θ_{K_2} = reaeration temperature coefficient, usually 1.0241.

Howe, however, proposed the following modified relationship for the temperature variation of K_2:

$$K_2 = \frac{(K_2)_{20}}{(\theta_{K_2})^{T-20}} \qquad \ldots (1.11)$$

in which θ_{K_2} is 1.0125.

In addition to the reareation by absorption of atmospheric oxygen, a small quantity of oxygen is introduced to the system by phyotosynthesis, and another portion is lost due to the respiration of aquatic plants and animals. It is convenient to combine the photosynthesis, respiration, and any other source or sink of oxygen in the stream, and express the rate of change of oxygen by the following relationship:

$$\frac{dC}{dt} = \pm S_r \qquad \ldots (1.12)$$

in which S_r = rate of change of dissolved oxygen concentration as a result of physical and biological forces, not included in Eqns. 1.2 and 1.7.

1.4.3 BOD and DO Profile

The rate of change of DO concentration and BOD along a stretch of a polluted stream, can be given by the following two one-dimensional DO-BOD equations, obtained by writing the material balance across an element of the stream water:

$$\frac{\partial C}{\partial t} + V\frac{\partial C}{\partial x} = \frac{1}{A}\frac{\partial}{\partial x}\left(AE\frac{\partial C}{\partial x}\right) - K_1 L + K_2 D \pm S_r \qquad \ldots (1.13)$$

and,

$$\frac{\partial L}{\partial t} + V\frac{\partial L}{\partial x} = \frac{1}{A}\frac{\partial}{\partial x}\left(AE\frac{\partial L}{\partial x}\right) - (K_1 + K_3)L \pm L_b \qquad \ldots (1.14)$$

in which, E = dispersion coefficient, usually m²/day ($L^2\, t^{-1}$).

It may be noted that, the second term in the L.H.S. of Eqns. 1.13 and 1.14, $V(\partial L/\partial x)$, represents the advective component of the flux across the cross section. The first term of the R.H.S. of the Eqns. 1.13 and 1.14 represents the longitudinal dispersion component of the flux. The remaining terms in the R.H.S. represent the sources and sinks, as discussed earlier in Sections 1.4.1 and 1.4.2.

However, in a stream the dispersion may be neglected without any serious error. And also it is convenient to express DO concentrations in terms of "DO saturation deficit", D. Again in a steady state condition, $\partial C/\partial t = 0$. Therefore, in a steady state condtion, and neglecting all sorts of longitudinal dispersion, the Eqns. 1.13 and 1.14 reduce to:

$$V\frac{dD}{dx} = -K_1L + K_2D \pm S_r \qquad \dots (1.15)$$

and

$$V\frac{dL}{dx} = -(K_1 + K_3)L + L_b \qquad \dots (1.16)$$

The solution of the Eqns. 1.15 and 1.16 yields the following relationship for the DO deficit at any distance x from the point of discharge:

$$D(x) = D_aF_2 + \frac{K_1}{K_2 - (K_1 + K_3)}\left[L_a - \frac{L_b}{K_1 + K_3}\right](F_1 + F_2)$$

$$+ \left[\frac{S_R}{K_2} + \frac{K_1L_b}{K_2(K_1 + K_3)}\right](1 - F_2) \qquad \dots (1.17)$$

in which, D_a = initial DO deficit, or D at $x = 0$,

$$F_2 = \exp\left(-K_2\frac{x}{V}\right), \qquad \dots (1.18)$$

$$F_1 = \exp\left[-(K_1 + K_3)\frac{x}{V}\right], \qquad \dots (1.19)$$

L_a = initial BOD$_L$, or L at $x = 0$,

and $\qquad$ V = Velocity of flow = Q/A.

A typical plot of the Eqn. 1.16 is shown in Fig. 1.1. The spoon shaped DO profile thus obtained is known as Oxygen Sag curve. The sag curve possesses two characteristic points:

(*i*) the critical point at a distance x_c, where the maximum DO deficit, D_c, occurs,

(*ii*) the point of inflection at a distance x_i,

where the rate of recovery is maximum.

By differentiation of Eqn. 1.17, and solving for x after equating the derivative to zero, the critical distance x_c may be obtained as follows:

$$x_c = \frac{V}{K_2 - (K_1 + K_3)} \ln \left[\frac{K_2}{K_1 + K_3} + \frac{K_2 - (K_1 + K_3)}{(K_1 + K_3)L_a - L_b} \right.$$

$$\left. \left\{ \frac{L_b}{K_1 + K_3} - \frac{K_2 D_a - S_r}{K_1} \right\} \right] \qquad \ldots (1.20)$$

To facilitate the analysis, as will be demonstrated in a following section, the Eqn. 1.17 is rearranged in the following form to give the initial 1st stage BOD, L_a:

$$L_a = \frac{D - F_3(1 - F_2) - D_a F_2}{F_4 \, (F_1 - F_2)} + \frac{L_b}{K_1 + K_3}, \qquad \ldots (1.21)$$

in which $\quad F_3 = \dfrac{S_r}{K_2} + \dfrac{K_1 L_b}{K_2 \, (K_1 + K_2)} \qquad \ldots (1.22)$

and $\quad F_4 = \dfrac{K_1}{K_2 - (K_1 + K_3)} \qquad \ldots (1.23)$

It may be noted that the values of F_3 and F_4 remain constant for a particular stretch of the stream.

It may however be noted that in many cases, K_3, L_b and S_r are negligible, and if neglected in the analysis, the Eqn. 1.17 reduces to the following form of well known Streeter and Phelps formula:

$$D = \frac{K_1 L_a}{K_2 - K_1}\left[\exp\left(-K_1 \frac{x}{V}\right) - \exp\left(-K_2 \frac{x}{V}\right)\right]$$

$$+ D_a \exp\left(-K_2 \frac{x}{V}\right) \qquad \ldots (1.24)$$

In such cases, x_c and D_c are given by the following two relationships

$$x_c \frac{V}{K_1 (f - 1)} \ln\left[f\left\{1 - (f - 1)\frac{D_a}{L_a}\right\}\right] \qquad \ldots (1.25a)$$

and $$D_c = \frac{L_a e^{-K_1 t_c}}{f}, \qquad \ldots (1.26)$$

in which, $\quad f = \dfrac{K_2}{K_1}$ = self-purification ratio $\qquad \ldots (1.27)$

Rearrangement of Eqn. 1.25a yields:

$$\left(\frac{L_a}{D_c f}\right)^{f-1} = f\left[1 - (f - 1)\frac{D_a}{L_a}\right] \qquad \ldots (1.25)$$

Eqn. 1.25 may be solved for L_a by trial and error methods.

1.4.4. Determination of K_1, K_2, K_3, L_b and S_r

In fact, K_1 is a reaeration rate constant of a biological process, and is to be determined in laboratory, testing samples from the stretch of the river under similar environmental conditions.

Once the value of K_1 is determined, the values of K_3 and L_b can be determined using the following relationship for BOD_L at any time x/V.

$$L_x = L_a \exp\left[-K_1\left(\frac{x}{V}\right)\right] \qquad \ldots (1.28)$$

in which, L_x = estimated theoretical value of BOD_L at a distance x from the point of discharge.

Now if L_x is greater than the actual value of L, it may be assumed that $L_b = 0$, and K_3 may be estimated by using the following relationship, which is obtained through integration of Eqn. 1.16, and substituting $L_b = 0$.

$$L = L_a \exp\left[-(K_1 + K_3)\frac{x}{V}\right] \qquad \dots (1.29)$$

When L_x is less than the actual value of L, it is assumed that $K_3 = 0$, and the value of L_b is calculated using the following relationship, obtained similarly from Eqn. 1.16 but substituting $K_3 = 0$.

$$L = L_a \exp(-K_1 x/V) + (L_b/K_1)[1 - \exp(-K_1 x/V)] \qquad \dots (1.30)$$

If L_x is equal to the actual value of L, both K_3 and L_b may be assumed to be equal to zero.

K_2 can best be determined by the Eqns. 1.8, 1.9 and 1.10. The remaining unknown S_r may now be calculated using Eqn. 1.17. S_r may be either positive or negative.

1.4.5 Allowable BOD Loading on Stream, L_a

To avoid anaerobic decomposition of the organic wastes, and also to provide adequate support to the aquatic life, the DO is not allowed to fall below about 3 mg/l. This information fixes the allowable maximum DO saturation deficit, D_{max}, in the stream.

For a given set of values of K_1, K_2, K_3, L_b, S_r, D_{max} and D_a, the maximum allowable BOD loading L_a can now be determined by an iterative process. The maximum allowable ultimate BOD of the treatment plant effluent may then be calculated from the stream flow and waste discharge data, assuming a complete mixing condition at the point of discharge.

The following procedure may be adopted for the estimation of allowable L_a:

(i) L_a is determined using the Eqn. 1.21 substituting $D = D_{max}$ and the maximum length of the stretch of the stream, x_{max}, as the value of x.

(*ii*) Using the value of L_a thus determined, the value of x_c is determined using the Eqn. 1.20.

The calculated value of x_c may be greater or less than the assumed value of $x = x_{max}$.

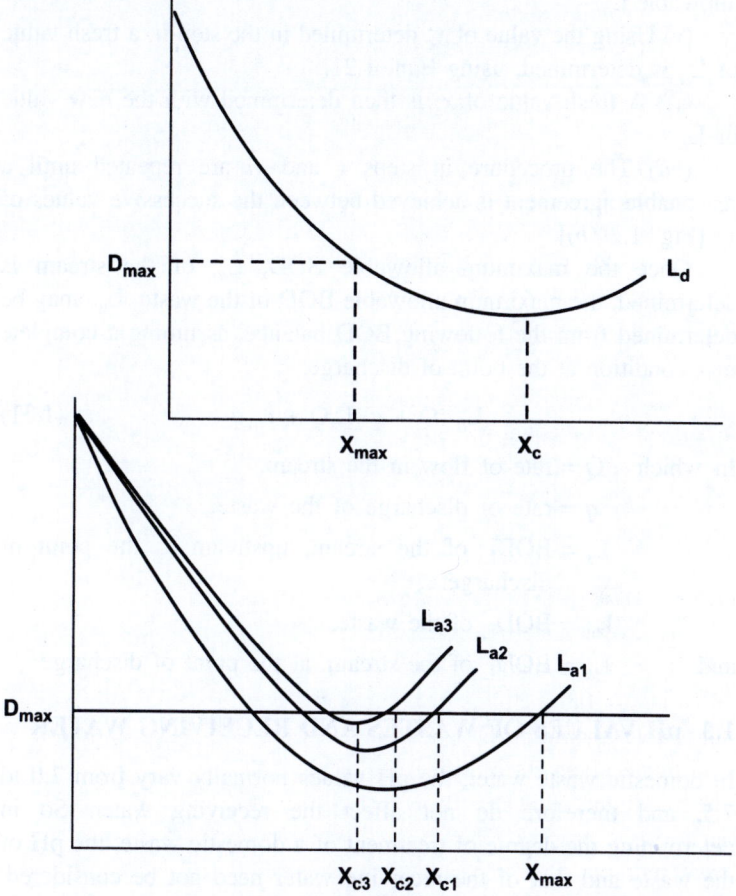

Figure 1.2: Definition Sketch of the Procedure Outlined in Section 1.4.5. (Loucks' Procedure).

(*iii*) if x_{max} is less than the value of x_c, calculated in step-*ii*, the estimated value of L_a in step-*i*, is not likely to result in a DO deficit greater than D_{max} within the stretch of the stream. And this L_a value may be accepted as a design value. [Fig. 1.2(*a*)].

(*iv*) But if x_{max} is greater than the value of x_c, calculated in step-*ii*, the calculated value of L_a in step-*i*, is likely to deplete the DO of the stream beyond the allowable limit. In such cases, following iterative procedure is to be adopted for the estimation of allowable L_a.

(*v*) Using the value of x_c determined in the step-*ii*, a fresh value of L_a is determined, using Eqn. 1.21.

(*vi*) A fresh value of x_c is then determined with the new value of L_a.

(*vii*) The procedure in steps *v* and *vi* are repeated until a reasonable agreement is achieved between the successive values of x_c [Fig. 1.2 (*b*)].

Once the maximum allowable BOD, L_a, of the stream is determined, the maximum allowable BOD of the waste, L_w, may be determined from the following BOD balance, assuming a complete mix condition at the point of discharge:

$$L_a (Q + q) L_s Q + L_w q \qquad \text{.... (1.31)}$$

In which Q = rate of flow in the stream,

q = rate of discharge of the waste,

L_s = BOD_L of the stream, upstream to the point of discharge.

L_w = BOD_L of the waste,

and L_a = BOD_L of the stream, at the point of discharge.

1.5 pH VALUES OF WASTES AND RECEIVING WATER

In domestic waste water, the pH values normally vary from 7.0 to 7.5, and therefore do not affect the receiving water. So in determining the degree of treatment of a domestic waste, the pH of the waste and that of the receiving water need not be considered. But a completely different picture may arise in the case of an industrial waste. Most of the industrial wastes are characteristically either acidic, or alkaline. Acidic wastes are corrosive to both metallic and concrete structures in the water courses, and are also toxic to the aquatic life. Moreover, when free acid reacts with the natural alkalinity of the water, it produces carbonate hardness, thus rendering it unfit for future laundry uses, or as boiler feed water.

Alkaline wastes, on the other hand, when discharged into the water courses, combine with the free carbon dioxide and further increase the alkalinity of the water. The degree of treatment of the acidic or alkaline waste is determined according to the pH of the receiving water after mixing, which can be estimated using the following approximate relationship.

For acidic wastes:

$$pH = 6.52 + \log\left(\frac{B-X}{50}\right) - \log\left(\frac{C}{22} - \frac{X}{50}\right) \qquad \dots \ (1.32)$$

and for alkaline waste:

$$pH = 6.52 + \log\left(\frac{B+X}{50}\right) - \log\left(\frac{C}{22} - \frac{X}{50}\right) \qquad \dots \ (1.33)$$

in which, B = alkalinity of water upstream to the point of discharge, in mg/l as $CaCO_3$.

X = concentration of acidic or alkaline waste after dilution with the water, in mg/l as $CaCO_3$.

and C = concentration of free CO_2 in mg/l.

The optimum range of pH for aquatic life is 6.8 to 9.0. Therefore, by substituting the limiting values of the pH (6.8 for acidic waste, and 9.0 for alkaline waste) in the Eqn. 1.32 or Eqn. 1.33, the maximum permissible value of X can be estimated. The permissible maximum concentration of the acidic or alkaline waste then can be determined, from an acid/alkali balance in the stream. It may however be noted that the Eqns. 1.32 and 1.33 are applicable for inorganic acids/alkalis only.

1.6 SUSPENDED SOLIDS

Organic suspended and dissolved solids undergo biodegradation and their pollution potentials are usually expressed in terms of BOD. The effects of these organic solids on the receiving water have been described earlier in Section 1.4. However, the fixed or inorganic solids and heavier organic solids settle quickly and form a sludge blanket near the point of discharge. The maximum concentration of the suspended solids (SS) in the receiving streams are often specified by the local authorities. The permissible concentration of SS in the waste effluent may be estimated as earlier, using the SS balance and neglecting the longitudinal dispersion; such a method assumes a complete mix condition at the point of discharge, which is far from the fact. For a more accurate method, the following relationship may be used:

$$S_w = (S_m - S_r)\left(a\frac{Q}{q}+1\right)+S_r \qquad \text{.... (1.34)}$$

in which, S_w = permissible SS in the waste effluent, mg/1,

S_m = permissible SS in the stream, at a particular cross-section, downstream to the point of discharge, mg/1.

S_r = SS in the stream, upstream to the point of discharge, mg/1,

a = degree of mixing, or, mixing coefficient for the said particular cross-section,

Q = rate of flow of receiving water,

and q = rate of discharge of the waste water.

The degree of mixing, a, in the Eqn. 1.35, is a function of flow rates, bed tortuosity, eddy diffusion, distance of the point of complete mixing (where $a = 1$) from the point of discharge, etc. Theoretically the mixing is complete (and $a = 1$) at a distance of infinity from the point of discharge. In practice it is convenient to assume the mixing to be sufficiently complete at a point where a is equal to 0.90 to 0.95. The distance of the cross-section of the polluted stream, l, where the degree of mixing equals to a particular value of a, may be estimated using the following relationship:

$$l = \left[\frac{2.3}{a}\log\frac{aQ+q}{(1-a)q}\right]^3 \qquad \text{.... (1.35)}$$

If the permissible concentration of SS in the receiving stream, S_m, is specified for a particular cross-section, at a distance of l

from the point of discharge, the degree of mixing, a, can be estimated by using the Eqn. 1.35; the substitution of this value of a in the Eqn. 1.34 will result in the permissible concentration of SS in the waste, S_w.

For all practical purposes, however, it may be assumed that the specified maximum permissible SS in the stream refers to that at the point of complete mixing. As such, the permissible SS in the waste effluent may be directly estimated using the Eqn. 1.34, assuming a equal to 0.90 or 0.95.

1.7 TOXIC SUBSTANCES

The toxic substances present in the industrial wastes not only disturb the ecological balance in the receiving water, but also may endanger the health of the animals and human beings who may have to use the polluted water for drinking purposes. The tolerance limits of the different toxic elements in the receiving water have been prescribed by different authorities. The permissible concentration of the toxic elements in the waste effluent may then be computed from the following relationship, obtained from the mass balance:

$$T_w = T_m\left(a\frac{Q}{q} + 1 \right) \qquad \text{.... (1.36)}$$

in which, T_w = allowable concentration of the toxic element in the waste effluent,

and T_m = permissible concentration of the toxic element in the receiving water, downstream to the point of discharge.

The permissible concentration of the toxic elements in the receiving water, T_m, refers to that at the point of complete mixing. In order to limit the concentration of toxic elements in the stream to the permissible value at any point in between the point of outfall and the point of complete mixing, the mixing coefficient, a, should be taken into considerations. In all other cases the value of a may be taken as unity.

When sufficient information about the individual toxicants, in regard to their concentration in the waste water and their allowable limits in the receiving water, are not available, the acute toxicity of the entire waste water is measured by means of bioassy tests, and are expressed in terms of "median tolerance limit", TL_m. The TL_m is defined as the concentration of waste in which just 50% of the test animals are able to survive for a specified period of exposure, usually 48–96 hours. Fish are usually used as test animals, and normally it is expected that a concentration of 1/10th of the 96-hr TL_m value will not cause any serious effect on the aquatic lives.

From the definitionz of TL_m, the water required for the dilution of the toxic waste, to reduce its toxicity to a value to cause 50% mortality among the test animals, is given by:

$$TL_m = \frac{q}{V + q} \times 100 \qquad \text{.... (1.37)}$$

in which, TL_m = median tolerance limit in percentage,
q = waste volume,
and V = dilution water volume.

The Eqn. 1.37 may be rewritten in the following form:

$$V = \frac{(100 - TL_m)}{TL_m} q \qquad \text{.... (1.38)}$$

In the stream pollution control, the stream flow necessary to dilute the waste sufficiently to a safe level is determined by multiplying the above dilution water volume by an application Factor, F_a, usually 10.

$$Q = V.F_a \qquad \text{.... (1.39)}$$

in which, Q = stream of flow rate,
F_a = application factor,
and V = dilution water flow rate.

1.8 ESTUARINE POLLUTION

When a certain amount of pollution is introduced continuously into a tidal stream, the tidal oscillation moves the polluted volume of water back and forth, and thereby causes a widely fluctuating water quality in the estuary. The prediction of water quality in terms of DO and BOD, or Nitrogenous nutrients, is based on the dynamic coupling between the hydrodynamic transport processes and the biochemical water quality change processes.

The biochemical processes which lead to the changes in the water quality in a stream has been discussed in the previous sections. The hydrodynamic transport processes include advection, mixing, and dispersion of the pollutants, as in a fresh water stream. But the problem becomes much more complicated in an estuary due to the longitudinal density gradient and the variation of the instantaneous longitudinal velocity of flow due to the tidal effect. Analysis of an estuary becomes further complicated, if the same is stratified due to the sea water intrusion near the bay. However, many dynamic estuaries, like Hooghly river, are vertically homogeneous. Where the estuary presents complex problems because of wide fluctuation of physico-chemical conditions within it, daily, seasonally and geographically, actual field studies are to be conducted, instead of depending on the mathematical models.

1.8.1 BOD and Do Profiles in An Estuary

The dynamic BOD-DO model in an estuary of varying and arbitrary cross-section consists of the following two equations, similar to Eqns. 1.13 and 1.14:

$$\frac{\partial C}{\partial t} + V_{xt} \frac{\partial C}{\partial x} = \frac{1}{A} \frac{\partial}{\partial x}\left(AE \frac{\partial C}{\partial x}\right) - K_1 L + K_2 D \pm S_r, \dots \quad (1.40)$$

and,

$$\frac{\partial L}{\partial t} + V_{xt} \frac{\partial L}{\partial x} = \frac{1}{A} \frac{\partial x}{\partial x}\left(AE \frac{\partial L}{\partial x}\right) - (K_1 + K_3)L + L_b \quad \dots (1.41)$$

in which, V_{xt} = instantaneous longitudinal velocity of flow, averaged over the cross-section, including tidal and fresh water components.

Main difficulty in the simultaneous solution of the Eqns. 1.40 and 1.41 arises from the unsteady hydraulic regime of the estuary. Moreover, all the terms in the above equations may vary with distance and time. Analytical solutions to Eqns. 1.40 and 1.41, as such, cannot be obtained. The available analytical solutions of the above two equations are based on many simplifying assumptions.

However, by means of numerical analysis method, using digital computers, it is possible to predict approximately the BOD and DO at any distance and time in the estuary for a given set of specific input data.

If L_a is the value of L at $x = 0$, for all general cases, L_a is usually given by:

$$L_a = \frac{W}{A\sqrt{V_{xt} + 4K_1E}} \qquad \dots \text{ (1.42)}$$

in which, W = BOD or pollution load in kg/day.

The Eqn. 1.42 indicates that the concentration peak will be created in the estuary during each slack water conditions.

If tidal height variations are considered, the cross-sectional area A is function of x and t. Usually it is sufficient to consider only the longitudinal variation of A, at the mean-tide level, obtained from the hydrographic data.

The velocity V_{xt} is also a function of x and t, and its determination is a hydrodynamic problem, and it depends on the geometry of the estuary, bed roughness, and the characteristics of the tide at the ocean entrance to the estuary. Finite difference formulation for such determination of V_{xt} is also available in the literature (Harleman and Lee). However, in a simplified estuary, the following variation of the instantaneous velocity with time may be assumed:

$$V_{xt} = V_f + V_{max} \sin\left[\frac{48\pi T}{12.5}\right], \qquad \dots \text{ (1.43)}$$

in which, V_{max} = maximum tidal velocity at any time T,

  T = time in days

  V_f = fresh water velocity at the head of the tide,

  $\dfrac{48\pi}{12.5}$ = frequency of the tide,

and 12.5 = tidal period

The dispersion coefficient, E, is usually assumed to vary only with x. The longitudinal variation of E is large in the region of salinity instrustion in the estuary. The variation of E may be estimated by writing mass balance equation for salinity, similar to the Eqns. 1.40 and 1.41, and from the actual field observations on salinity variation. However, in the upstream fresh water zone, the value of E is fairly small, and depends mainly on tidal velocity, and is given by the following equation:

$$E = 63 \ n\text{VR}^{5/6} \qquad \qquad \text{.... (1.44)}$$

in which, n = Mannings coefficient of roughness of the estuary,
R = hydraulic radius in meters,
V = local tidal velocity in meters/sec,
and E = longitudinal dispersion coefficient in m^2/sec.

For the finite difference formulation of the numerical method of solution of the Eqns. 1.40 and 1.41, the readers may refer to the literature listed at the end of this chapter.

In a simplified estuary, where the terms S_r, L_b and K_3 in the Eqns. 1.40 and 1.41, are neglected, all other terms except the velocity are assumed to be constant, and assumed that the average velocity of flow in a tidal cycle is zero, the steady state solutions of the Eqns. 1.40 and 1.41, are given by the following two equations (8):

$$L(x) = L_a \exp \left[\ \pm x(K_1/E)^{\frac{1}{2}} \ \right], \qquad \text{.... (1.45)}$$

and,

$$C(x) = C_3 - \frac{K_1 L_a}{K_2 - K_1} \left[\exp \left\{ \ \pm x(K_1 / E)^{\frac{1}{2}} \right\} \right.$$

$$\left. -(K_1 / K_2)^{\frac{1}{2}} \exp \left\{ \pm x(K_1 / E)^{\frac{1}{2}} \right\} \right] \qquad \text{.... (1.46)}$$

in which, $L_a = \dfrac{W}{2A\sqrt{K_1 E}}$, the value of L at $x = 0$,

and W = input pollution load (BOD) in kg/day.

The BOD and DO values, averaged over a tidal cycle, obtained through a numerical analysis, assuming the velocity of flow to vary as given in the Eqn. 1.43 and all other terms in the Eqns. 1.40 and 1.41 to be constant, closely approximate to the values obtained for a steady state condition values, as given by the Eqns. 1.45 and 1.46.

Example 1.1: A city discharges 115 MLD of sewage into a

stream whose minimum rate of flow is 8500 litres/sec. The velocity of the stream is about 3.2 Kmph. The temperature of the sewage is 20°C, while that of the stream is 15°C. The 20°C 5 day BOD of the sewage is 200 mg/1, while that of the stream is 1.0 mg/1. The sewage contains no DO, but the stream is 90% saturated at the upstream of the discharge. At 20°C, K_1 is estimated to be 0.3 per day, while K_2 is 0.7 per day. Determine the critical oxygen deficit and its location. Also estimate the 20°C 5 day BOD of a sample taken at the critical point. Use temperature coefficient of 1.047 per °C for K_1 and 1.024 per °C for K_2. Neglect K_3, L_b and S_r.

Solution :

1. Saturation concentration of Oxygen at 15°C = 10.2 mg/1.
   Do in the stream = 0.9 × 10.2 = 9.18 mg/1.

2. Discharge of sewage = 115 MLD = 1330 litres/sec.

3. Temperature of stream after mixing

$$= \frac{1330 \times 20 + 8500 \times 15}{1330 + 8500} = 15.7°C$$

4. Do of the mixture

$$= \frac{1330 \times 0 + 8500 \times 9.18}{1330 + 8500} = 8.0 \text{ mg}/1$$

5. 5 day 20°C BOD of the mixture

$$= \frac{1330 \times 200 + 8500 \times 1}{1330 + 8500} = 27.7 \text{ mg}/1$$

6. Assuming BOD rate constant K = 0.3 per day at 20°C, initial first stage BOD, L_a

$$= \frac{27.7}{1 - e^{-0.3 \times 5}} = 35.7 \text{ mg/1}$$

7. Now, K_1 at 15.7°C
   $$= 0.3 \times (1.047)^{15.7-20} = 0.247 \text{ per day}$$
   and K_2 at 15.7°C
   $$= 0.7 \times (1.024)^{15.7-20} = 0.63 \text{ per day}$$

8. Saturation concentration of DO at 15.7°C = 10.1 mg/l

Initial DO deficit, D_a = 10.1 – 8.0 = 2.1 mg/1

Self-purification ratio, $f = K_2/K_1$

$$= 0.63/0.247$$

$$= 2.55$$

9. Therefore, from Eqn. 1.25,

$$x_c = \frac{2.3 \times 24 \times 3.2}{0.247(2.55-1)} \cdot \log\left[2.55\left\{1-(2.55-1)\frac{2.1}{35.7}\right\}\right]$$

$$= 169 \text{ Km}$$

$$t_c = 169/(24 \times 3.2) = 2.2 \text{ days.}$$

10. Critical deficit, D_c

$$= \frac{35.7 \times e^{-0.247 \times 2.2}}{2.55}$$

$$= 8.15 \text{ mg/1}$$

Minimum DO = 10.1 – 8.15 = 1.95 mg/1

11. Now first stage BOD at critical point, L_c

$$= L_a \, e^{-K_1 t_c} = 35.7 \times e^{-0.247 \times 2.2}$$

$$= 20.8 \text{ mg/1}$$

∴ 5 day 20°C BOD at 169 Km downstream to the point of discharge

$$= 20.8 (1 - e^{-0.3 \times 5})$$

$$= 16.2 \text{ mg/1}$$

QUESTIONS

1. Draw a typical oxygen sag curve in a stream subjected to pollution.

2. How do you determine the reaeration rate constant of a biological process?

3. Write down the equations that depict the dynamic BOD-DO Model in an estuary.

2 | Primary Treatment of Waste Water

2.0 SCOPE

A complete treatment of waste water involves many unit operations and processes. In this chapter the design of the principal unit opeartions used in the primary treatment of the waste water are discussed, along with relevant background information.

2.1 BAR SCREENS

The first step in the treatment of waste water is the removal of larger particles of floating and suspended matter by coarse screening. In most of the plants this is accomplished by a set of inclined parallel bars, fixed at certain distance apart in a channel. These could have been properly called as "Racks"; instead these are referred to as "Bar Screens".

In large plants at least two screens in parallel are to be provided. If only one unit is installed, it is absolutely necessary to provide a bye pass channel. The screens are normally accomodated in one screen chamber. The cross-sectional area of this chamber is always greater than that of inflowing sewer. The length of this channel should be sufficiently long to prevent eddies around the screen.

2.1.1. The dimensions of the channel, and the size of the openings between the bars, and their slopes are determined conforming to the following hydraulic requirements and specifications:

1. Approach velocity of the waste water in the screening channel shall not fall below a self cleansing velocity of 37.5

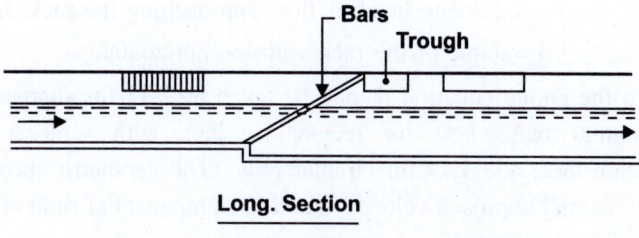

Long. Section

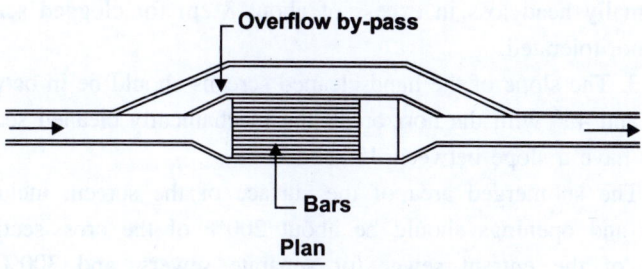

Plan

Figure 2.1: Bar Screen (Hand Cleaned).

cm/sec, or rise to a magnitude at which screenings will be dislodged from the bars.

The suggested approach velocity in the screening channel is 60 cm/sec to 75 cm/sec, for all the grit bearing waste water. The slope of the floor of the channel may be adjusted to maintain this velocity.

The suggested maximum velocity through the screen is 30 cm/sec at average rate of the flow for hand cleaned bar screens; and 75 cm/sec at the normal maximum flow for mechanically cleaned screens.

2. Head losses resulting from the screening operation must be controlled so that back water will not cause the entrant sewer to operate under pressure. Head loss can be calculated from the following empirical relationship given by Kirschmer:

$$h = \beta(w/b)^{4/3} h_v \sin \theta \qquad \ldots (2.1)$$

in which, h = head loss, in m,

β = a bar shape factor,

w = width of the bars facing the flow.

b = clear spacing between the bars,

h_υ = velocity head of flow approaching the rack, in m,

and $\qquad$ θ = angle of the rack with the horizontal.

In the above equation β may be taken as 2.42 for sharpedged rectangular bars, 1.83 for rectangular bars with semi-circular upstream face, and 1.79 for circular rods. (The geometric mean of the horizontal approach velocity υ and its component at right angles to the bars $\upsilon \sin θ$, i.e., $\upsilon\sqrt{\sin θ}$ is taken as the effective velocity). Normally head loss in excess of about 8 cm for clogged screens are not tolerated.

3. The slope of the hand cleaned screens should be in between 30° and 45° with the horizontal; the mechanically cleaned screens may have a slope between 45° and 80°.

The submerged area of the surface of the screen, including bars and openings should be about 200% of the cross-sectional area of the entrant sewer for separate sewers, and 300% for combined sewers.

The net submerged area of the opening of the screen should be about 500 cm^2 per MLD of flow for separate sewerage and 750 cm^2 for MLD of flow for combined sewerage system.

4. Clear spacing of bars may be from 25 to 50 mm for hand cleaned bar screens; this may range from 15 mm to 75 mm in case of mechanically cleaned bar screens. The width of the bars facing the flow may be from 8 mm to 15 mm, and the depth may vary from 25 mm to 75 mm; but the sizes less than 15 mm × 50 mm are normally not used. They are welded together at the rear face. Usual Bar sizes and opening are given in table 2.1.

All the above mentioned parameters in the design are so inter-related that, assumption of the magnitude of one or two will fix the magnitude of the others. So, in the design of bar screen, the controlling parameters are assumed and the dependent parameters are computed; if these are not within prescribed limits, new assumptions are made concerning the controlling parameters and depedent parameters are computed until a satisfactory design is attained.

TABLE 2.1: Usual Bar Sizes and Openings.

Dimension of the Bar facing flow, mm	Clear spacing between bars, mm	Area of the opening/ gross surface area of the screen, %
6	18	75
6	24	80
6	30	83.3
6	36	85.6
9	18	66.7
9	24	72.8
9	30	77
9	36	80
12	18	60
12	24	66.7
12	30	71.5
12	36	75

Example 2.1: Design a bar screen chamber through which maximum, average and minimum rates of flow are respectively 12 MLD, 10 MLD and 6 MLD (which corresponds to flows of 300 1/capita/day, 250 1/capita/day and 150 1/capita/day respectively for a population of 40,000).

Solution:

1. Average rate of flow: 10 MLD

$$= \frac{10 \times 10^6 \times 10^3}{10^6 \times 24 \times 60 \times 60}$$

$$= 0.116 \ m^3/sec.$$

   Maximum rate of flow = 0.139 m^3/sec.

   Minimum rate of flow = 0.0695 m^3/sec.

2. It is assumed that the screen will be cleaned manually; so an inclination of the bars = 30° is provided. Assumed size of the bar is 50 mm × 6 mm, 6 mm being the dimension facing the flow.

   A clear spacing of 30 mm between the bars is provided.

3. Assumed velocity of flow normal to the screen is 30 cm/sec at average rate of flow.

4. Net submerged area of the screen opening required

$$= \frac{0.116 \ m^3 / \ sec}{0.3 \ m / \ sec} = 0.387 \ m^2$$

(which is equivalent to 387 cm² per MLD, and is less than that specified).

Assuming velocity of flow normal to the screen is 75 cm/sec at maximum rate of flow, net submerged area of the screen opening

$$= \frac{0.139}{0.75} = 0.185 \ m^2$$

Provided a net submerged area of 0.387 m².

5. Gross submerged area of the screen

$$= \frac{0.387}{0.833} \quad \text{(from table 2.1)}$$

$$= 0.467 \ m^2$$

6. Submerged cross-sectional area of the screen

$$= 0.467 \ m^2 \times \sin 30° = 0.2335 \ m^2$$

$\therefore$ Velocity of flow in the screen chamber$= \dfrac{0.116}{0.2335}$

$$= 0.497 \ m/sec.$$

So velocity in the screen chamber is more than the self cleansing velocity of 39.5 cm/sec.

Now providing 20 nos. of bars, the gross width of the screen, and thus of the chamber.

$$= 20 \times 0.006 + 21 \times 0.03 = 0.75 \ m$$

$\therefore$ Liquid depth $= \dfrac{0.2335}{0.75} = 0.312 \ m$

$\therefore$ Now providing a free board = 0.25 m, total depth of the channel = 0.312 + 0.25 = 0.60 m.

$\therefore$ Size of the channel provided: depth = 60 cm

width = 75 cm

Now the liquid depth found above corresponds to the average rate of flow. Under this condition the hydraulic radius,

$$R = \frac{0.312 \times 0.75}{2 \times 0.312 + 0.75} = \frac{0.2335}{1.376} = 0.17 \ m$$

But in Mannings' Formula, $v = \frac{1}{n} R^{\frac{2}{3}} S^{\frac{1}{2}}$

or

$$S^{\frac{1}{2}} = \frac{v.n}{R^{\frac{2}{3}}} = \frac{0.013 \times 0.497}{0.17^{\frac{2}{3}}} = 21 \times 10^{-3}$$

$$\therefore \ S = 420 \times 10^{-6} = 0.42 \times 10^{-3} = 1 \ in \ 2400$$

Now the head loss through the screen, h

$$= \beta \left(\frac{w}{b} \right)^{\frac{4}{3}} h_v \sin \theta$$

$$= 2.42 \left(\frac{6}{30} \right)^{\frac{4}{3}} \frac{(0.497)^2}{2 \times 9.81} \sin 30°$$

$$= \frac{2.42}{8.55} \times \frac{0.247}{2 \times 9.81} \times \frac{1}{2} = 1.785 \times 10^{-3} m$$

$$= 0.1785 \ cm \ (\text{when clean})$$

For half clogged screen, the width of the opening is assumed to be $b/2$ or 15 mm.

Under this condition, the head loss h,

$$= 0.1785 \times (2)^{4/3} = 0.45 \ cm.$$

So, head loss under the half clogged condition is also within the limit of 8 cm and hence the design is safe.

Here, liquid depth, $h = \dfrac{\text{Flow}}{\text{width} \times \text{max velocity}} = \dfrac{0.139}{0.75 \times v_m}$

$$\therefore \text{ Hydraulic radius } = \left[\cfrac{\cfrac{0.75 \times 0.139}{0.75 \times \upsilon_m}}{0.75 + 2\cfrac{0.139}{0.75 \times \upsilon_m}}\right]$$

$$= \cfrac{0.139 / \upsilon_m}{0.75 + \cfrac{0.37}{\upsilon_m}}$$

$$= \frac{0.139}{\upsilon_m} \times \frac{\upsilon_m}{0.37 + 0.75\, \upsilon_m}$$

$$= \frac{0.139}{0.37 + 0.75\, \upsilon_m}$$

Now from Mannings's formula: $v = \dfrac{1}{n} R^{\frac{2}{3}} S^{\frac{1}{2}}$

$$\therefore \upsilon_m = \frac{1}{0.013}\left[\frac{0.139}{0.37 + 0.75\, \upsilon_m}\right]^{\frac{2}{3}} \times 21 \times 10^{-3}$$

or $\dfrac{13}{21}\upsilon_m = \left[\dfrac{0.139}{0.37 + 0.75\, \upsilon_m}\right]^{\frac{2}{3}}$

Now this equation has to be solved by Trial and Error method.
1st Trial assume $\upsilon_m = 0.6$ m/sec.

$$\therefore \text{LHS} = 0.371$$

$$\text{RHS} = \left[\frac{0.139}{0.37 + 0.45}\right]^{\frac{2}{3}} = \left[\frac{0.139}{0.82}\right]^{\frac{2}{3}} = [0.1695]^{\frac{2}{3}}$$

2nd Trial assumed $\upsilon_m = 0.55$ m/sec.

$$\therefore \text{LHS} = 0.34$$

$$\text{RHS} = \left[\frac{0.139}{0.37 + 0.412}\right]^{\frac{2}{3}} = 0.315$$

3rd Trial assumed $\upsilon_m = 0.5$ m/sec.

$$\therefore \text{LHS} = 0.3095$$

$$\text{RHS} = \left[\frac{0.139}{0.37 + 0.375}\right]^{\frac{2}{3}} = 0.326$$

4th Trial assumed $\upsilon_m = 0.52$ m/sec.

$$\therefore \text{LHS} = 0.322$$

$$\text{RHS} = \left[\frac{0.139}{0.37 + 0.39}\right]^{\frac{2}{3}} = 0.32$$

$\therefore \upsilon_m$ may be taken as equal to 0.52 m/sec, which is also within the limit.

Check for the head loss under the maximum rate of flow:

$$h_m = 2.42 \left(\frac{6}{30}\right)^4 \frac{(0.52)^2}{2 \times 9.81} \times \frac{1}{2}$$

$$= 0.195 \text{ cm (when clean)}$$

and $h_m = 0.195 \ (2)^{4/3}$

$$= 0.49 \text{ cm when half clogged.}$$

So both are within limits. Hence the design is safe.

2.2 GRIT CHAMBER

Grit chambers are designed to remove grit, consisting of sand, gravel, cinders, or other heavy solid materials that have specific gravities much greater than those of the organic putrescible solids in the waste water. This chamber is assumed to be one in which particles settle as individual entities, and where there is no

38

significant interaction with the neighbouring particles. This type of settling is referred to as Type I settling or free settling.

2.2.1 Theory of Type I Settling

An idealised rectangular continuous horizontal flow settling tank of this type may be divided into four zones as shown in the Fig. 2.2.

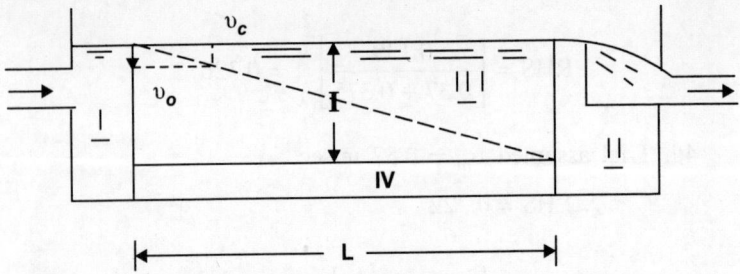

Figure 2.2: Idealized Rectancular Settling Tank.

In the above figure, Zone I is the inlet zone, which distributes the flow uniformly over the cross-section; Zone II is the outlet zone, which collects the clear water through overflow weirs; Zone III is the settling zone where settling occurs; and Zone IV is the sludge zone where sludge accumulates. L is the length of settling zone, H is the depth of the settling zone, υ is horizontal velocity of waste water through the tank, υ_o is settling velocity of the particles of specified specific gravity and size.

If υ_s = settling velocity of any particle, from the geometry, we can conclude that (1) particles having settling velocities $\upsilon_s \geq \upsilon_o$, will be fully removed, and (2) particles having settling velocities $\upsilon_s < \upsilon_o$ will be partially removed.

In the design of settling tank of this type, generally υ_o is specified, or is calculated for the given size of the particle and its specific gravity. In fact, υ_o should be the settling velocity of the particles smallest in size. This velocity can be expressed as flow or discharge per unit surface area of the tank, and is usually called as "Overflow Rate" or "Surface Settling Velocity" in the design of Settling Tanks.

From the above figure again, in order to have 100% removal of the particles, having settling velocity $\geq \upsilon_o$, we must maintain the following geometric relationship.

$$\frac{L}{H} = \frac{v}{v_o} \qquad \dots (2.2)$$

Now, to prevent the scouring of the deposited particles, the magnitude of v should not exceed a value, beyond which it will lift the deposited particles. If this magnitude of v is given by v_c, the equation 2.2 can be modified as:

$$\frac{L}{H} = \frac{v_c}{v_o} \qquad \dots (2.3)$$

Calculation of v_c: the horizontal velocity of flow within the tank or chamber, to start the motion of the particle, v_c, can be given by the following equation:

$$v_c = \sqrt{\frac{8\beta}{f} g\,(S-1)\,D} \qquad \dots (2.4)$$

in which $\quad \beta$ = constant factor, 0.04 for unigranular sand, and 0.06 or more for non-uniform sticky material,

$\quad f$ = Weisbach—Darcy friction factor, which depends on the characteristics of the surface, over which flow is taking place, and Reynold number; may be taken as 0.03 for settling tanks and grit chambers,

$\quad g$ = Acceleration due to gravity,

$\quad S$ = specific gravity of the particle to be removed,

and $\quad D$ = Diameter of the particle.

Now normally it is assumed that the smallest size of the particle that will be removed, will have a diameter of 0.2 mm and specific gravity 2.65; corresponding to these values, v_c = 22.7 cm/sec.

Calculation of v_s: Settling velocity of any discrete particle depends on its individual characteristics and also on the characteristics of the fluid. Assuming particles to be spherical, the settling velocity of any particle, v_s, can be given by the following formula:

$$v_s = \sqrt{\frac{4}{3}\frac{q}{C_D}\,(S-1)\,D} \qquad \dots (2.5)$$

in which C_D = Newton's Drag Coefficient

$$= \frac{24}{R} + \frac{3}{\sqrt{R}} + 0.34, \text{ for } 1 < R < 10^4$$

$$= 24/R, \text{ when } R < \frac{1}{2},$$

and $\quad v$ = Kinematic viscosity of the fluid.

For the value of $R < \frac{1}{2}$, $C_D = \frac{24}{R}$, the equation 2.5 reduces to:

$$v_s = \frac{g}{18}\left[\frac{S-1}{\rho}\right]D^2 \qquad \qquad \text{... (2.6)}$$

For the value of $R > \frac{1}{2}$, we have to find v_s by trial and error method.

2.2.2 Control of Velocity Through the Grit Chamber

For proper functioning of the grit chamber, the velocity through the grit chamber should not be allowed to change inspite of the change in flow. One of the most satisfactory types of automatic control is achieved by providing a proportional Weir at the outlet. The shape of the opening between the plates of a Proportional Weir is made in such a way that the chamber depth will vary directly as the discharge, as a result of which the chamber velocity will remain constant for all flow conditions. The Fig. 2.3. shows the end view of the grit chamber with the above mentioned velocity control device:

The flow through this weir, Q, can be given by the following formula:

$$Q = C.b\sqrt{2ag}\left(H - \frac{a}{3}\right) \qquad \qquad \text{... (2.7)}$$

in which, C = constant, 0.61 for symmetrical sharp edged weirs; a and b are the dimensions of the weir as shown in the above figure, and H = depth of flow from the level as shown in figure.

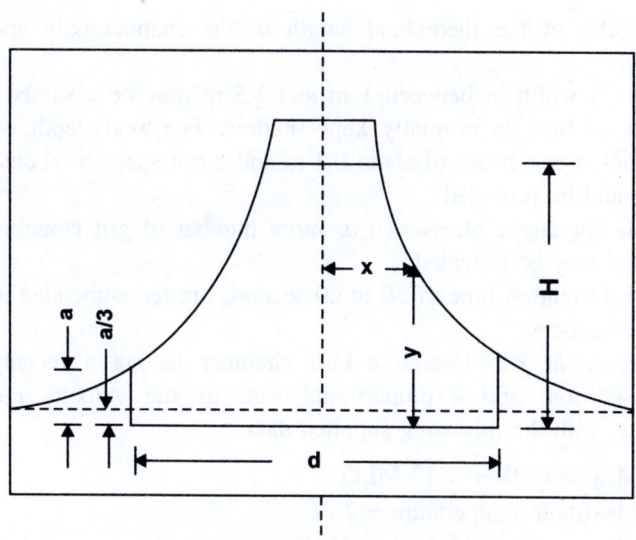

Figure 2.3: Proportional Flow Weir.

The dimension of 'a' may be from 25 mm to 50 mm; a suggested value of 'a' is 35 mm. As all other parameters except 'b' in the equation 2.7 is fixed, 'b' can be computed from this equation.

The equation of the curve forming the edge o f the weir is given by the following formula:

$$x = \frac{b}{2}\left(1 - \frac{2}{\pi} \tan^{-1}\sqrt{\frac{y}{a} - 1}\right) \qquad \ldots (2.8)$$

in which x and y are the coordinates of the points as defined in the above figure.

2.2.3 Design Criteria

1. The design of the horizontal-flow grit chambers should be such that under the most adverse conditions, all the grit particles of size 0.2 mm or more in diameter will reach the bed of the channel prior to their reaching the outlet end.

2. The length of the channel will be governed by the depth required which is again governed by the settling velocity. A minimum allowance of approximately twice the maximum depth should be given for inlet and outlet zones. An allowance of

20%-50% of the theoretical length of the channel may also be given.

3. A width in between 1 m and 1.5 m may be assumed. The depth of flow is normally kept shallow. For total depth of the channel, a free board of about 0.3 m and a grit space of about 0.25 m should be provided.

4. For larger plants two or more number of grit chambers in parallel may be provided.

5. Detention time of 30 to 60 seconds are recommended in the grit chambers.

Example 2.2: Design a Grit chamber having a rectangular cross-section, and a proportional weir as the velocity control device, with the following supplied data:

Maximum flow = 12 MLD

Maximum temperature = 26°C

Minimum termperature = 15°C

Specific gravity of grit particles = 2.65

Diameter of the smallest grit particles
to be removed = 0.02 cm

Solution: The settling velocity will be minimum at the minimum temperature. At the minimum temperature of 15°C, kinematic viscosity is 1.14×10^{-2} cm²/sec.

Assuming R = 0.5 in the first trial for v_s, from Eqn. 2.6,

$$v_s = \frac{gD^2}{18} \frac{(S-1)}{v} = \frac{981 (0.02)^2 \, 1.65}{18 \times 1.14 \times 10^{-2}}$$

$$= 3.15 \text{ cm/sec.}$$

$\therefore$ Reynolds Number, $R = \dfrac{vD}{v} = \dfrac{3.15 \times 0.02}{1.14 \times 10^{-2}}$

$$= 5.53 > 0.5$$

$\therefore v_s \neq 3.15$ cm/sec, as Eqn. 2.6 is valid for R < 0.5 2nd trial: assumed, $v_s = 2.4$ cm/sec.

$$\therefore \qquad R = \frac{2.4 \times 0.02}{1.14 \times 10^{-2}} = 4.21$$

$$\therefore \qquad C_D = \frac{24}{4.21} + \frac{3}{\sqrt{4.21}} + 0.34 = 7.50$$

$\therefore$ form Eqn. 2.5,

$$v_o = v_s = \sqrt{\frac{4}{3} \times \frac{981}{7.50} \times 1.65 \times 0.02} = 2.4 \text{ cm/sec.}$$

Hence O.K.

$\therefore v_s = 2.4$ cm/sec.

Now for $\beta = 0.06$, $f = 0.03$, and $D = 0.02$, from Equ. 2.4, v_c = 22.7 cm/sec.

Now $Q = 12$ MLD $= 12 \times 10^9$ cm^3/day $= 0.139$ m^3/sec.

$\therefore$ Cross sectional area, $A = \dfrac{0.139}{22.7 \times 10^{-2}} = 0.6125$ m^2

If a width of 1 m is provided, the liquid depth required, $H = 0.6125$ m.

Total depth of the chamber $= 0.6125 + 0.30$ free board $+ 0.25$ space for sludge accumulation $= 1.1625$ m $= 1.2$ m

Now $\dfrac{v_o}{v_c} = \dfrac{H}{L} = \dfrac{2.4}{22.7}$

$\therefore$ Theoretical length required, $L = \dfrac{22.7 \times 0.6125}{2.4}$

$$= 5.8 \text{ m}$$

Providing 25% allowance for inlet and outlet, total length of the chamber $= 1.25 \times 5.8 = 7.25$ m.

Proportional Weir:

Assuming $C = 0.61$, and $a = 0.035$ m in Eqn. 2.7,

$$b = \frac{0.139}{0.61 \sqrt{2 \times 0.035 \times 9.81} \times \left(0.6125 - \dfrac{0.035}{3}\right)}$$

$$= 0.46 \text{ m.}$$

The coordinates of the curve forming the edge of the weir may now be calculated using Eqn. 2.8 as shown in Table 2.2.

TABLE 2.2 Design of Proportional Weir

	y/a	$y/a-1$	$\sqrt{y/a-1}$	$\tan^{-1}\sqrt{y/a-1}$	$\frac{2}{\pi}\tan^{-1}\sqrt{y/a-1}$	$1-\frac{2}{\pi}\tan^{-1}\sqrt{y/a-1}$	$x/b=\frac{1}{2}(1-\frac{2}{\pi}\tan^{-1}\sqrt{y/a-1})$	b	a	x	y
1	0	0	0	0	0	1	0.5	.46	0.035	.23	.035
2	1	1	1	0.785	.5	.5	.25			.115	.07
3	2	2	1.415	0.955	.61	.39	.195			.0896	.105
4	3	3	1.732	1.047	.666	.334	.167			.0767	.14
5	4	4	2.00	1.107	.704	.296	.148			.068	.175
6	5	5	2.236	1.15	.73	.27	.135			.062	.21
7	6	6	2.449	1.183	.752	.248	.124			.057	.245
8	7	7	2.646	1.209	.77	.23	.115			.053	.28
9	8	8	2.828	1.231	.784	.216	.108			.0496	.315
10	9	9	3.00	1.249	.796	.204	.102			.0469	.35
11	10	10	3.162	1.265	.802	.198	.099			.0455	.385
12	11	11	3.317	1.278	.812	.188	.094			.0432	.42
13	12	12	3.469	1.29	.82	.18	.09			.0414	.455
14	13	13	3.606	1.3	.828	.172	.086			.0395	.49
15	14	14	3.742	1.31	.832	.168	.084			.0386	.525
16	15	15	3.873	1.318	.84	.16	.08			.0368	.56
17	16	16	4.00	1.326	.844	.156	.078			.0359	.595
18	17	17	4.123	1.333	.85	.15	.075			.0345	.63
19	18	18	4.243	1.34	.852	.148	.074			.034	.665
20	19	19	4.359	1.345	.856	.144	.072			.0331	.70
21	20	20	4.472	1.351	.86	.14	.07			.0322	.735

2.2.4 Alternative Device for Velocity Control in the Grit Chamber

Instead of retangular cross-section of grit chamber with proportional weir, we can use a parabolic cross-section of the grit chamber with fixed width control section. In practice of course, the parabolic cross-section can satisfactorily be approximated by a trapezoidal cross-section as shown in the fig. 2.4.

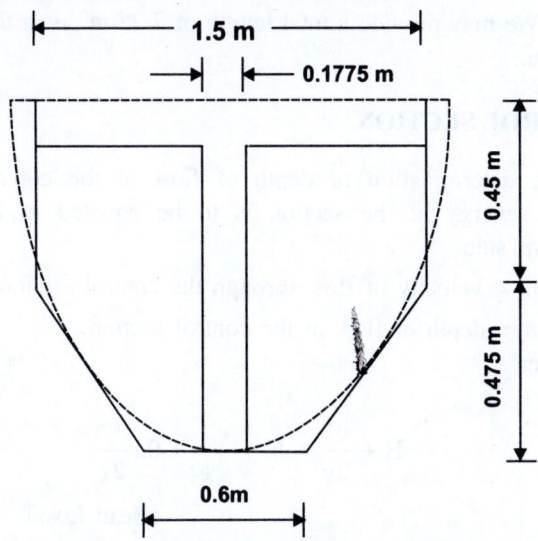

Figure 2.4: Grit Chamber of Parabolic Cross Section.

For the purpose of design of this type of grit chamber it is assumed that the head loss through the control section is 10% of the velocity head.

The procedure for the design is explained by the following solution of the previous example. (Example 2.2)

Solution:

Settling velocity of the particles v_s = 2.4 cm/sec.

The flow through velocity, v_c = 22.7 cm/sec.

Normal maximum flow = 0.139 m³/sec.

Cross sectional area required = 0.6125 m².

Now in a parabilic section, if H is the depth of flow and T is the top width, then area A = 2/3 H.T.

Let us provide T = 1.5 m

$$\therefore \ H = \frac{0.6125 \times 3}{2 \times 1.5} = 0.6125 \ m$$

∴ Theoretical length required,

$$L = \frac{22.7}{2.4} \times 0.6125 = 5.8 \ m$$

∴ We may provide a total length of 7.25 m, as in the previous example.

CONTROL SECTION

For the determination of depth of flow at the control section, specific energy at the section is to be equated to that in the upstream side.

If v' = Velocity of flow through the control section,

h' = depth of flow at the control section,

Then

$$H + \frac{v_c^2}{2g} = h' + \frac{v'^2}{2g} + 0.1 \frac{v'^2}{2g}$$

(Head Loss)

Now, all the control sections are critical sections, and in the

critical section, $h = 2 \dfrac{v^2}{2g}$

$$\therefore \ h' = 2.\frac{v'^2}{2g}$$

$$\therefore \ 0.6125 + \frac{(0.227)^2}{2 \times 9.81} = \frac{v'^2}{2g} \left(2 + 1 + 0.1 \right)$$

$$= \frac{3.1}{2g} \left(v' \right)^2$$

or $\dfrac{(v')^2}{2g} = 0.1985$ m

∴ depth in control section = 2 × 0.1985 = 0.397 m
∴ Velocity through control section, v'

$$= \sqrt{2 \times 9.81 \times 0.1985} \text{ m/sec.}$$

$$= 1.97 \text{ m/sec.}$$

∴ If w is the width of the control section,
$$Q = (w \times 0.397) \, 1.97 = 0.139 \text{ m}^3/\text{sec.}$$

$$w = \dfrac{0.139}{0.397 \times 1.97} = 0.1775 \text{ m}$$

Check for average flow condition:

Now as in critical section $h' = 2\dfrac{(v')^2}{2g}$, and $v' = \dfrac{Q}{wh'}$, the depth of flow at any condition can be given by:

$$h' = \sqrt[3]{\dfrac{Q^2}{w^2 g}}$$

∴ At average flow (0.116 m³/sec), depth of flow at the control section,

$$h' = \sqrt[3]{\dfrac{0.116 \times 0.116}{0.1775 \times 9.81 \times 0.1775}}$$

$$= 0.35 \text{ m}$$

Assuming velocity head to be very small in the channel upstream to the section, depth H of flow in the grit chamber

$$= \dfrac{3.1}{2} \times 0.35 = 0.544 \text{ m.}$$

Width of flow in the chamber = $Q/(v_c \frac{2}{3} H)$

$$= \frac{0.116 \times 3}{0.227 \times 2 \times 0.544}$$

$$= 1.41 \text{ m.}$$

∴ The adopted cross-section as shown in the Fig. 2.4 is safe.

2.3 Primary Sedimentation Tank (PST)

Effluent of the grit chamber, containing mainly light weight organic matter, is settled in the primary sedimentation tanks. In the addition

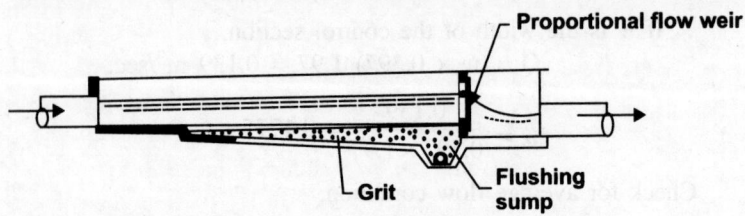

Figure 2.5: Grit Chamber.

to the settling characteristics, some other characteristics also interfere in the process of sedimentation. The organic matter tends to agglometrate during settling. After agglomeration they settle at a rate higher than that of the parent particles. As depth of liquid

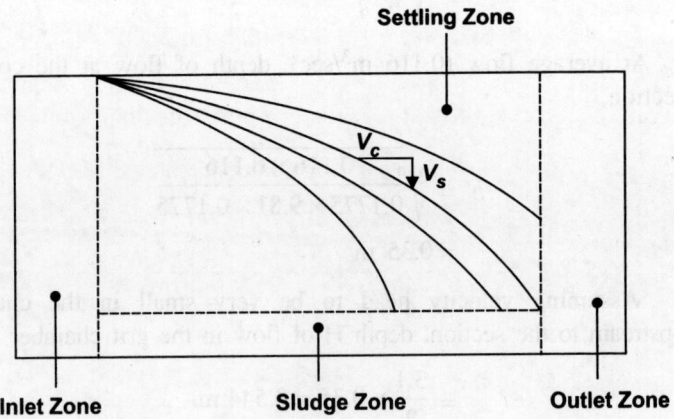

Figure 2.6: Settling Path of Particles in an Idealized Settling Tank, in Type-II Settling.

increases, chances of contact between the particles, and thus the size of the agglomerated particles increases which in turn makes the process of settling more rapid. This is true for any settlement of particles in a dilute suspension. The entire process is dependent on the range of particle sizes, the concentration of particles, mean velocity gradient in the system and may other known and unknown parameters. As such, the classical laws of sedimentation, which were briefly described in the last setion on the grit chamber design, will not apply to the sedimentation of organic solids in PST. Moreover, due to the involvement of many unknown parameters in the process of this type of settling (referrred to as Type II settling), no satisfactory formulation of this process is possible. Fig. 2.6 shows the typical plot of the settling path of particles subjected to the Type II settling, in an indealized settling tank.

In the design of the primary sedimentation tank, therefore, only an experimental settling velocity analysis will give reliable information, on the basis of which design can be done. In the absence of any actual tests, result, however, the conventional magnitude of the parameters, obtained through previous operational experience, may be assumed.

2.3.1 Experimental settling velocity anaysis

The test is usually conducted in test cyclinder of diameter not less 100 mm and of sufficient height, preferably more than the proposed height of sedimentation tank. The test cylinder should have at least three sampling ports for collection of samples at different time. The test is started with an uniformly mixed

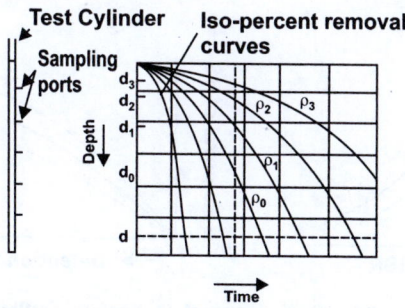

Figure 2.7: Experimental Settling Velocity Analysis.

suspension and is allowed to settle in quiescent condition. Samples are withdrawn from the ports at different time and are analysed for suspended solids concentration. The percent removal of suspended solids at different time and depth of cylinder are then calculated and plotted in a depth-time grid as shown in Fig. 2.7. "Iso-precent removal curves" are then drawn on the above grid.

Using the above curves, for a given detention time (t), over all suspended solids removal, in a clarifier of any given depth (d), may be computed as follows:

$$R = P \; p_0 + \frac{(d_o + d_1)}{2d_o} \Delta p + \frac{(d_1 + d_2)}{2d_o} \Delta p + \ldots\ldots$$

where, R = overall % suspended solids removal,

$$\Delta p = p_0 - p_1 = p_1 - p_2 = p_2 - p_s = \ldots\ldots\ldots$$

p_0, p_1, p_2, etc. = % removals as defined in Fig. 2.7.

d_0, d_1, d_2 etc. = depths as defined in Fig. 2.7.

(*Note:* Iso-percent removal curve, which is just above the point of intersection of horizontal line through d and vertical line through t, is designated by p_0).

Now d and t defines the overflow rate or surface settling rate (SSR). For different combination of d and t, i.e. for different SSR, % removal of suspended solids may be calculated and plotted as in Fig. 2.8(a).

In settlng tank design, the effective settling velocity or the surface settling rate, for a given percent removal, is determined by

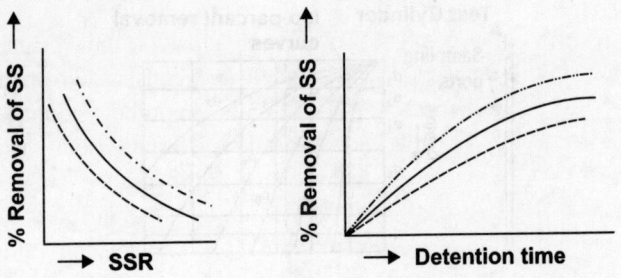

Figure 2.8: (a) % Removal vs Surface Settling Rate Curves
(b) % Removal vs Detention Time Curves.

dividing a depth of 1.8 m by the time required to attain such a removal at that depth. The firm line in the Fig. 2.8(a) shows a typical plot of % removal vs. surface settling rate curve. Similar plot of % removal vs. time of settling curve may also be obtained for the specific depth of 1.8 m, as shown by the firm line in Fig. 2.8(b).

Now, the flocculating effects of the suspended matter varies as the initial concentration in type II settling, and hence a series of similar tests are to be conducted with samples of different concentrations in the anticipated range of concentration of suspended solids of the waste water. Therefore, a set of curve similiar to that shown in Fig. 2.8(a and b) will be obtained. From this analysis a range of percent removal of suspended solids will be established corresponding to a specific surface settling rate and detention time. Usually a suspended solids removal of 50 to 60% in the primary sedimentation tanks is assumed.

The plots as shown in Fig. 2.8 give the surface settling rates and detention time for different percentage removals in a batch process. For the application of these values in a continuous flow settling tank, corresponding to the desired degree of treatment, the surface settling rates are to be reduced by a factor of 1.25 to 1.75, and the detention time is to be increased by a factor of 1.5 to 2.0.

When tank performances are expressed in terms of percent removal of BOD and COD, sets of curves similar to Fig. 2.8 may be obtained for BOD or COD, for the same laboratory batch process.

2.3.2 Design Criteria

(1) In the absence of any test reuslts, the primary sedimentation tanks in India are designed and constructed on the basis of the performance of these units in other countries, particularly in the USA. The Fig. 2.9 based on US data usually used to predict the performance of the primary sedimentaiton tanks in BOD removal corresponding to different surface settling rates. The primary sedimentation tanks which are followed by biological treatments are usually designed on the basis of 35% BOD removal in the primary treatment.

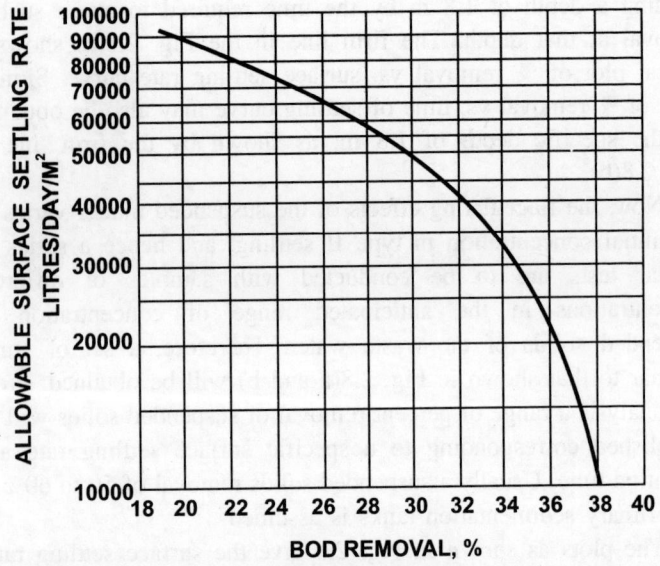

Figure 2.9: Efficiency of Primary Clarifiers in BOD Removal (Based on US Data)

However, the observations under Indian conditions (in Delhi, Punbjab and West Bengal) indicate that the efficiences of the units under similar surface settling rates are much higher, probably due to the nature of the suspended organic matter and the BOD contributed by them in domestic waste water in India. The expected efficiencies of primary sedimentation tanks under Indian conditions is shown graphically in Fig. 2.10, and may be used in selecting the surface settling rates. It may be noted that at a surface settling rate of 40,000 $1/m^2/day$ which corresponds to 32% BOD removal in U.S. conditions, the SS and BOD removals are 80% and 56.5% respectively under Indian conditions.

(2) The detention time and the flow through velocity are interrelated parameters. The effect of flow through velocity in settleable solid concentration in the effluent is shown in Fig. 2.11. As such, the flow through velocity of 3 cm/sec. is practically the maximum limit for the waste considered in the said figure. However, in normal plant practices, a flow through velocity of 1 cm/sec. at average flow is used. The primary sedimentation tanks

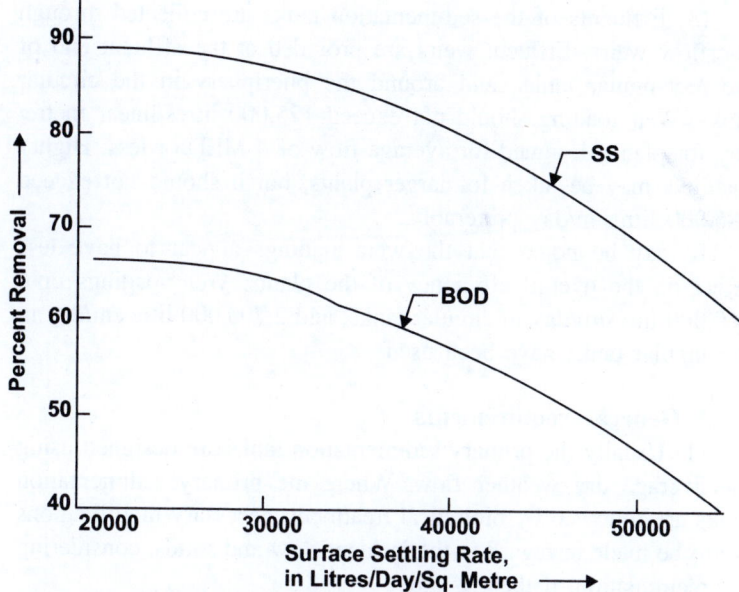

Figure 2.10: Efficiencies of Primary Clarifiers in BOD and Suspended
Solids Removal (Bhaskaran and Mathur)

may also be designed assuming a detention period in the range of
90 to 150 minutes at the average rate of flow.

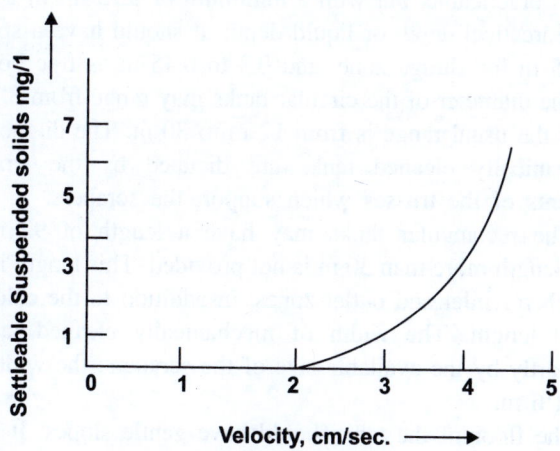

Figure 2.11: The Effect of Flow-Through velocity in the Settleable Solids
Concentration in the Effluent.

(3) Effluents of the sedimentation tanks are collected through overflow weirs. Effluent weirs are provided at the effluent end of the rectangular tanks, and around the pheriphery in the circular tanks. Weir loading should not exceed 125,000 litres/linear metre/ day for plants designed for average flow of 4 MLD or less. Higher loadings may be taken for larger plants, but it should not exceed 185,000 litres/m/day preferably.

It may be noted that the weir loadings appear to have less effect on the overall efficiency of the plants. Weir loadings upto 860,000 litres/m/day in circular tanks, and 2,700,000 litres/m/day in rectangular tanks have been used.

2.3.3 General requirements

(1) Usually the primary sedimentation tanks are designed using the average dry weather flow. Where the primary sedimentation tanks are followed by biological treatment, necessary modifications are to be made in regard to the influent flow and solids, considering the recirculation, if there is any.

(2) The number of tanks is determined mainly by the limitations of tank size. Normally all the treatment plants should have at least two tanks in parallel.

(3) The depth of mechanically cleaned tanks should be as shallow as practicable, but with a minimum of 2.15 m. In addition to this theorectical depth or liquid depth, it should have a space of about 0.25 m for sludge zone, and 0.3 to 0.45 m as free board.

(4) The diameter of the circular tanks may range from 3.7 m to 60 m, but the usual range is from 12 m to 30 m. The diameters of the mechanically cleaned tanks are dictated by the structural requirements of the trusses which support the scrapers.

(5) The rectangular tanks may have a length of 92 m, but usually a length more than 30 m is not provided. This length inludes 1.3 m each for inlet and outlet zones, in addition to the calculated theoretical length. The width of mechanically cleaned tanks is dictated partly by the available size of the scrapers; the width may be around 6 m.

(6) The floor of the tank should have gentle slopes. It should be around 1% and 8% respectively for ciruclar and rectangular tanks.

(7) The scrapers are attached to the rotating arms in the case of circular tanks, and to the endless chain in the case of rectangular tanks. These scrapers collect the solids in a central sump, and the solids are withdrawn regularly in the circular tanks. In rectangular tanks the solids are collected in the sludge hoppers at the influent end, and are withdrawn at fixed intervals. The volume of sludge produced in the tank will govern the design of facilities for removal of sludge.

In rectangular tanks, the minimum slope of side walls of the sludge hoppers shall be 1.7 vertical to 1 horizontal; the hopper floors shall have a maximum dimension of 0.6 m.

(8) Inlets for both rectangular and circular tanks are to be designed to distribute the flow equally across the cross-section, or, in all directions. There are several alternatives.

(9) The scum removal arrangement also is to be made ahead of the effluent weir on all primary settling tanks.

Example 2.3: Design a primary settling tank to handle an average rate of flow of 10 MLD.

Solution:
We assume, desired BOD removal in primary treatment units as 32%.

∴ From figure 2.9 surface settling rate = 40,000 1/m²/day.

∴ Required top surface area of the tank

$$= \frac{\text{flow}}{40,000} = \frac{10 \times 10^6}{40,000} = 250 \text{ m}^2$$

Let us assume, width = 6 m

∴ required theoretical length $= \dfrac{250}{6}$ m

The above theoretical length will make a total length of (41.66 + 1.3 + 1.3) m or 44.26 m, which is too large; also it is adviced to have at least two tanks in parallel.

∴ Proving two tanks in parallel, total length of each

$$= \frac{41.65}{2} + 1.3 + 1.3 = 20.83 + 2.6 = 23.43 \sim 23.5 \text{ m}$$

Now, $\dfrac{\text{Flow}}{\text{Surface area}} = \dfrac{\text{depth}}{\text{detention time}} = $ Surface settling rate

Let us provide a detention time of 1.5 hr.

$\therefore$ Liquid depth required $= 40,000 \times 10^{-3} \times \dfrac{1.5}{24}$

$$= 2.5 \text{ m.}$$

$\therefore$ Flow through velocity $= \dfrac{0.116 \text{ m}^3/\text{sec}}{2.5 \text{ m} \times 6 \text{ m}}$

$$= 0.773 \text{ cm/sec. Hence O.K.}$$

Total depth provided $= 2.5$ m $+ 0.3$ m as free board $+ 0.25$ m as the space for sludge accumulation $= 3.05$ m.

Effluent weir

Let us assume an weir loading of 185,000 1/m/day.

$\therefore$ total length of the weir required $= \dfrac{10 \times 10^6}{2 \times 185,000}$

$$= 27 \text{ m.}$$

Proving the side channel effluent weir, as shown in the figure 2.12(a) and with a channel width of 0.6 m, the dimension of *b* in the figure is given by:

$$6 + 2 (6 - 0.6 \times 2) + 2b = 27 \text{ m}$$

Solving $b = 4.65$ m

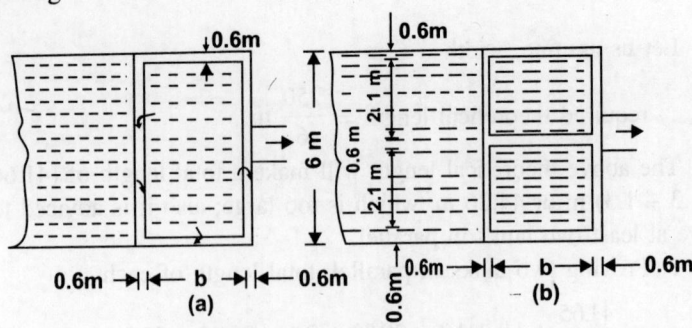

Figure 2.12: Effluent Weirs.

Alternatively the side channel effluent weirs can be arranged in the following way, as shown in Fig. 2.12(b):

Assuming channel width = 0.6 m, we get:

$6 + 4 \times 2.1 + 4b = 27$ m

solving, $b = 3.15$ m

Either of these arrangements may be adopted.

Design of sludge hopper

Providing two hoppers in each tank, the top width of each hopper = 3 m. Let us provide a slope of hopper: 2 vertical in 1 horizontal, as shown in Fig. 2.13, and 0.6 m × 0.6 m size of hopper floor.

∴ Depth of the hoppers

$$= 2 \left(\frac{3}{2} - \frac{0.6}{2} \right) = 2.4 \text{ m}$$

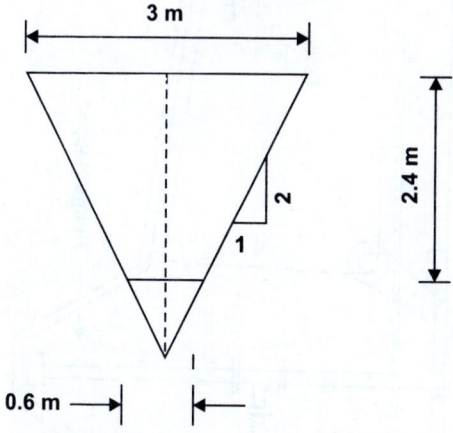

Figure 2.13: Sludge Hoppers.

Volume of each hopper

$$= 1/3 \left[(3)^3 - (0.6)^3 \right] = 8.93 \text{ m}^3$$

∴ volume of hopper per tank = 17.86 m³

58

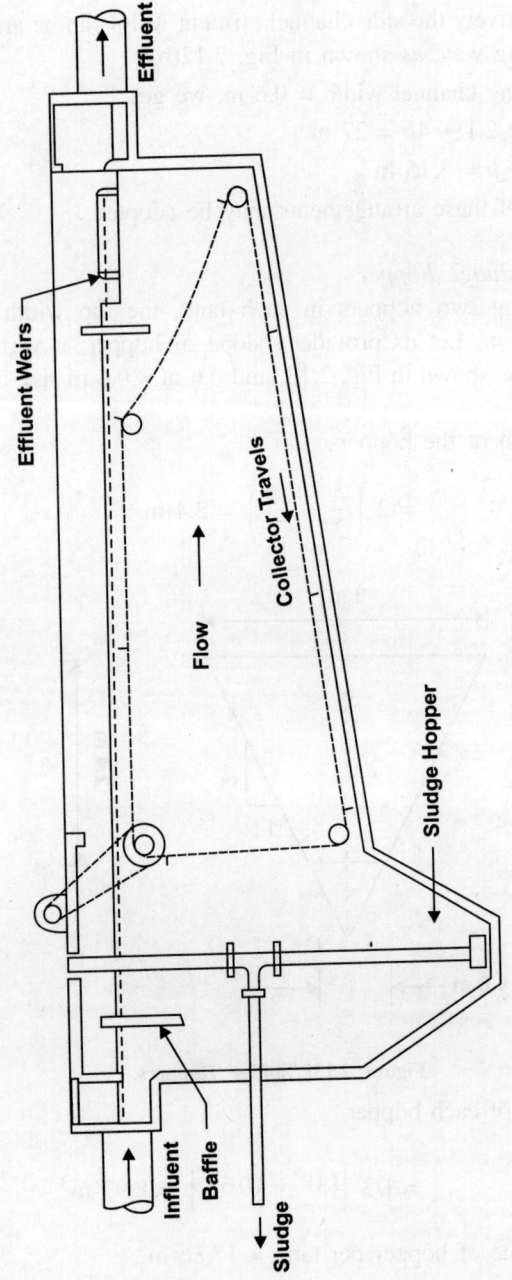

Figure 2.14: Primary Sedimentation Tank.

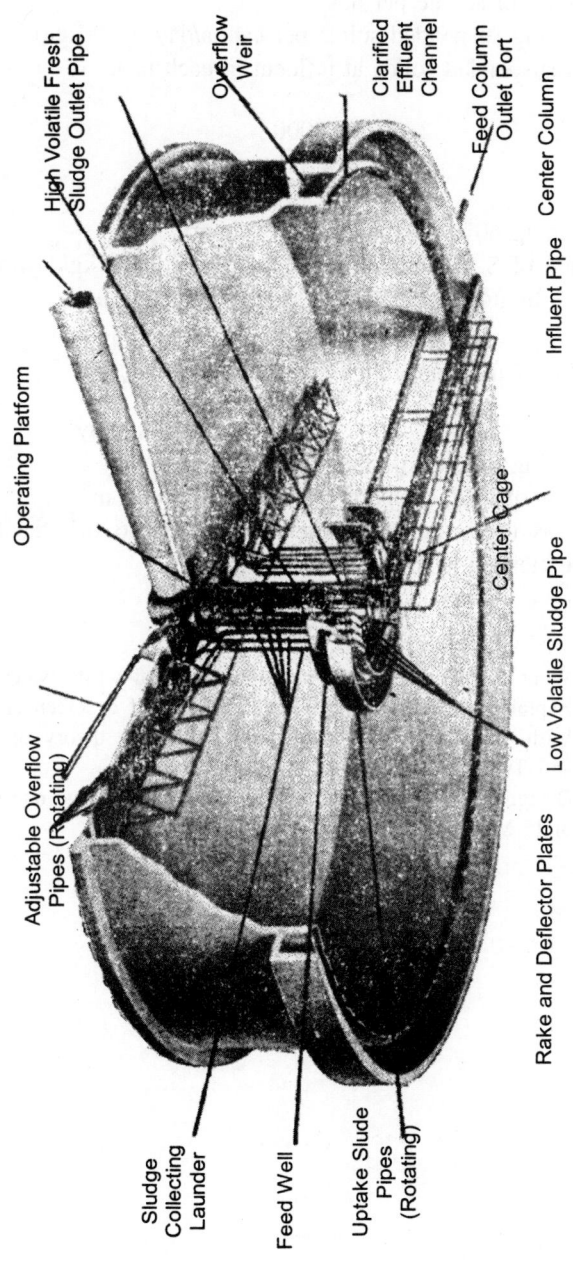

Figure 2.15: Dorr-Oliver Round Clarifier.

High Volatile Fresh Sludge Outlet Pipe

Overflow Weir

Clarified Effluent Channel

Feed Column Outlet Port

Influent Pipe

Center Column

Operating Platform

Center Cage

Low Volatile Sludge Pipe

Rake and Deflector Plates

Adjustable Overflow Pipes (Rotating)

Sludge Collecting Launder

Feed Well

Uptake Sludge Pipes (Rotating)

Quantity of sludge per day:

Assuming suspended solids per capita/day = 100 gm,

Total suspended solids at influent in each tank

$$= \frac{100 \times 40000}{10^3 \times 2}$$

Assuming 60% S.S. removal in primary,

Weight of S.S. removed = 0.6 × 2000 = 1200 kg/day/tank

Let us assume that sludge contains 96% moisture.

$$\therefore \text{ Wt. of the sludge/tank } = \frac{1200 \times 100}{(100 - 96)}$$

$$= 30000 \text{ kg/day.}$$

Assuming density of sludge = 1.0,

Volume of sludge/tank/day = 3000, litre = 30 m^3

Now as the volume of hopper/tank = 17.86 m^3, the hopper should be cleaned at the interval of 12 hrs.

QUESTIONS

1. What are the steps in primary treatment of waste water ?
2. Explain the steps involved in the design of a screen chamber.
3. With the help of neat sketches, explain the theory of Type I and Type II settling.
4. Design a primary settling tank to handle an average rate of flow of 15 MLD.

3 | Design Principles in Biological Treatment

3.0 INTRODUCTION

Biologically degradable organics may exist in waste water in soluble, colloidal or suspended form. Biological removal of degradable organics involves a sequence of steps including mass transfer, adsorption, absorption and biochemical enzymic reactions. An understanding of the microbiological metabolism and growth kinetics is therefore necessary for a rational design of the biological treatment processes.

3.1 MICROBIOLOGICAL METABOLISM IN AQUATIC ENVIRONMENT

Stabilization of organic substances by micro-organisms in a natural aquatic environment, or in a controlled environment of biological treatment process, is accomplished by two distinct metabolic processes: respiration and synthesis.

Respiration is a microbiological process in which a portion of the available organic or inorganic substrate is oxidized by the biochemical reactions, being catalyzed by large protein molecules known as enzymes produced by the micro-organisms, to liberate energy. The oxidation or dehydrogenation can take place both in aerobic and anaerobic conditions. Under aerobic conditions, the oxygen acts as the final hydrogen-acceptor for the oxidation; on the other hand, sulfates, nitrites, nitrates, carbon dioxide and even organic compounds act as hydrogen-acceptor in anaerobic conditions. Metabolic end products of the respiration are true inorganics like carbon dioxide, water, ammonia, methane, hydrogen sulphide etc. Fig. 3.1.

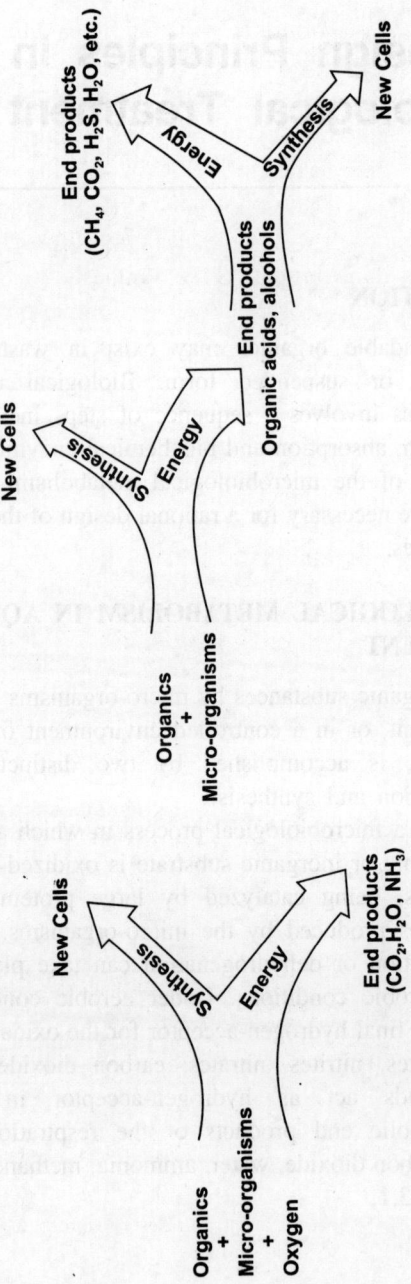

Figure 3.1: Metabolism of Organic Substrates.

The energy derived from the respiration, is utilized by the micro-organisms to synthesize new protoplasms through another set of enzyme catalyzed reactions, from the remaining portion of the substrate. The heterotrophic micro-organisms derive the energy required for cell synthesis exclusively through the oxidation of organic substances; on the other hand the autotrophic micro-organisms derive the energy for synthesis either from the inorganic substances or from light through photo synthesis.

The energy is also required by the micro-organisms for maintenance of their life activities. In the absence of any suitable external substrate, the micro-organisms derive this energy through the oxidation of their own protoplasms. Such a process is known as endogenous respiration. The metabolic end product of the endogenous respiration is same as that in primary respiration.

The complete process of liberation of energy and its utilization in the synthesis of new cells by the heterotrophic micro-organisms is shown schematically in Fig. 3.2.

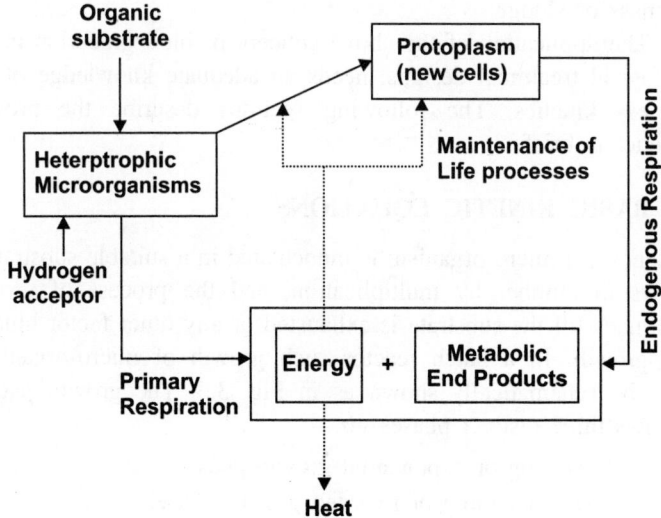

Figure 3.2: Schematic Representation of Heterotrophic Microbiological Metabolism.

64

Though the metabolic processes in both aerobic and anaerobic processes are almost similar, the yield of energy in an aerobic process, using oxygen as hydrogen acceptor, is much more than that in an anaerobic condition. This is the reason why the aerobic systems liberate more energy and thus produce more new cells than the anaerobic systems.

3.2 PRINCIPLES OF WASTE TREATMENT

The natural process of microbiological metabolism in aquatic environment is capitalized in the biological treatment of waste water. Under proper environmental conditions, the soluble organic substances of the waste are completely destroyed by biological oxidation, part of it is oxidized while rest are converted into biological mass, in the biological reactors. The end products of the metabolism are either gas or liquid; on the other hand, the synthesised biological mass can flocculate easily, particularly with increasing mean age of the cells, and are separated out in a clarifier. Therefore, the biological treatment system usually consists of (1) a biological reactor, and (2) a settling tank to remove the produced biomass or sludge.

The application of the above concept of biodegradation in the biological treatment designs, needs an adequate knowledge of the process kinetics. The following sections describe the process kinetics in brief.

3.3 BASIC KINETIC EQUATIONS

Whenever a micro-organism is innoculated in a suitable substrate it grows in number by multiplication, and the process of growth continues till the substrate is exhausted or any other factor hinders the growth. In a batch reactor such growth of micro-organisms may be schematically shown as in Fig. 3.3. The growth pattern follows three distinct phases *viz.*

      (*i*) log or exponential growth phase,
      (*ii*) declining or retarded growth phase,
and   (*iii*) endogenous growth phase or death phase.

In the log growth phase, the supply of the substrate is always adequate and the rate of metabolism is only dependent on the ability

65

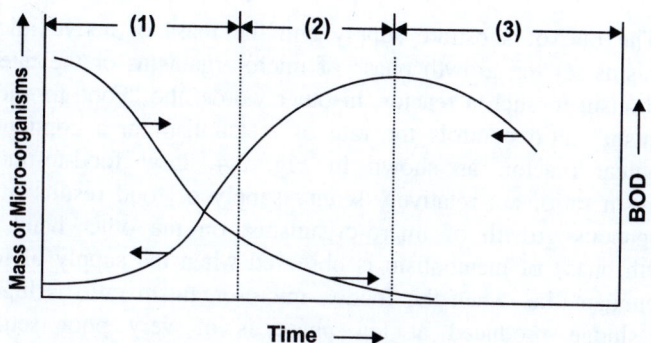

Figure 3.3: Microorganism Growth Pattern in Batch Reactors:
(1) Log Growth Phase, (2) Declining Growth Phase, (3) Endogenous
Growth Phase.

of the micro-organism to utilize the substrate. In the declining growth phase the rate of metabolism decreases due to the limitations in substrate supply. In the endogenous growth phase, the micro-organisms are forced to oxidize their own protoplasms for energy (endogenous respiration) and thereby decrease in number.

The growth pattern as shown in Fig. 3.3 is not applicable in a continuous biological reactor, where the substrate or food for the micro-organism is continuously supplied.

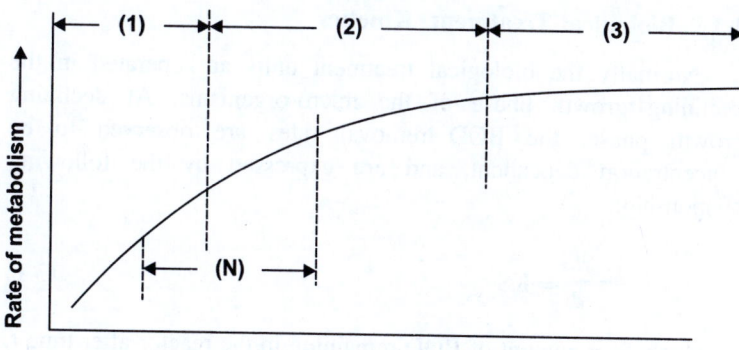

Figure 3.4: Rate of Microbiological Metabolism in Continuous
Biological Reactors: (1) Endogenous Growth Phase, (2) Declining
Growth Phase, (3) Log Growth Phase, (N) Normal Range of Operations
in Most of the Biological Reactors.

The rate of substrate supply and the mass of active micro-organisms set the growth phase of micro-organisms or the rate of metabolism in such a reactor. In other words, the "Food-to-micro-organism" ratio controls the rate of metabolism in a continuous biological reactor, as shown in Fig. 3.4. Low food-to-micro-organism ratio, i.e. relatively scanty supply of food results in an endogenous growth of micro-organisms; on the other hand log growth phase of metabolism is observed when the supply of food is abundant, i.e. when the food-to-micro-organism ratio is higher. The sludge produced at log phase is of very poor settling characteristics, and that in the endogenous phase not only settles well but it is also more stable in nature. As such, in all biological reactors, the system is so adjusted as to create a rate of metabolism ranging in between endogenous and declining growth phase, as shown in Fig. 3.4.

As evident from Fig 3.3, the formulation of a biological process is possible either in terms of changes in the substrate concentration or in terms of microbiological growth. In the analyses and design of a biological treatment system, any such formulations can be relied upon, depending on the availability of relevant data. However, the microbiological growth kinetics offer a more rational method of designing the biological treatment units.

3.3.1 Biological-Treatment Kinetics

Normally the biological treatment units are operated in the declining growth phase of the micro-organisms. At declining growth phase, the BOD removal rates are observed to be concentration dependent, and are expressed by the following relationship:

$$-\frac{dS}{dt} = KS \qquad \qquad \dots (3.1)$$

in which S = amount of BOD remaining in the reactor after time t, mg/l (ML^{-3}),

   t = contact time, days (T),

and   K = 1st order BOD removal rate constant, per day (T^{-1}).

It may be noted that the value of K may be based on either overall BOD_5 or only soluble BOD_5. Soluble BOD_5 removal rate constants are always higher than the overall BOD_5 removal rate constants. The value of K may vary with different types of wastes and environmental conditions. But for normal domestic waste it may be assumed to be equal to 0.23/day.

Integration of the Eqn. 3.1 between the limits $S = S_0$ at $t = 0$, and $S = S_1$ at $t = t$ yields:

$$\frac{S_1}{S_0} = e^{-kt} \qquad \qquad .... (3.2)$$

in which S_0 = initial concentration of BOD and S_1 = final concentration of BOD.

The Eqn 3.2 is a normal 1st order BOD removal rate equation and is applicable for a batch operation only.

The Eqn 3.1 is sometimes written in the following form, which takes into account of the mass of micro-organisms in the batch reactor:

$$-\frac{dS}{dt} = kXS \qquad \qquad .... (3.3)$$

in which, k = specific BOD (overall or soluble, as the case may be) removal coefficient, 1/mg/day, which equals to K/X, and   X= micro-organism concentration in the reactor, mg/1.

In most of the biological reactor designs, the concentration of the volatile suspended solids (VSS) in the reactor is taken as the concentration of the microbial mass. This assumption is true only when the waste under treatment is soluble in nature; but the estimate with such assumption may be two fold to five fold in an aerobic treatment, and upto two fold in anaerobic treatment, with domestic waste water, because of a large amount of inactive (i.e. non-biomass) volatile suspended solids present in it. However, since only the settled waste waters are treated, in most of the biological treatment units, the above approximation may not give rise to an identifiable error.

The value of k and K are generally reported for a standard reactor temperature of 20°C. These values at other temperatures may be calculated using the following relationship.

$$k_T = k_{20} \ \theta^{T-20} \qquad \qquad .... \ (3.4)$$

in which T = temperature in °C,

and θ = temperature coefficient.

Different values of θ have been reported by different workers for different temperature ranges. Unless otherwise stated, for temperatures not below 20°C, θ may be assumed as 1.047, in batch processes. In continuous flow biological reactors, however, the following values may be assumed for θ :

(i) Activated Sludge Process   —   1.00–1.03
(ii) Trickling Filter   —   1.02–1.04
(iii) Aerated Lagoon   —   1.06–1.09

3.3.2. Microbiological Growth Kinetics

In any low-substrate-concentration system, under proper environmental conditions, the mass of micro-organisms will tend to increase due to cell synthesis and decrease due to the endogenous respiration; the net rate of growth of bio-mass, dX/dt, may be given by the following relationship:

$$\frac{dX}{dt} = Y\frac{dF}{dt} - k_d X \qquad \qquad .... \ (3.5)$$

in which   X = concentration of the micro-organisms in the reactor, mg/1 (ML^{-3}),

  t = time of contact in the reactor, days (T),

  dX/dt = growth of micro-organisms in unit time per unit volume, $(ML^{-3} \ T^{-1})$, mg/1 /day.

  $\dfrac{dF}{dt}$ = rate of substrate utilization, mg/l/day $(ML^{-3} \ T^{-1})$.

  Y = growth yield coefficient, mg/mg,

and k_d = micro-organism decay coefficient, per day, (T^{-1})

Now, if S is the concentration of a soluble substrate in the reactor, for a normal biological reactor, where the substrate removal is mediated through the micro-organisms only, the rate of reduction of substrate concentration can closely be approximated to the rate of substrate utilization,

$$i.e, \qquad \frac{dS}{dt} = \frac{dF}{dt} \qquad \qquad .... (3.6)$$

The Eqn. 3.5, may be modified in the following form:

$$\frac{dX/dt}{X} = Y \frac{dF/dt}{X} = -k_a \qquad ... (3.7)$$

In Eqn. 3.7, the term $(dX/dt)/X$ is referred, to as the "specific growth rate" and is often symbolized as μ, and the term $dF/dt/X$ is referred to as the "specific substrate utilization rate".

Now in a batch process, if the substrate is supplied to the micro-organisms in excess, the specific substrate utilization rate remains constant under a particular set of substrate type, micro-organisms, and environmental conditions. However, when the concentration of the substrate becomes growth limiting, i.e. when the substrate supply falls short, the specific utilization and hence specific growth of micro-organism declines. The relationship between the substrate concentration and the specific utilization rate is usually given by the following continuous hyperbolic function, and is shown graphically in Fig. 3.5.

$$\frac{dF/dt}{X} = \frac{k_m S}{K_s + S} = \frac{dS/dt}{X} \qquad .... (3.8)$$

in which k_m = maximum rate of specific substrate utilization, per day (T^{-1}),

and K_s = substrate concentration at the utilization rate of $k_m/2$, mg/l (ML^{-3})

Now in normal biological reactors where the substrate concentration is usually low, i.e. $S \ll K_s$, the specific utilization rate approaches to $k_m S/K_s$

$$i.e. \qquad \frac{dF/dt}{X} = \frac{dS/dt}{X} = \frac{k_m S}{K_s}$$

$$or \qquad \frac{dS}{dt} = \frac{k_m}{K_s} XS \qquad .... (3.9)$$

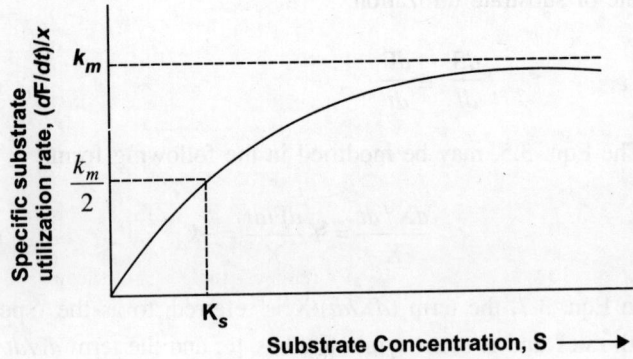

Figure 3.5: Effect of the Substrate Concentration on the Specific Substrate Utilization Rate.

Hence, comparing Eqns. 3.3 and 3.9, when the substrate concentrations are given by soluble BOD, in a low-BOD-process,

$$k = \frac{k_m}{K_s} \qquad \qquad .... \ (3.10)$$

For any general case, using Eqn. 3.8, the Eqn. 3.7 may be rewritten as follows:

$$\frac{dX/dt}{X} = \frac{Yk_m S}{K_s + S} - ka \qquad \qquad .... \ (3.11a)$$

$$= \frac{\mu_m S}{K_s + S} - ka \qquad \qquad .... \ (3.11b)$$

in which μ_m = maximum specific growth rate (μ) = $Y\,k_m$.

For the analysis of any biological process any one of the Eqns. 3.5, 3.7, and 3.11 may be used.

3.4 CONTINUOUS FLOW TREATMENT MODELS

The batch reactor kinetics, discussed earlier, are not directly applicable in the normal situations encountered in the waste treatment plant. However, these kinetic equations give an idea of the

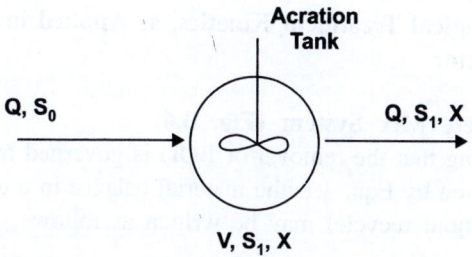

Figure 3.6: Complete Mix Continuous Biological Reactor.

pattern of substrate removal or microbial growth. Normally these equations are combined with the equations representing the hydraulic flow conditions of the reactor, to establish relationships between the process parameters and variables.

Depending on the hydraulic flow patterns within the continuous reactor, the reactors are classified as:

(*i*) Complete-mix type,

and (*ii*) Plug-flow type,

(*iii*) Partially mixed, or Arbitrary flow type.

In the complete-mix continuous reactors the influent substrate completely intermixes with the microbial content of the reactor, thereby producing an uniform composition throughout the reactor; as such, the composition of the effluent liquid is also the same as that in the reactor.

In plug-flow, the liquid flow leaves the reactor in the same order, in which it enters, i.e., each element of the influent liquid simply traverses along the length of the reactor without losing its identity, and behaves just like a moving unit batch reactor. In the arbitrary flow or partially mixed system, the substrate is uniformly dispersed, and some sort of mixing occurs in the reactor.

It may be noted that, the mathematical modelling of the arbitrary flow reactors is difficult, and most of the reactors in practice are analysed assuming either plug flow or complete mix condition. Moreover, most of the biological reactors are operated with some sort of recycling, and this is also to be considered while formulating a biological process in the reactor.

As in batch process, the continuous biological processes can also be formulated either in terms of biological treatment kinetics, or in terms of microbiological growth kinetics.

3.4.1 Biological Treatment Kinetics, as Applied in Continuous Reactor

(a) Complete Mix System (Fig. 3.6)

Assuming that the removal of BOD is governed by a 1st order equation given by Eqn. 3.1, the material balance in a complete-mix reactor (without recycle) may be written as follows.

$$\begin{bmatrix} \text{Rate of change of BOD} \\ \text{in the reactor} \end{bmatrix} = \begin{bmatrix} \text{Rate of BOD} \\ \text{inflow} \end{bmatrix} - \begin{bmatrix} \text{Rate of BOD} \\ \text{outflow} \end{bmatrix}$$
$$- \begin{bmatrix} \text{Rate of BOD} \\ \text{removal} \end{bmatrix}$$

or, $\quad \dfrac{dS}{dt} V = QS_0 - QS_1 - KS_1 V \qquad \qquad$ .... (3.12)

in which, $\quad Q$ = waste flow rate,

$\qquad \qquad V$ = volume of the reactor,

$\qquad \qquad S_0$ = influent BOD (Substrate) concentration,

$\qquad \qquad S_1$ = effluent BOD concentration,

and, $\qquad K$ = 1st order BOD removal rate constant,

Now in steady state, $dS/dt = 0$.

$\therefore$ From Equ. 3.12, $\quad \dfrac{S_1}{S_0} = \dfrac{1}{1 + K(V/Q)} \qquad$ .... (3.13)

In terms of specific BOD removal rate coefficient, k, the Eqn. 3.13 may be rewritten as follows:

$$\frac{S_1}{S_0} = \frac{1}{1 + kX(V/Q)} \qquad \qquad \text{.... (3.13a)}$$

in which X = microbial mass concentration in the reactor.

(b) Partially Mixed System

In practice, the hydraulic regime of most of the reactors is neither complete mix type nor plug flow type, but assumes an intermediate condition between the two. The hydraulic regime in the reactors of this type may be considered as plug flow with certain degree of dispersion. Wehner and Wilhem derived the following equation for the partially mixed chemical reactors:

$$\frac{S_1}{S_0} = \frac{4ae^{1/2d}}{(1+a)^2 e^{a/2d} - (1-a)^2 e^{-2a/2d}} \qquad \text{.... (3.14)}$$

in which, $a = \sqrt{1 + 4 K(V/Q) d}$,

d = dispersion factor (dimensionless) = D/UL
= D (V/Q) L^2,

U = fluid velocity,

D = axial dispersion coefficient,

and   L = characteristic length of travel path of a typical particle in the tank.

The above equation, though seems to be complicated may be used in the design of partially mixed Biological Waste Treatment Systems, with the aid of the design chart prepared by Thirumurthi (Fig. 3.7). The curves in the Fig. 3.7 give the value of the product K(V/Q), corresponding to the % BOD remaining (S_1/S_0), for different values of the dispersion factor, d.

However, as an approximation, the second term in the denominator of the Eqn. 3.14, which is small may be neglected, in which case the Eqn. 3.14 is simplified as:

$$\frac{S_1}{S_0} = \frac{4ae^{(1-a)2d}}{(1+a)^2} \qquad \text{.... (3.14a)}$$

It may be noted that the dispersion factor for the idealized complete mix system is infinite, while that for idealized plug flow system is zero. Dispersion factor in the long aeration tanks in the conventional activated sludge process ranges from zero to 0.2; the same in the complete mix aeration tanks ranges from 4 to infinity; and it varies in between 0.1 to 1.0 in most of the waste stabilization ponds.

3.4.2. Microbiological Growth Kinetics, as Applied in Continuous Reactor

(a) Complete-Mix-No Recycle System (Fig. 3.6)

In the analysis of these type of biological reactor, the following

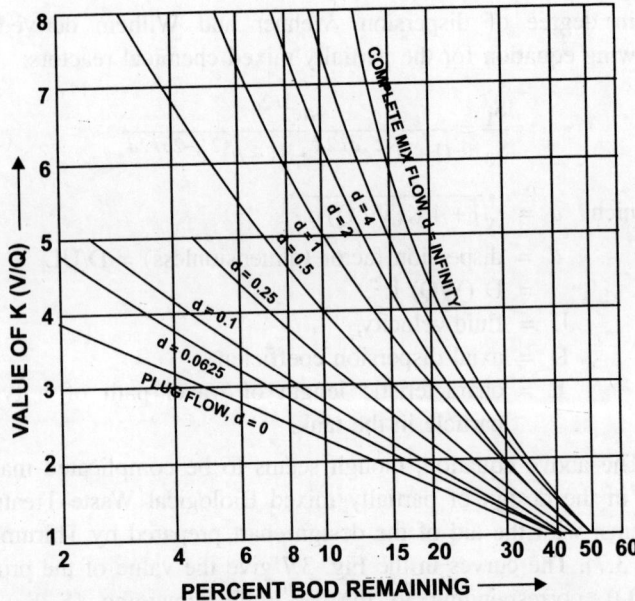

Figure 3.7: Design Formula Chart for Partially Mixed Reactors (Thirumurthi)

assumptions are made:

(*i*) Liquid waste flow into the reactor at a constant rate Q, and is instantaneously and homogenously mixed with the contents of the reactor.

(*ii*) the mixed liquor of the reactor is withdrawn at a rate equal to the rate of inflow Q,

(*iii*) influent does not contain any active microorganisms, and

(*iv*) the microorganism concentration in the effluent and that in the reactor are equal.

Under these conditions, the material mass balance across the system may be written as follows:

$$\begin{bmatrix} \text{Rate of Chage} \\ \text{of biomass in reactor} \end{bmatrix} = \begin{bmatrix} \text{Net Growth rate} \\ \text{of microbial mass in the reactor.} \end{bmatrix}$$

$$= \begin{bmatrix} \text{Rate of microorganisms} \\ \text{outflow from the reactor} \end{bmatrix}$$

or
$$\frac{dX}{dt} V = \left(Y \frac{dF}{dt} - k_d X \right) V - QX \qquad \text{.... (3.15)}$$

But as in a steady state, the microbial mass reaches a constant value,

i.e.,
$$\frac{dX}{dt} = 0$$

∴ the eqn. 3.15 reduces to:

$$\left(Y \frac{dF}{dt} - k_d X \right) V = QX$$

or
$$Y \frac{dF/dt}{X} - k_a = \frac{Q}{V} \qquad \text{.... (3.16)}$$

Now, the mean retention time of the microbial mass in the reactor, or "mean cell residence time", θ_c, is given by:

$$\theta_c = \frac{VX}{QX} = \frac{V}{Q} = \theta \qquad \text{.... (3.17)}$$

in which, θ = hydraulic retention time,

θ_c = mean cell residence time.

So, using Eqn 3.17, the Eqn 3.16 may be rewritten as follows:

$$\frac{1}{\theta_c} = Y \frac{dF/dt}{X} - k_a \qquad \text{.... (3.18)}$$

The specific substrate utilization rate, $(dF/dt)/X$ represents the "Food-to-micro-organisms ratio", U, or, the "process loading factor", when considered on a finite mass and time basis.

$$U = \frac{\Delta F / \Delta t}{X_M} \qquad \text{.... (3.19)}$$

in which $\Delta F / \Delta t$ = amount of food utilized "per unit time of Δt,

X_M = mass of the active micro-organisms in the reactor.

∴ Using Eqn. 3.19, Eqn. 3.18 reduces to the following form:

$$\frac{1}{\theta_c} = Y.U - k_d \qquad \text{.... (3.20)}$$

The Eqn. 3.20 is the basic or the controlling equation in any biological reactor of this type. Hence the process can be controlled by regulating either θ_c or U.

Using Eqn. 3.8, and solving Eqn. 3.18 for the effluent substrate concentration, S_1, the following relationship is obtained:

$$S_1 = \frac{K_s(1 + k_a\theta_c)}{\theta_c(Yk_m - k_d) - 1} \qquad \text{.... (3.21)}$$

The Eqn 3.21 completely defines the value of the effluent substrate concentration S_1, for a particular type of substrate with characteristic values of k_m, K_s, Y and k_d, and for a given value of control process parameter, θ_c.

Now in a complete-mix-no-recycle system assuming that the substrate removal is mediated only through the micro-organisms, dF/dt may be closely approximated to the substrate removal per unit volume of the reactor.

$$\frac{dF}{dt} = \frac{Q(S_0 - S_1)}{V} \qquad \text{.... (3.22)}$$

So using Eqn 3.22, and putting θ_c = V/Q in Eqn. 3.18, the following expression for X may be derived:

$$X = \frac{Y(S_0 - S_1)}{1 + k_d\theta_c} \qquad \text{.... (3.23)}$$

The Eqn 3.23 correlates the microbial mass concentration, X, with the mean cell residence time, θ_c, for the specific characteristic values of Y and k_d of a particular type of substrate, and desired degree of treatment. The mixed liquor volatile suspended solid concentration (MLVSS) may be taken as the microbial mass concentration, with the reservations as stated earlier in Section 3.3.1.

(b) Complete Mix-cellular Recycle System (Figs. 3.8 and 3.9)

The system consists of a biological reactor, followed by a Clarifier-Thickener unit, and a sludge recycle line, as shown in Figs. 3.8 and 3.9

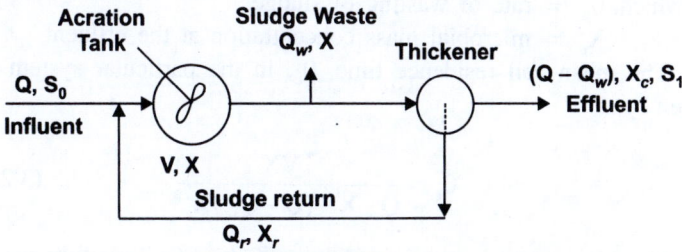

Figure 3.8: Complete Mix-Cellular Recycle System: Wasting is Accomplished from the Reactor.

The following additional assumptions are made in the analysis of the reactor of this type:

(i) all waste utilization occurs in the reactor only,

(ii) total microbial mass in the system is equal to the microbial mass in the reactor (i.e., the volume of the settling tank is small and the recycling is continuous).

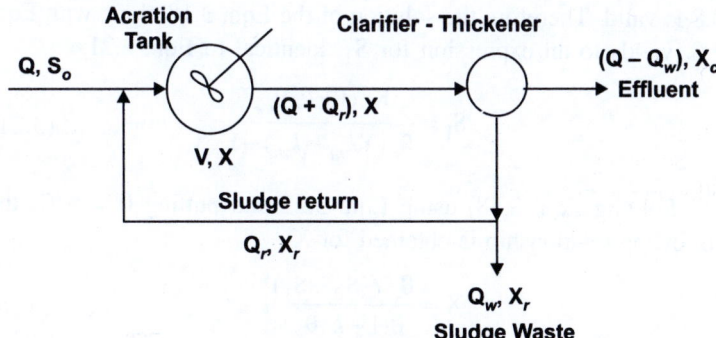

Figure 3.9: Complete Mix-Cellular Recycle-System: Wasting is Accomplished from the Recycle Line.

In the system as shown in Fig. 3.8, the mass balance of the microbial mass across the entire system is given by the following equation:

$$V\left(\frac{dX}{dt}\right) = \left(Y\frac{dF}{dt} - k_d X\right)V - (Q - Q_w)X_e - Q_w X \qquad .... (3.24)$$

in which, θ_w = rate of wasting of sludge,

X_e = microbial mass concentration at the effluent.

The mean cell residence time, θ_c, in the particular system is given by:

$$\theta_c = \frac{\overset{\bullet}{V}X}{Q_w.X + (Q - Q_w)X_e} \qquad .... (3.25)$$

Now for a steady state condition $dX/dt = 0$. Therefore using Eqn. 3.25, the Eqn. 3.24 reduces to the following form, which are identical to equations derived for complete-mix-no recycle system:

$$\frac{1}{\theta_c} = Y\frac{dF/dt}{X} - k_d \qquad .... (3.18)$$

or

$$\frac{1}{\theta_c} = YU - k_d \qquad .... (3.20)$$

Now even in a complete mix cellular recycle system, the Eqn. 3.8 is valid. Therefore the solution of the Eqn. 3.18 along with Eqn. 3.8, yields to an expression for S_1, identical to Eqn. 3.21.

$$S_1 = \frac{K_s(1 + k_d\theta_c)}{\theta_c(Yk_m - k_d) - 1} \qquad ....(3.21)$$

Solving Eqn. 3.18, using Eqn. 3.22 and putting $\theta = V/Q$, the following relationship is obtained for X:

$$X = \frac{\theta_c Y(S_0 - S_1)}{\theta(1 + k_d\theta_c)} \qquad .... (3.26)$$

It may be noted that, in the system as shown in Fig. 3.9, where the wasting is done from the recycle line, the mean cell residence, time, θ_c, is given by:

$$\theta_c = \frac{VX}{Q_w.X_r + (Q - Q_w)X_c} \qquad .... (3.27)$$

in which X_r = microbial mass concentration in the recycle line.

Q_w = rate of wasting of sludge from the recycle line.

The steady state equations, and the expressions for S_1 and X, remain unchanged in this case also.

Again, as the microbial mass concentration at the effluent, X_e, is very small compared to X, the Eqns. 3.25 and 3.27 may be reduced to the following simplified form, particularly for plant controls:

$$\theta_c = \frac{V}{Q_w},$$ when the wasting is done from the reactor directly,

.... (3.25a)

and $\theta_c = \dfrac{VX}{Q_w X_r}$, when the wasting is done from the recycle line

.... (3.27a)

The treatment process fails or becomes ineffective when the substrate concentration at the effluent becomes equal to that in the inlet, i.e. when $S_1 = S_0$. The critical mean cell residence time, $(\theta_c)_c$, required to make a treatment process effective, i.e. to prohibit the complete wash out, may be determined from Eqn. 3.21, after setting $S_1 = S_0$, and is given by:

$$\frac{1}{(\theta_c)_c} = \frac{Y k_m S_0}{K_s + S_0} - k_d \qquad \text{.... (3.28)}$$

(c) Plug Flow — Cellular Recycle System (Fig. 3.10)

The analysis of a plug flow-cellular recycle system becomes difficult, because of the varying nature of the flow and the microbial mass concentration along the reactor length. A plug flow reactor model assumes that no longitudinal mixing occurs between the contents of the reactor; as stated earlier, each element of the mixed liquor simply traverses along the length of the reactor without losing its identity, and behaves just like a moving unit batch reactor. In such a reactor, while the waste concentration decreases along the length, from influent towards effluent, the microbial mass concentration increases due to the utilization of the substrate by the micro organisms. But as the substrate removal and the microbial growth are interdependent, it is not possible to obtain an explicit

80

analytical solution for the system. However, the operational observations indicate that the increase of microbial mass in the reactor effluent over that in the influent is not very significant; this leads to a simplifying assumption that the microbial mass concentration remains constant (as long as $\theta_c/\theta > 5$). The rate of decrease of substrate concentration is assumed to follow a relationship similar to the rate of waste utilisation, given by Eqn. 3.8.

$$-\frac{dS}{dt} = \frac{dF}{dt} = \frac{k_m . S . \overline{X}}{K_s + S} \qquad \text{.... (3.29)}$$

in which $\overline{X}$ = average concentration of microbial mass in the reactor.

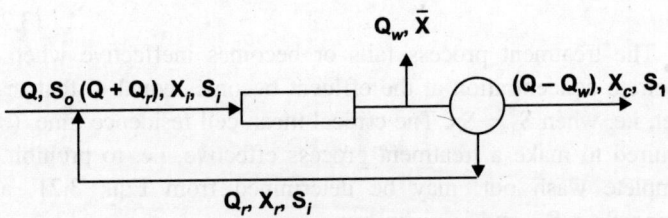

Figure 3.10: Plug flow-cellular recycle system.

Integrating Eqn 3.29, within limits $S = S_i$ at $t = 0$, and $S = S_1$ at $t = \theta$, the following relationship is obtained:

$$k_m \overline{X}\theta = (S_i - S_1) + K_s \ln(S_i / S_1) \qquad ...(3.30)$$

in which S_i = substrate concentration of the liquid after mixing the influent flow with recycled flow,

and θ = hydraulic retention time = $V/(Q + Q_r)$.

The S_i may be calculated using the following relationship;

$$S_i = \frac{QS_0 + Q_r S_1}{Q + Q_r}$$

The material mass balance across the entire system would yield an expression similar to that in the complete mix cellular recycle system, as given by Eqn 3.15. Now noting that the quantity dF/dt, on a finite time and mass basis, is equal to $Q(S_0 - S_1)/V$, the above expression is rewritten as follows:

$$\frac{1}{\theta} = \frac{YQ(S_0 - S_1)}{XV} - k_d \qquad \dots (3.31)$$

Now, substituting the value of θ, as given by $V/(Q+Q_r)$, in the Eqn 3.30, solving for $\overline{XV}$, and substituting in Eqn. 3.31 yield the following equation, for the value of $Q_r/Q \gg 1$:

$$\frac{1}{\theta_c} = \frac{Yk_m(S_0 - S_1)}{(S_0 - S_1) + K_s \ln(S_0 / S_1)} - k_d \qquad \dots (3.32)$$

As can be seen from Eqn. 3.32, the θ_c is a function of both S_0 and S_1 in a plug flow cellular recycle system. On the other hand, as apparent from Eqn. 3.21 the θ_c is function of S_1 only in the complete mix system (with or without cellular recycle).

It can be shown that the relationships between Q_r and θ_c, and that between $\overline{X}$ and θ_c are identical to those in complete mix system with recycle, where X only is to be replaced by $\overline{X}$.

$$\theta_c = \frac{V\overline{X}}{Q_w\overline{X} + (Q - Q_w)X_e},$$

and

$$\overline{X} = \frac{\theta_c}{\theta} \frac{Y(S_0 - S_1)}{1 + k_d\theta_c}$$

3.4.3 Comparison Between Complete Mix and Plug Flow Systems

It has been shown that the plug flow system is more efficient in the waste water treatment, than the complete mix system. On the other hand, while complete mix system can easily absorb shock loads, the other cannot sustain such type of loading. In practice, however, true plug flow conditions cannot exist; some degree of dispersion or back mixing occurs in the reactor. When a high degree of back mixing occurs in the tank, the condition

TABLE 3.1 : Summary of Steady State Relationships in Biological Reactors.

Parameters	Complete Mix System		Plug flow system with recycle
	without recycle	with recycle	
Effuent Substrate Concentration	$S_1 = \dfrac{K_s(1+k_d\theta_c)}{\theta_c(Yk_m - k_d)-1}$	$S_1 = \dfrac{K_s(1+k_d\theta_c)}{\theta_c(Yk_m - k_d)-1}$	(no explicit solution)
Microbial Mass Concentration	$X = \dfrac{Y(S_0-S_1)}{1+k_d\theta_c}$	$X = \dfrac{\theta_c\,Y(S_0-S_1)}{\theta(1+k_d\theta_c)}$	$X = \dfrac{\theta_c}{\theta}\,\dfrac{Y(S_0-S_1)}{1+k_d\theta_e}$
Excess Microbial Mass Concentration	$\dfrac{QY(S_0-S_1)}{1+k_d\theta_c}$		$\dfrac{QY(S_0-S_1)}{1+k_d\theta_e}$
Mean Cell Residence Time, θ_c	$\dfrac{1}{\theta_c} = \dfrac{Yk_m S_1}{K_s+S_1}-k_d$ or $\dfrac{1}{\theta_c} = YU - k_d$	$\dfrac{1}{\theta_c} = \dfrac{Yk_m S_1}{K_s+S_1}-k_d$ or $\dfrac{1}{\theta_c} = YU - k_d$	$\dfrac{1}{\theta_c} = \dfrac{Yk_m(S_0-S_1)}{(S_0-S_1)+\ln(S_0/S_1)}-k_d$

approaches a complete-mix reactor. Attempts have been made to analyse a dispersed-plug flow model. But observations from several studies suggest that equations for a complete mix system may be used for plug flow system without significant difference; this would be a conservative procedure, as any deviation from the complete mix towards plug flow would result in a somewhat higher effluent quality (Lawrence and McCarty).

3.5 OXYGEN REQUIREMENT IN AEROBIC PROCESSES

The BOD removal in a biological reactor is accomplished not entirely by oxidation—a part of the BOD is removed from the system, without being oxidized, in the form of wastage of excess sludge (synthesized biomass) from the system.

So, the oxygen requirement will be equal to the amount that would have been required, if all the BOD was removed by oxidation only (i.e. the ultimate BOD removed), less, a credit for the fraction of BOD removed by sludge wasting. Therefore one can write:

(Rate of oxygen usage) = (Rate of BOD_L removal)—f (Rate of excess sludge wasted),

in which, f is a function that denotes the amount of oxygen "saved" per unit weight of sludge removed from the system. Based on the stoichiometry of cell oxidation, it has been shown that f has a constant value of 1.42 kg of O_2 per kg of cells.

Therefore, on a finite time and mass basis:

O_2 requirement/day = Ultimate BOD removed/day — 1.42 (Excess sludge wasted per day) .... (3.33)

Now assuming BOD rate constant = 0.23/day at 20°C,

$$\frac{\text{Ultimate BOD}}{5\,\text{day } 20^\circ\text{C BOD}} = \frac{1}{1-e^{0.23 \times 5}} = 1.47 \qquad .... (3.34)$$

Using Eqn. 3.34, the Eqn. 3.33 may be modified in the following form:

Weight of oxygen required/day
= (5 day 20°C BOD removed/day) 1.47 − 1.42 (excess sludge wasted/day)
= 1.47 Q $(S_0 - S_1)$ − 1.42 V (X/θ_c) .... (3.35)

in which   Q = rate of inflow of waste,

S_0, S_1 = Substrate concentration at the inlet and outlet of the system respectively, BOD_5 in mg/l.

V = Volume of the reactor,

X = microorganism concentration in the reactor,

and θ_c = Mean cell residence time.

The Eqn. 3.35 may be used directly to calculate the theoretical oxygen requirement of the system. It may be noted that, the above expression is not materially different from that given by Eckenfelder and O'Connor, as presented below:

$$\text{Oxygen requirement/day} = Q(S_0 - S_1)\,a' + VXb' \qquad \text{.... (3.36)}$$

The Eqn 3.35 may be rewritten in the following form:
Weight of oxygen required/day.

$$= 1.47\ Q(S_0-S_1) - 1.42\ V\left(Y\frac{dF}{dt} - k_d X\right)$$

$$= Q(S_0-S_1)\,(1.47 - 1.42Y) + 1.42\ k_d XV, \qquad \text{.... (3.37)}$$

in which Y is expressed in mg of cell per mg of 5 day BOD. The terms (1.47-1.42 Y) and 1.42 k_d of the Eqn 3.37 are referred to as a' and b' respectively in the Eqn 3.36.

3.6 PRODUCTION OF SLUDGE

The rate of production of sludge in any reactor may be computed either directly from the Eqn 3.5, or using the concept of "mean cell residence time", as given below:

$$\frac{dX}{dt} = \frac{X}{\theta_c} \qquad \text{.... (3.38)}$$

in which dX/dt = net rate of change of microbial mass,

X = microbial mass concentration,

and θ_c = mean cell residence time.

Therefore, the quantity of microbial mass produced per unit time is given by:

$$\frac{dX}{dt}V = \frac{XV}{\theta_c} \qquad \text{.... (3.39)}$$

Now for a complete mix no recycle system, using Eqn. 3.23,

$$\text{Daily production of microbial mass} = \frac{QY(S_0 - S_1)}{1 + k_d \theta_c} \quad ....(3.40)$$

Similarly, for the complete mix-cellular recycle system, and the plug flow cellular recycle system, using Eqn. 3.26,

$$\text{Daily production of microbial mass} = \frac{QY(S_0 - S_1)}{1 + k_d \theta_c} \quad ... (3.41)$$

The above expression of microbial mass production represents the volatile suspended solids (VSS) production in the tank, from which suspended solid production and hence sludge production can be estimated.

The sludge produced, thus determined, is to be removed regularly to maintain the steady state condition. In conventional and high rate activated sludge process, the removal of excess sludge or wasting is done continuously. In extended aeration system, the wasting is done either continuously or at a time in each day. Again in Waste Stabilization Ponds and Aerated Lagoons the accumulated sludges are cleaned only after a long time.

QUESTIONS

1. Draw a sketch to describe microbiological metabolism in an aerobic process.
2. Compare complete mix system with complete mix-cellular recycle system.
3. Compare the steady state relationships in biological reactors.
4. How do you determine oxygen requirement in an aerobic process?

4 | Design of Conventional Biological Treatment Facilities

4.0 INTRODUCTION

Conventional biological treatment of waste water under aerobic conditions, include Activated Sludge Process and Trickling filter. The activated sludge process consists of a fluid bed reactor known as aeration tank, where the organic wastes are stabilized, a secondary settling tank where the biological mass is separated from the effluent, and the recycle line. On the other hand, the trickling filter consists of a fixed-film biological reactor, which *is* followed by a secondary settling tank. While recycling is essential for ASP, the trickling filters may or may not require the same. In this chapter, both the processes and their design methods will be discussed briefly.

4.1 ACTIVATED SLUDGE PROCESS

As stated earlier, the waste water is stabilized in this process biologically in the Aeration Tank, under aerobic conditions. The aerobic conditions are achieved by the use of diffused or mechanical aeration. The contents of the tank are referred to as Mixed Liquor. The biological mass, which is produced in this tank, is settled in the secondary settling tank. A portion of the settled biological mass, containing active micro organisms is recycled; the rest is wasted. The wasting of the sludge controls the mean cell residence time. A better control on the mean cell residence time can be achieved if wasting is done directly from the aeration tank.

For effective and quick separation of the biological mass, it is also important that the microbiological growth forms a satisfactory floc. It has been seen that increased mean cell residence time and adequate aeration enhances the settling characteristics of the biological floes.

The conventional activated sludge process consists of an aeration tank, a secondary or final settling tank, and a sludge recycle line. The aeration tank is having air diffusers on one side of the tank bottom to provide aeration and mixing. Settled raw sewage and the returned sludge enter the head of the tank, and cross the tank following a spiral flow pattern. The air supply is tappered along the length of the tank to match the quantity of air supplied to the oxygen demand exerted due to the microbiological activities.The effluent mixed liquor is settled in the settling tank and the 'sludge is returned at a desired rate. [Fig. 4.1(a)].

In the complete mix or high rate activated-sludge process, the influent settled sewage and return sludge flow are introduced at several points in the aeration tank from a central channel. The mixed liquor is aerated and completely mixed by diffused or mechanical aeration. The treated liquor is collected through effluent side channels and is settled in the secondary settling tank. [Fig. 4.1(b)]. In the complete mix process volumetric loading is normally very high and is approximately three times of that in conventional process ; the hydraulic retention time is also kept low in this process.

The contact stabilization, a modification of the ASP, provides an economical method of waste treatment, when a high percentage of BOD can be removed by "biosorption", within a short time. In this process, the first two phases in the biological treatment, viz., the biosorption and the stabilization are separated and made to occur in two separate tanks. In the first tank, as shown in Fig. 4.1(c), the waste is aerated with stabilized return sludge for a short contact period of 30-60 minutes. During this period the waste organics get adsorbed on the microbiological mass. The mixed liquor is then thickened in a secondary settling tank. The settled sludge is then transported to a sludge stabilizer, where the same is aerated, for a longer time (usually 3 to 6 hrs). The required stabilization period depends on the initial BOD removal and on the detention period of the contact tank. A portion of the stabilized

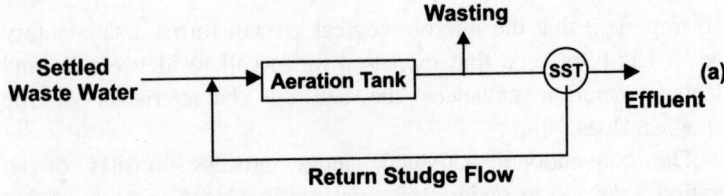

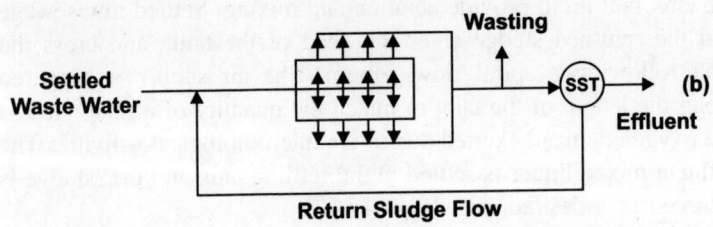

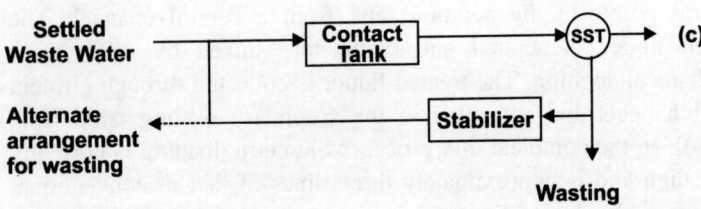

Figure 4.1(a): Conventional Activated Sludge Process, (b) High Rate Activated Sludge Process, (c) Contact Stabilization Process.

sludge is recycled; the rest is wasted.

Other modifications of activated sludge process, such as Extended Aeration Process, and Oxidation Ditch will be described in the next chapter.

4.1.2 Process Design Considerations

The effluent quality or the treatment efficiency can be controlled by controlling either the mean cell residence time, θ_c or the food-to-microorganism ratio (loading factor), U. In any system with recycle, both θ_c and U are independent of the hydraulic retention time of the tank and that of the system. Hence any degree of treatment can be achieved without altering the hydraulic retention time. However, it is easier to measure and control the value of θ_c, and as such the same is used in many rational designs, as a controlling process parameter.

In any system, θ_c can be given by the Eqn. 3.25(a) when the wasting is accomplished from the reactor, and by the Eqn. 3.27(a) when the wasting is done from the recycle line:

$$\theta_c \simeq \frac{V}{Q_w} \qquad \text{.... (3.25a)}$$

$$\theta_c \simeq \frac{VX}{Q_w X_r} \qquad \text{.... (3.27a)}$$

Therefore, the wasting of sludge containing microbial cells provides a direct method of control of θ_c.

The loading factor, U, which is also very often used, is directly related to θ_c, and their relationship can be given by the following equation:

$$\frac{1}{\theta_c} = YU - k_d \qquad \text{.... (3.20)}$$

The volume of the reactor, V, can be obtained by the following modification of the Eqn. 3.26, for a complete mix system:

$$XV = \frac{Q.\theta_c(S_0 - S_1)Y}{1 + k_d\theta_c} \qquad \text{.... (4.1)}$$

For a conventional aeration system (plug flow), X in the above equation is to be replaced by $\overline{X}$. It should be noted that, as the kinetic growth equations are applicable for soluble substrate only, S_0 and S_1 do not represent the total BOD of the influent and effluent respectively, but represent only the portion of the total BOD which is soluble. However, in the activated sludge process which treats settled waste only, total influent BOD may be considered as representating the influent soluble BOD. The BOD of effluent is attributed to that due to biological cells and that due to soluble organics, escaping the treatment. Therefore the effluent soluble BOD may be estimated after giving a credit to that due to biological cells. The ultimate BOD (BOD_L) of the waste cells may be taken as 1.42 mg/l.

The rate of production of excess volatile suspended solids (given by the mass of micro-organisms) is given by the following equation:

$$\frac{dX}{dt} = Y\frac{dF}{dt} - k_d X \qquad \text{.... (3.5)}$$

or directly by:

$$\frac{dX}{dt} = \frac{X}{\theta_c} \qquad \text{.... (3.38)}$$

∴ Total excess volatile suspended solids (VSS) production per unit time

$$= \frac{dX}{dt}.V = \frac{XV}{\theta_c} \qquad \text{.... (3.39)}$$

The above equation gives the mass of excess volatile suspended solids production, from which the total suspended solids production and hence the sludge production can be estimated.

The rate of return sludge flow, Q_r, depends on the volatile suspended solids concentration in the secondary settling tank underflow, X_r, and the mixed liquor volatile suspended solids (MLVSS), X, and is given by the following equation:

$$(Q + Q_r)X = Q_r . X_r, \qquad \text{.... (4.2)}$$

The oxygen requirement of the system can be estimated using the Eqn. 3.33:

wt. of O_2 required/day = (ultimate BOD consumed/day)
              −1.42 (Production of VSS/day).

The actual quantity of air to be supplied is estimated considering the fraction of oxygen in air, and the oxygen transfer efficiency of the aerators.

Volume of theoretical air requirement

$$= \frac{\text{(Weight of oxygen required)}}{\text{(Specific weight of air at standard temperature)} \times \text{(fraction of oxygen in air, by weight)}} \qquad \text{... (4.3)}$$

The specific weight of air, at Mean Sea Level, is 1.2 kg/m^3 at 20°C, and 1.16 kg/m^3 at 30°C. The fraction of oxygen in air is 23.2%.

Now, the volume of actual air requirement

$$= \frac{\text{(theoretical requirement)}}{\text{(oxygen transfer efficiency)}} \quad \text{.... (4.4)}$$

For porous tube diffusers, used in conventional activated sludge units, oxygen transfer efficiency may be assumed as 8%. The oxygen transfer efficiency for coarse bubble diffusers is around 6%.

4.1.2.1 Process parameters

The kinetic growth coefficients and the controlling process parameters for a particular waste are to be determined prior to any design of the treatment unit. However, in the absence of any actual test result, the Table 4.1 may be consulted for all normal domestic waste waters.

4.1.2.2 General requirements

(*i*) More than one tank is to be provided if total tank volume exceeds 150 m^3.

(*ii*) Normally liquid depth should be between 3 m and 4.3 m; a free board of 0.3-0.6 m is also to be provided.

(*iii*) Width to depth ratio may vary from 1 to 1, to 1 to 2.2.

(*iv*) Length may go upto 150 m. In a diffused air aeration-conventional system, the length is dictated by the air flow requirement, to some extent.

(*v*) A minimum air flow of approximately 0.3 m^3/min/metre length of tank is required for adequate mixing velocities and to avoid deposition of solids. (3 porous diffuser tubes can deliver a volume of air of 0.114-0.425 m^3/min/unit).

(*vi*) Air supplied should not be less than 62.50 m^3/kg of BOD removed.

4.1.3 Sludge Bulking

The biological sludge which does not settle well or settles and compacts poorly in the secondary settling tank, leaving a small amount of clear supermatant at the top, is called the bulking sludge. The bulking may be due to either (*i*) the growth of filamentous micro-organisms, which do not allow desirable compaction, or (*ii*) the production of non-filamentous highly hydrated biomass. There are many reasons for such bulking. The presence of toxic substances, insufficient aeration, frequent shock loadings and excessive rate of supply of substrate may also cause sludge bulking. In general, however, adequate supply of oxygen, and a proper design of the unit, so that the micro-organisms remain in the endogenous growth phase of their metabolism, is not expected to produce bulking sludges. Bulking of sludge due to the presence of toxic substances in the influent can be controlled by a suitable chemical treatment. In an activated sludge plant, treating an Antibiotic Plant waste, the bulking of sludge due to the filamentous growth were successfully controlled by treatment with bordex mixture (a mixture of copper sulphate and lime in the ratio 1: 1 by weight) at a dose of 10 to 20 mg/l.

Table 4.1: Design Specifications for Activated Sludge Process Systems

Parameters	Conventional ASP	Complete Mix ASP
1. Y, $\dfrac{\text{mg of VSS}}{\text{mg of BOD}_5}$	0.5-0.67	0.5-0.67
2. k_d, day^{-1}	0.056-0.01	0.055-0.07
3. θ_c, days	5-15	5-15
4. U, $\dfrac{\text{kg of BOD}_5}{\text{kg ML VSS}}$ per day	0.2-0.4	0.2-0.6
5. Volumetric loading, kg BOD$_5$ /1000m^3	320-640	800-1925
6. MLSS, mg/1 (MLVSS =80% of MLSS)	1500-3000	3000-6000
7. Hydraulic retention time, t, hr	4-8 (higher value for lower rate of flow)	3-5
8. Recirculation ration	0.25-0.5	0.25-1

Example 4.1: Design the Aeration Tank in a conventional activated sludge process using the same data as given in the previous example.

Solution:

Population = 40,000; Av. rate of flow = 10 MLD

$$= 0.116 \ m^3/sec.$$

Raw sewage suspended solids = 100 gms/capita/day.

$$= 4,000 \ kg/day.$$

Assuming 60% removal of S. Solids in primary,

Suspended Solids in Aeration Tank influent = 1,600 kg/day.

Raw sewage 5-day BOD = 76 gms/day/capita.

Assuming 32% BOD removal in primary, BOD in the

Aeration Tank influent = $0.68 \times 76 \times 40000/1000$

$$= 2065 \ kg/day.$$

Influent BOD concentration = 2065 × 1000000/

$$(10 \times 1000000) = 206.5 \ mg/1,$$

and influent suspended solids concentration = 160 mg/1.

Let overall efficiency of the plant desired be 95% and 90% respectively in the removal of SS and BOD.

$$\therefore \ \text{Effluent SS} = \frac{0.05 \times 100 \times 40000}{10000} = 20 \ mg/1$$

and effluent BOD = $0.1 \times 76 \times 40000 \times 10^{-4} = 30.2 \ mg/1$.

Assuming VSS in effluent = 80%, 5-day BOD of the effluent waste cells = (0.8 × 20) 1.42/1.47= 15.45 mg/1.

$\therefore$ 5-day BOD of the influent waste escaping treatment (or soluble BOD_5) = 30.2–15.45 = 14.75 mg/1.

Tank volume:

Assumptions:

  (*i*) Growth yield coeff. = 0.6

  (*ii*) Microorganism decay coeff. = 0.06,

  (*iii*) MLSS = 2000 mg/1; MLVSS

$$= 0.8 \times 2000 = 1600 \ mg/1,$$

  (*iv*) Mean cell residence time = 10 days.

$$\therefore \ \text{Tank volume} = \frac{YQ \ \theta_c (S_0 - S)}{\overline{X}(1 + k_d \ \theta_c)}$$

$$= \frac{0.6 \times 10 \times 10^6 \times 10(206.5 - 14.75)}{1600 \times (1 + 0.06 \times 10)} \text{ litres}$$

$$= 4500 \text{ cu. metres.}$$

Provide two units of tanks.

Total production of VSS per day $= (\overline{X}V)/\theta_c$

$$= \frac{1600 \times 4.5 \times 10^6}{10} \text{ mg/day}$$

$$= 720 \text{ kg/day}$$

Production of SS per day $= 720/0.8 = 900$ kg/day.

Wt of Oxygen required per day = (Ultimate BOD removal)—
(ultimate BOD of the waste cells)

$$= 1.47 \times (206.5 - 14.75) \times 10 - 1.42 \times 720$$

$$= 1920 \text{ kg/day}$$

Assuming specific wt of air = 1.15 kg/cu. metre, and oxygen % of the air = 23%,

Volume of air (theoretical) required $= \dfrac{1920}{1.15 \times 0.23}$

$$= 7275 \text{ m}^3/\text{day}$$

Assuming oxygen transfer efficiency of the porous diffuser tubes = 8%, Actual air requirement = 7275/0.08

$$= 91{,}000 \text{ m}^3/\text{day}$$

The supply of air will be tapered along the length of the tank.

50% of the air will be supplied in the 1st $\frac{1}{3}$rd zone, and 30% and 20% of the requirement in the next two equal zones.

Requirement of air in the last $\frac{1}{3}$rd zone/tank

$$= (0.2 \times 91{,}000) \, \frac{1}{2} \, \text{m}^3/\text{day} = 6.35 \text{ m}^3/\text{min.}$$

$\therefore$ Assuming the minimum rate of air flow in the last $\frac{1}{3}$rd zone, i.e., a rate of 0.3 m^3/min/meter length, total length of each tank required

$$= 3 \times 6.35/0.3 = 63.5 \text{ m.}$$

$\therefore$ Provide a length of 64 m

$\therefore$ Cross-sectional area of each tank

$$= 4500/2 \times 64 = 35 \text{ m}^2$$

Provide a liquid depth of 4.0 m and width of 8.75 m. Total depth of tank provided = 4.0 m + 0.5 m free board = 4.5 m.

If the wasting is accomplished from the tank effluent pipe,

The rate of sludge wasting $= \dfrac{V}{\theta_c} = \dfrac{4500 \text{ m}^3}{10} = 450 \text{ m}^3/\text{day.}$

If the wasting is accomplished from the return sludge line,

the rate of sludge wasting $= \dfrac{VX}{\theta_c X_r} = \dfrac{4500 \times 1600}{10 \times 8,000}$

$$= 90 \text{ m}^3/\text{day}$$

Assuming SS in return sludge =1%,

VSS in the return sludge = 0.8% = 8000 mg/1

Rate of return sludge flow, Q_r:

$1600(10 + Q_r) = 8000 \times Q_r$

or $\qquad Q_r = \dfrac{16000}{6400} = 2.5 \text{ MLD}$

$\therefore$ Recirculation ratio R = Q_r/Q = 0.25

Hydraulic retention time = t = V/Q = $\dfrac{4500 \times 10^3}{10 \times 10^6}$

= 0.45 days = 10.8 hours, which, is a bit high

The design can be revised taking a lower magnitude for θ_c.

Check for U, Volumetric loading and air requirement:

$$U = \frac{(206.5 - 14.75) \text{ mg} / 1 \times 10 \times 10^6}{1600 \text{ mg} / 1 \times 4.5 \times 10^6} = 0.266$$

$$\text{Volumetric loading} = \frac{191.75 \times 10 \text{kg}}{4500 \, m^3}$$

$$= 430 \text{ kg of BOD/1000 m}^3 \text{ of tank volume}$$

$$\text{Air requirement} = \frac{128,000 \text{ m}^3 / \text{day}}{191.75 \times 10 \text{ kg} / \text{day}}$$

$$= 66 \text{ m}^3/\text{day per kg of BOD removed.}$$

Hence the design is OK but for the higher hydraulic retention time, which can be brought down within limits, by revising the design with a lower value of mean cell residence time.

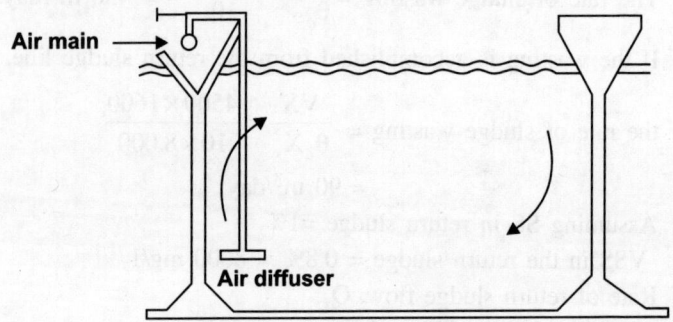

Air main →

Air diffuser

Figure. 4.2: Cross Section of a Typical Aeration Tank in Conventional Activated Sludge Process.

4.1.4 Secondary Settling Tank for Activated Sludge Plants

The final or secondary settling tank serves the dual function of clarification of liquid overflow, and the thickening of the sludge underflow. The allowable surface settling rate in such a tank has to be determined considering both the factors of clarification and thickening. The conventional design usually overlooks the requirement for the thickening. But it can be shown from the following equation (which is a modified form of Eqn 4.2) that the MLSS concentration is entirely dependent on the concentration of the solids in the underflow of the settling tank:

$$X = X_r \frac{Q_r}{Q + Q_r} \qquad \qquad \dots \ (4.5)$$

It may be noted that both the MLSS and the underflow concentration, C_u, are the functions of X and X_r respectively. So any failure on the part of the settling tank in the production of desired underflow concentration will lead to an unsatisfactory performance of the entire activated sludge process.

The settling characteristics of a concentrated suspension (like the mixed liquor of activated sludge process) are somewhat different from those of dilute suspensions. The type of settling which takes place in a secondary settling tank is referred to as "zone settling" and "compression". In zone settling the solid-liquid interface visibly subsides in the settling tank at a uniform rate, till the concentration at the interface approaches a "critical concentration". At this stage, the rate of subsidence of interface decreases due to the increased density and viscosity of the suspension, and a transition zone occurs. A compression zone occurs after that, when the solid particles come into physical contact with each other, and get compacted due to the weight of the sludge in the upper layers.

The behaviour of a particular sludge in both clarification and thickening can best be estimated through a laboratory batch settling test in a cyclinder.

4.1.4.1. Laboratory Batch Settling Test:

The test is started with a uniformly mixed suspension of height H_o and concentration C_o. Time required by the solid-liquid interface to cross each centimeter mark in the column are recorded, and are plotted in a height-time graph, as shown in Fig. 4.3.

The portion AB of the curve indicates hindered settling, BD represents transition zone, and DE represents compression. The transition zone is characterised by a point of compression C. This point can be estimated by bisecting the angle formed by the tangents at the hindered settling zone and compression zone.

It is assumed that the compaction of solids starts from the point of compression. Now it has been shown that the interface concentration at the "point of concentration" is a critical concentration, and that the solid flux (product of concentration and rate of subsidence) is minimum at this point. So this critical concentration and the corresponding rate of subsidence are the

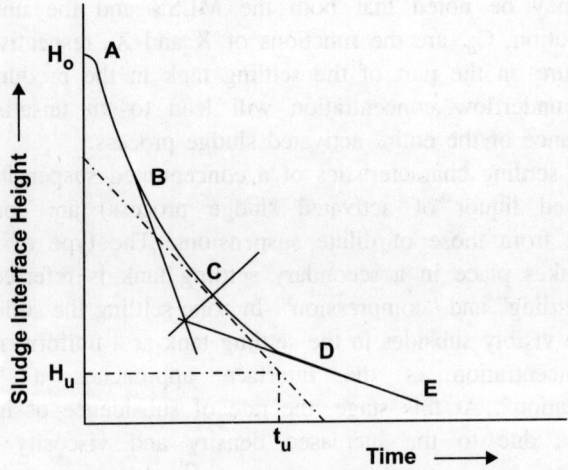

Figure. 4.3: Laboratory Batch Settling Test.

controlling factors in the estimation of the sizes of tanks required for thickening.

The surface settling rate required for hindered settling could be directly computed from the portion AB of the curve in Fig. 4.3.

The surface settling rate required for the desired thickening i.e., for attaining a desired underflow concentration, C_u, is computed from the following relationship:

$$\text{Surface settling rate} = \frac{Q}{A} = \frac{H_o}{t_u} \qquad .... \ (4.6)$$

in which Q = volumetric flow rate,

           A = required surface area,

           H_o = initial interface height in the test cylinder,

and t_u = time required to attain the desired underflow concentration, C_u.

The value of t_u can be determined by a graphical method given by Talmadge and Fitch. A height H_u is first determined from the following relationship, which gives the height of the sludge interface if all the solids in the system were at an underflow concentration of C_u:

$$C_o\, H_o = C_u\, H_u \qquad .... \ (4.7)$$

Figure. 4.4(a): Dorr-Oliver Low Pressure Aeration System.

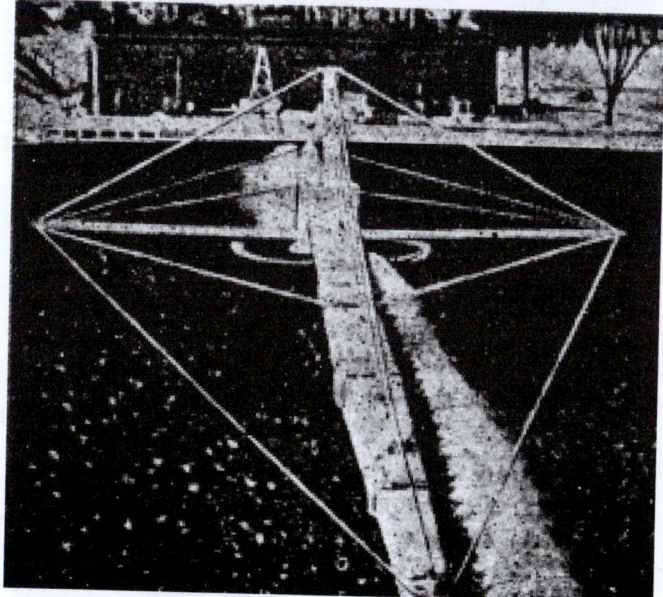

Figure. 4.4(a): Dorr Trickling Filter.

Now the intersection of the tangent drawn at the point of compression C (in Fig. 4.3), with the horizontal line at the height H_u establishes the value of t_u.

If the surface settling rate required for clarification is found to be lesser than that required for thickening, the clarification governs the design, and vice versa. At this stage this must be mentioned that, a flocculation effect in settling is observed with some activated sludges and hence this is to be taken into account while estimating the parameters for the design of clarifiers.

Since, the above process parameters are determined from a laboratory batch test, some scale-up factors will have to be used to convert these criteria to the continuous processes. Computed surface settling rate for hindered settling is reduced by a factor of 1.2 to 2.0, while that for thickening is reduced by a factor of 1.0 to 1.5.

In the absence of any actual test data the overflow rates (based on only the plant flow), as given in the Table 4.2 may be used for the design:

TABLE 4.2: Overflow Rates for Different MLSS and Recirculation Ratio

MLSS	Overflow rate, $l/m^2/day$ for recirculation ratio of	
	0.25	0.50
500	57000	57000
1000	57000	57000
1500	49000	49000
2000	49000	43000
2500	47000	39000
3000	39000	32500
3500	33500	28000

The following are the other requirements in a secondary settling tank of Activated sludge, process:

In small tanks the weir loading should not exceed 125,000 1/m/day at average flow or 250,00 1/m/day at maximum flow. For large tanks the loading should not exceed 370,000 1/m/days, at maximum flow, when the weir troughs are placed away from the upturn zone of the density current of the mixed liquour flow; when troughs are placed near the upturn zone the rate should not exceed 250,000 1/m/day, at maximum flow. Optimum position of the circular weir troughs to intercept well clarified effluent is at a distance of $\frac{2}{3}$ to $\frac{3}{4}$ of the radial distance from the centre.

Depth of the liquid below the overflow weirs should not be less than 3.65 m.

The flow-through velocity in a rectangular tank should not exceed 30.5 m/hr.

In circular tanks, the tank radius should not preferably exceed five times the side water depth. In rectangular tank the maximum length should not exceed ten times the depth.

4.2 TRICKLING FILTER

Trickling filters consists of a highly permeable filter bed, made up of rocks of 25 mm to 100 mm size, over which the sewage is uniformly sprayed. The depth of the bed varies from 1 to 2.5 m. Diameters are dictated to some extent by the mechanical equipments used for spaying of the sewage. Usually the sewage is sprinkled over the surface of the bed by means of the rotary distributors. The treated sewage is collected from the bed through an underdrainage system and is settled in the Final Settling Tank.

The filter media gets coated with' a biological film or slime layer, soon after its exposure to the sewage flow.

As the sewage is allowed to trickle over the surface of the filter media, the organic matter is adsorbed on the slime layer and is degraded by the aerobic microorganisms. But as the process continues the biological mass and hence the thickness of the slime layer increases; as a result, very near to the surface of the media, the condition in the layer becomes anaerobic; and due to the shortage of supply of food, the microorganisms there, reach . state of starvation. At this stage the microorganisms in contact with the media lose their ability to cling to the surface of the media; the flowing liquid then washes the slime off the media, and a new layer of slime starts to form. The phenomenon of losing this slime layer is refered to as "sloughing". The cycle of slime formation and its consequent sloughing is a function of the hydraulic and organic loading of the bed.

The overall performance of a Trickling filter will depend not only on its hydraulic and organic loading, but also on influent waste composition, pH, temperature, availablity of air, and many other factors. But as the biological slim layer is very unstable, and also it is difficult to predict the hydraulic characteristics, it is very difficult to derive a relationship between the different process variables and the filter performance.

Based on hydraulic and organic loadings, the trickling filters may be divided into two classes: (1) Low rate trickling filters and (2) High rate trickling filters. Their basic difference in respect to some important factors are tabulated in the Table. 4.3.

TABLE 4.3. Comparison between Low Rate and High Rate Trickling Filters

	Parameters	Low rate filter	High rate filter
(i)	Hydraulic loading, mL/hectars/day	9.35-37.5	93.5-37.5
(ii)	Organic loading, 5 day BOD in kg/m³/day	0.11-0.37	0.37-1.85
(iii)	Depth	1.8-3.0	0.9-2.4
(iv)	Recirculation	None	1: 1 to 4: 1
(v)	Volume of bed	5 times	1
(vi)	Sloughing	Intermittent	Continuous
(vii)	Effluent characteristics	Nitrified fully	Nitrification only at low loading.
(viii)	Nuisance potential	Many flies and chances of odour	Less odour and flies.

$$(1 \text{ mL/ha/day} = 10^2 \text{ Lit/m}^2\text{/day})$$

As observed from the Table 4.3 the recirculation is not essential in a trickling filter, but it is employed in high-rate filters to improve their efficiency. Though the recirculation helps in seeding the filter bed, the primary purpose of the recirculation is to dilute a strong waste and to create further opportunity of treatment of the filter effluent. This may also be employed to maintain a satisfactory operating rate for the rotary distributor during periods of low flow. Some times the recirculation is combined with secondary sludge removal, where the recirculated flow is returned to the primary inlet.

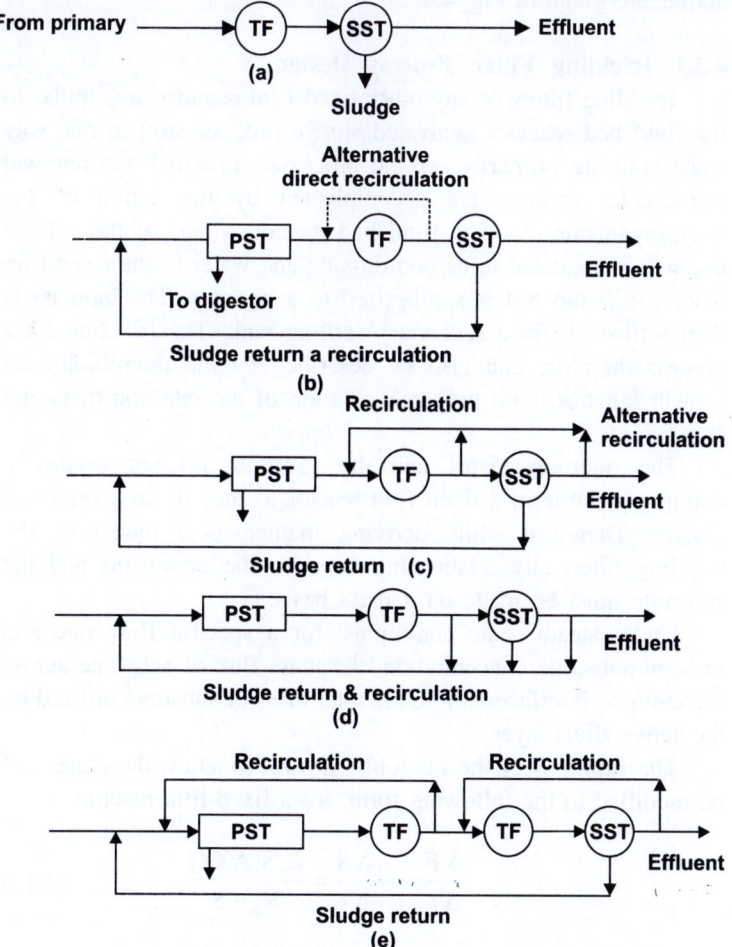

Figure. 4.4(a) Low Rate Trickling Filters, (b) High Rate Trickling Filter (for Lower Recirculation Ratio), (c) High Rate Trickling Filter (for Higher Recirculation Ratio), (d) High Rate Trickling Filter, (e) Two Stage Trickling Filters.

For the treatment of very strong wastes, some times two high rate trickling filters in series with or without an intermediate clarifier, are adopted. These are called two stage filters. But it has been found that the same degree of treatment can be obtained with a single stage activated sludge process, at a much lower cost.

Flow diagrams of different types of trickling filters, discussed above, are given in Fig. 4.4.

4.2.1 Trickling Filter Process Design

Trickling filters or any other fixed film reactors are similar to the fluid bed reactors (activated sludge process, say) in one way since both are primarily aerobic processes, in which the removal of soluble organics are accomplished by the action of the microorganisms. As in a fluid bed reactor, a part of the soluble organics is oxidized to carbon dioxide and water in the fixed film reactor, and the rest is synthesized to a biomass. The biomass is then settled out in a secondary settling tank. The trickling filter process therefore can also be described by the microbiological growth kinetics, with proper evaluation of the rate constants and coefficients.

The microorganisms and the substrate do not occupy a common volume in a fixed film reactor as they do in a fluid bed reactor. Therefore while deriving mathematical model of the trickling filter, any relationship between the organisms and the substrate must be made on a mass basis.

Under steady state conditions, for a specific flow rate and influent substrate concentration, the mass flux of substrate across slime-liquid interface will be the total mass of substrate utilized by the active slime layer.

The Eqn. 3.8 of the microbial growth kinetics, therefore, can be modified in the following form, for a fixed-film reactor:

$$\frac{\Delta F}{\Delta t} = -\frac{\Delta S}{\Delta t} = \frac{k_m S (AXd)}{K_s + S} \qquad \text{.... (4.13)}$$

in which $\Delta S / \Delta t$ = substrate removed per unit time and volume of reactor mg/day/unit volume,

k_m = maximum rate of specific substrate utilization,

$$\text{per day} = \mu_m/Y,$$

K_s = substrate concentration at the utilization rate of $k_m/2$, mg/1,

S = substrate concentration, mg/1,

A = specific surface area of the filter medium (area per unit volume of reactor),

X = concentration of the microorganism in the slime layer,

and $\quad d$ = thickness of the active slime layer.

It may be noted that, the entire mass of the microorganisms attached to the filter surface is not active in the removal of soluble organics. The effective film thickness, d, remains constant for a specified flow rate at steady state condition. The substrate removal is therefore directly related to the surface area of the filter media.

It should also be noted that for a specific applied organic loading, and increase in flow rate increases the mass flux of oxygen across the air-liquid interface resulting in the increased mass flux of oxygen across the liquid-slime interface, which again causes a greater depth of penetration of oxygen in slime layer, as a result there is an increase in the volume of active slime per unit surface area, and hence an increased rate of substrate utilization.

As pointed out earlier, the substrate removal is directly related to the surface area of the media. Therefore the mass flux of the substrate (across the slime-liquid interface) per unit area may be taken as a process parameter for comparisons between different systems. When the mass flux of the substrate is plotted against the substrate concentration, a family of curves is obtained for different flow rates, as shown in Fig. 4.5. The curves are different at different flow rates because of the dependence of the substrate utilization on the rate of flow, as mentioned earlier.

Now the maximum ordinate of the curve, for a specified flow rate, in Fig. 4.5 is the magnitude of $k_m Xd$. The magnitude of K_s, is obtained from the same plot, corresponding to the ordinate of $k_m Xd/2$. Thus, $k_m Xd$ and K_s, corresponding to a specified flow rate, for a particular type of substrate may be determined experimentally, and it will eliminate the determination of microorganism concentration in slime layer, X, and the thickness of the active slime layer, d.

Now the hydraulic regime in a fixed film reactor is true plug flow type. The substrate concentration in the liquid interface vary along the depth of the reactor. This variation may be linear in higher hydraulic loadings and nonlinear in lower hydraulic loadings. Again, the rate of substrate removal varies from zero-order at high concentration in the top layers to first-order at low concentration in deep layers. However, the continuous functions, as given in Eqn. 4.13, encompass all types of removal rate, which may occur in the fixed film reactor.

By integrating the Eqn. 4.13, within limits $S = S_0$ at $t = 0$,

and $S = S_1$ at $t = HD/Q$, the following relationship is obtained for the plug flow fixed-film reactor, without recirculation:

$$k_m (AXd) \frac{HD}{Q} = (S_0 - S_1) + K_s \ln (S_0 - S_1) \quad .... \quad (4.14)$$

or, $$(k_m Xd) \frac{AH}{Q} D = (S_0 - S_1) + K_s \ln (S_0 - S_1) \quad ..... \quad (4.15)$$

in which H = cross-sectional area of the filter bed,
        D = depth of the filter bed,
and   Q = hydraulic loading or rate of flow in the bed.

The Eqn. 4.14 or 4.15 provides a suitable relationship for the prediction of the trickling filter performance. However, it is assumed that the end product of the biological reactions within the reactor will not affect the rate of substrate utilization at the increased filter depths.

It may be noted that the term $k_m Xd$ in the LHS of the Eqn. 4.15 is the maximum mass flux of the substrate (mass removal per unit surface area of the media), the inverse of the term A/Q represents the rate of flow per unit specific surface area of the media, the inverse of the term AH/Q represents the rate of flow per unit specific surface per cross-sectional area of the bed (which may also be defined as the flow per unit wetted perimeter), and the inverse of the term AHD/Q represents the rate of flow per unit surface area. So, the rates of flow, as indicated in Fig. 4.5 may be expressed in terms of either rate of flow per unit specific surface area or rate of flow per unit wetted perimeter, or rate of flow per unit surface area.

In normal loading and depth ranges of trickling filters, there exists a continuous and almost linear function relating the substrate concentration and filter depth as will be obtained from the plot of Eqn. 4.15. In such cases, it is convenient to express the hydraulic loading in terms of either flow per unit surface area, Q/AHD, or flow per unit specific surface area of the media, Q/A. However, for greater depths, lower flow rates, and lower influent substrate concentration, the above continuous function may become non-linear. In such cases it is convenient to leave the term D apart as a dependent variable, and express the hydraulic loading in terms of the rate of flow per Unit wetted perimeter, Q/(AH).

As such, the rates of flow, in Fig. 4.5, are expressed in terms of flow per unit wetted'perimeter, as a general case. Where the efficiency of the process does not depend on depth, as in biological discs, the flow rates may conveniently be expressed as flow per unit surface area of the media.

In the rational design of a trickling filter it is therefore necessary to evaluate the values of K_s and $k_m Xd$ from the "mass

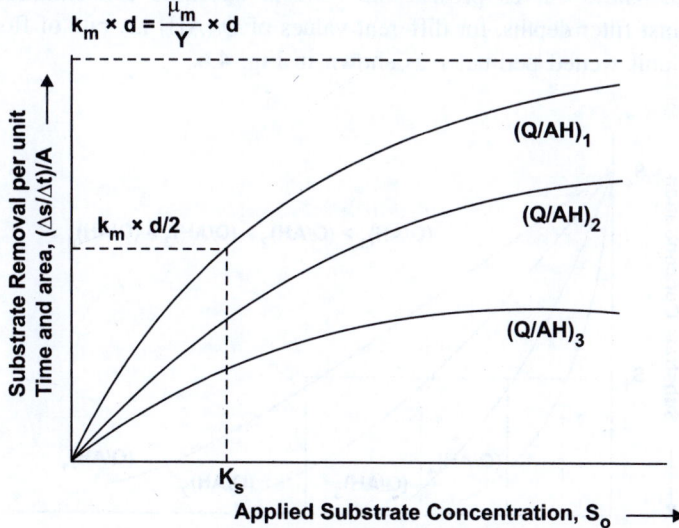

Figure. 4.5 Effect of substrate concentration on the mass flux of the substrate, at different flow rates.
[$(Q/AH)_1 > (Q/AH)_2 > (Q/AH)_3$]

108

flux of substrate Vs substrate concentration" curves, obtained experimentally in a laboratory scale trickling filter for different flow rates, as in Fig. 4.5. Usually, in the above curves the flow rates are expressed in terms of flow per unit time per unit wetted perimeter. Therefore, along with K_s, and $k_m X d$, the quantity $Q/(AH)$ gets fixed for a given system. But, for a particular system, the specific area, A, and the Q are constant. Thus the cross-sectional area can be determined directly from the assumed rate of flow per unit wetted perimeter. The remaining process variable, the depth of the bed, D, may be determined by a substitution of the already determined quantities, the values of influent substrate concentration, So, and the desfred effluent substrate concentration, S_1, in the Eqn. 4.15. However, under similar influent and effluent substrate conditions, and hydraulic loading, an alternative design can be made assuming different flow rates per unit wetted perimeter. It will give different values of K_s, and $k_m X d$ required for the design.

In order to have a freedom in the choice of the dimensions of the filter bed, under a specific influent substrate concentration and flow, a family of curves may be plotted using Fig. 4.5 and Eqn. 4.15. These curves present the effluent substrate concentrations against filter depths, for different values of $Q/(AH)$ i.e. rate of flow per unit wetted perimeter as shown in Fig. 4.6.

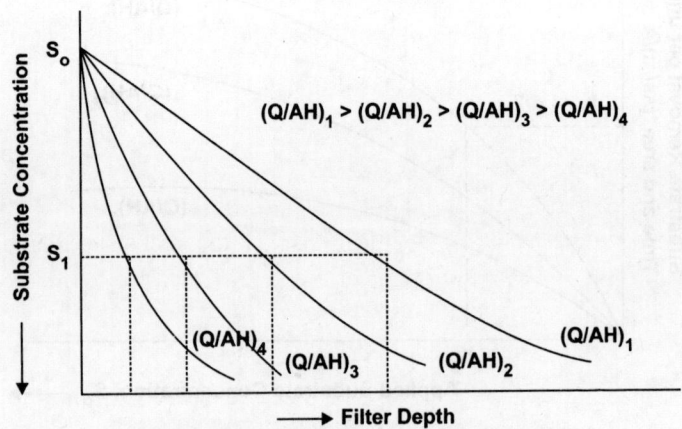

Figure 4.6 Relationship between Substrate Concentration and Filter Depth at Steady State.

With such a plot, for a given effluent substrate concentration, the selection of particular combination of Q/H and D will complete the design, since A is constant in a given system.

The design procedure discussed above may be extended to a high rate trickling filter, with recirculation of filter effluent. The flow rate and influent organic loading, however, are to be calculated accordingly.

It may be noted that the improvement in the efficiency of the process by recirculation is due to the initial dilution and more chances of contact of soluble organics, and probably due to the removal of a part of BOD in the flowing liquid phase.

The biological processes are usually controlled by means of either food-to-microorganism ratio (or process loading factor), U, or mean cell residence time, θ_c . But such a method of control in a fixed film biological reactor is not possible as both the above process parameters cannot be controlled externally. The wasting of microbial mass in a fixed film reactor is accomplished automatically and continuously in the form of sloughing. The food to microorganism ratio is dependent on the flow rate, as both the slime layer thickness and the mass of microorganism vary with the rate of flow under similar influent conditions.

In fact, for a particular influent substrate concentration, the selection of a particular combination of hydraulic loading, Q/H, and a corresponding filter depth, D, plays a key role in the control of the process in a fixed film reactor.

The specific surface area of the filter bed may be calculated using the following relationship:

$$A = \frac{6(1-f)}{\Psi d} \qquad \qquad .... \ (4.16)$$

in which, A = specific surface in m^2/m^3,
  d = geometric mean size of the rocks, used in filter bed in m,
  f = porosity of the bed,
and Ψ = spherecity of the rocks.

With normal range of the size of rocks in the bed (63 mm to 100 mm; with geometric mean size of 80.467 mm) and assuming 40% porosity, the surface area per unit volume of bed comes to about 50 m^2 per m^3.

4.2.2 Empirical Method of Design

The National Research Council of USA (NRC) has developed an empirical equation for trickling filter performance, based on some extensive studies of the past records. The NRC equation for a first stage or single stage filter is modified for SI units as follows:

$$VF = \frac{W}{5.08}\left(\frac{E_1}{1-E_1}\right)^2 \qquad \dots (4.17)$$

in which E_1 = fractional efficiency of BOD removal of the process, including recirculation and sedimentation,

W = BOD loading in kg/day,

V = Volume of the filter media in m^3,

F = recirculation factor,

$$= \frac{1+R}{\left(1+\dfrac{R}{10}\right)^2},$$

and R = recirculation ratio = $\dfrac{Q_r}{Q}$

For the second stage filter, the NRC equation can be written in SI units as follows:

$$VF = \frac{W'}{5.08}\left[\frac{E_2}{(1-E_1)(1-E_2)}\right]^2 \qquad \dots (4.18)$$

in which E_2 = fractional efficiency of BOD removal of the second stage filter,

and W' = BOD loading to second stage filter in kg/day.

The suggested values for different other process variables are given in the Table 4.3.

The diameter of the filter is dictated by the available size of the rotary distributors; but usually it does not exceed 30 m.

Many other empirical equations are also available for the design of trickling filters. One such equation as suggested by Eckenfelder is as follows:

$$\frac{S_e}{S_0} = \exp\left(-K_o' \frac{D}{Q_a{}^n}\right) \qquad \text{... (4.19)}$$

where S_e = effluent BOD, mg/l

S_o = raw water BOD, mg/l

K_o' = treatability factor, per day

D = depth of filter bed, m

Q_a = surface loading or liquid flow rate per unit surface area of media, m³/m²/day.

n = characteristics of particular filter media.

In high rate trickling filters, with recirculation ratio R, the above equation is modified as follows:

$$\frac{S_e}{S_o} = \frac{\exp(-K_o'D/Q_a{}^n)}{1+R-R\exp(-K_o'D/Q_a{}^n)} \qquad \text{... (4.20)}$$

Now the constants K_o' and n in Eqns. 4.19 and 4.20 can be determined by running a laboratory scale trickling filter for different surface loadings (Q_a) and plotting "ln S_e/S_0 vs D" curves for each Q_a values. Eqn 4.19 suggests that different straight lines will be obtained for different Q_a values, and slopes of these lines will be equal to $K_o' Q_a{}^{-n}$. A subsequent plot of "ln (slope) vs ln Q_a" curves, which are also straight lines, yields the value of n and K_0'. Reported values of n for 25 mm rock, 25 mm rounded gravel, 63 mm rock and 63 mm rounded gravel are 2.36, 3.00, 3.80 and 5.40 respectively; those for K_o' (at 20°C) are 0.74, 0.625, 0.645 and 0.57 respectively.

4.2.3 Secondary Settling Tanks for Trickling Filters

The function of secondary settling tanks in the trickling filters is to produce a clarified liquid. These tanks may be either rectangular or circular. The design of these tanks is similar to the design of primary settling tanks, except that the over-flow rate at peak flow should not exceed 49,000 l/m²/day. The design flow may be taken as plant flow plus recirculated flow minus underflow (often neglected), where the recirculation is made from the secondary tank.

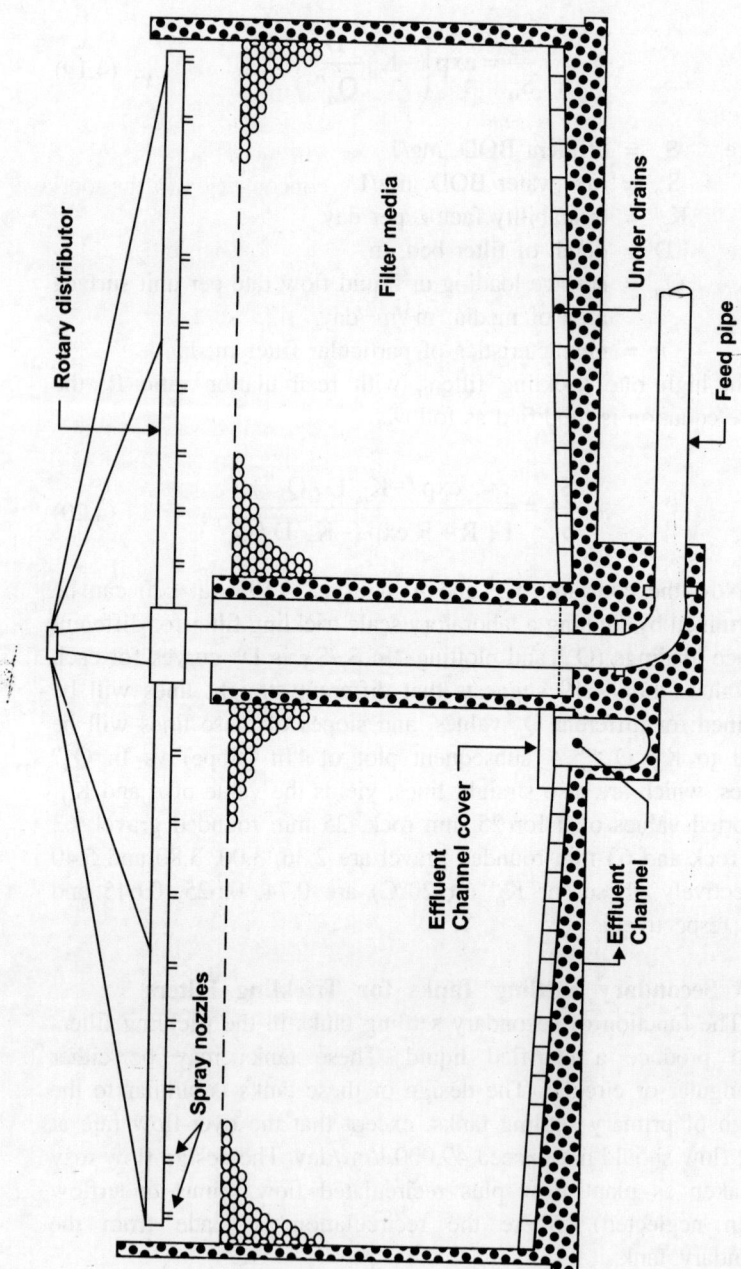

Figure 4.7: Trickling Filter.

QUESTIONS

1. Draw flowsheets for (a) conventional activated sludge process, (b) high rate ASP, (c) contact stabilization process.
2. In a secondary settling tank, what is the recommended OFR for a MLSS value of 3000 mg/l and recirculation ratio of 0.25?
3. Compare the construction, operation and maintenance of low rate and high rate trickling filters.
4. Explain the empirical method of design of trickling filters.

5 | Sludge Treatment and Disposal

5.0 INTRODUCTION

Most of the suspended solids and some of the dissolved organic solids, in a converted form, are separated from the main flow of the waste water, in the form of sludge. The quantity and the characteristics of the sludge depend upon its origin. In general, the sludge contains 92-98% of moisture, and is loaded mostly with putrescible organic substances. The high organic content of the sludge demands further treatment prior to its final disposal either on land or into the sea. In this chapter, all the sludge treatment processes will be described in brief. The design methods of the commonly used municipal sludge treatment processes will be dealt with in subsequent sections.

5.1 SLUDGE TREATMENT PROCESSES

Sludge treatment may include all or a combination of the following unit operations and processes: (1) Thickening, (2) Digestion, (3) Conditioning, (4) Dewatering, and (5) Incineration.

Thickening is meant for the reduction of moisture content of the sludge. Both gravity thickeners and floatation thickeners are employed for this purpose.

Digestion is a biological method of treatment and is meant for the reduction of organic content of the sludge. The digestion may be accomplished under either anaerobic or aerobic conditions.

Conditioning improves the drainability of the digested sludge, and may be accomplished by any of the following processes like Elutriation, Chemical conditioning, Heat treatment, Freezing etc. Chemical conditioning consists of addition of some coagulants like ferric chloride, lime, alum, and organic polymers. Elutriation, which involves mixing of digested sludge with water and resettling, may be used alone to improve the drainability of the sludge, or may be used prior to chemical conditioning. In the later case it reduces the chemical requirement. Elutriation may be used in advance to chemical conditioning wherever cost is justified. Chemical conditioning of sludge, with or without elutriation is necessary when the dewatering of sludge is accomplished by vacuum filtration.

Dewatering is accomplished either by air drying in Sand Drying Beds, or by mechanical means using Vacuum Filtration, Centrifugation, heat drying etc.

Dewatered sludge may some times require further treatment such as Incineration, to reduce the volume and the organic content of the sludge. Incineration involves the combustion of the sludge in a reactor under high temperature, along with auxiliary fuels, if needed.

The ultimate disposal of dewatered sludge or the ash after incineration may be onto the land or into the sea.

At this stage it should be made clear that out of all the processes described above, the digestion and auxiliary processes like incineration constitute the treatment part of the sludge processing while the remaining processes are meant for the reduction of its moisture only, prior to either digestion, or final, disposal.

5.1.1 Gravity Thickeners

Gravity Thickeners are deep circular tanks, provided with a slow moving deep truss or deep arm equipped with vertical members or pickets. The sludge is continuously stirred by the pickets and is concentrated and compressed at the bottom. The clear water overflows and is sent back to the primary inlet.

Separate sludge thickeners are not frequently used in the municipal sewage treatment plants. But it can be shown that certain benefits can be realised from its use, particularly when handling an activated sludge. Primary sludge usually contains about 4 to 6% of suspended solids. On the other hand, the activated sludge characteristically contains about 0.5% suspended solids, which

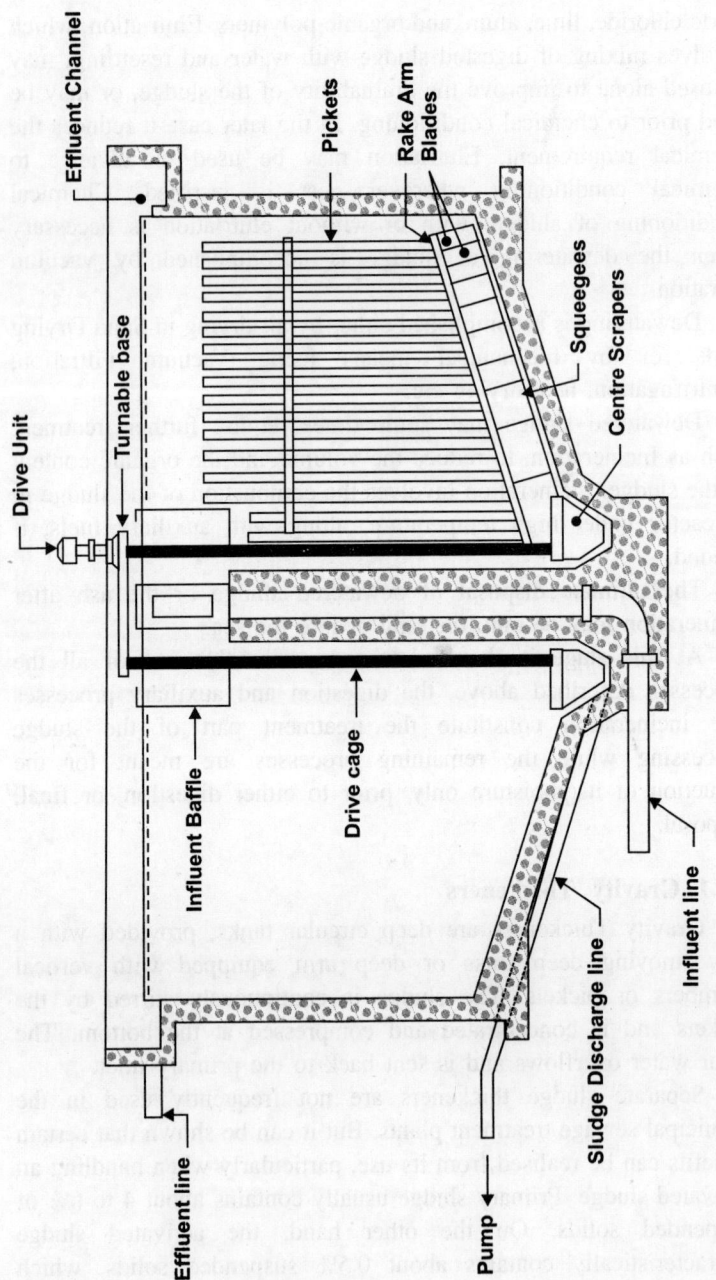

Figure 5.1: Gravity Thickener.

after settling increases to about 2%. It is apparent that any increase in the solid concentration by thickening, would result in the reduction of the volume of the sludge. This in turn would allow a decrease in the size of the Digesters, pumps, and other equipments handling the sludge.

The conventional clarifiers are designed to achieve the maximum removal of solids. The underflow concentration there usually never exceed 2%. On the other hand the thickeners are designed to get as thick a sludge as possible. Any attempt to achieve both the objectives in one unit, would require longer detention time and hence a larger size of the settling tank. Therefore, it is always desirable to have separate clarifier and thickener units, so that each can be designed and operated under the optimum conditions.

In the sewage treatment plants, employing activated sludge process, the thickeners could be used with another advantage. Wasting of sludge directly from the aeration tank, followed by the thickening, instead of the wasting from the recycle line, offers better control over the mean cell residence time and hence on the performance of the process.

The Gravity thickeners are to be designed for the desired underflow concentration on the basis of laboratory settling studies, described in the previous chapter.

In the absence of any test result, the gravity thickener units may be designed using a surface settling rate of 36500 $l/m^2/day$, and the solid loading as suggested in the Table 5.1.

5.1.2 Floatation Thickeners

The thickening of sludge, particularly the activated sludge, can also be accomplished in a floatation unit, instead of gravity thickeners. In the current floatation practice, certain amount of clarified waste is pressurized to about 3-4 kg/cm^2 in the presence of sufficient air within it. This air-saturated pressurized liquid is then mixed with the sludge, which results in the release of minute air bubbles in the solution. The sludge flocs are floated by these air bubbles and skimmed off from the surface. A typical rectangular floatation unit is shown in Fig. 5.2.

The process parameters in the design of floatation units include

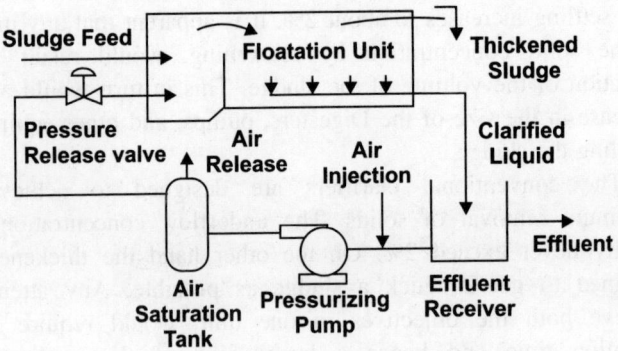

Figure 5.2: Dissolved Air Floatation Unit.

the detention time, the surface-rise-rate in the floatation unit, and the dimensionless air-to-solid ratio. The optimum air-to-solid ratio, in kg of air per kg of solid treated, should be determined from a pilot-plant study and may vary from 0.03 to 0.1, in the case of activated sludge. This ratio determines the amount of recycle to be pressurized, the pressure to be applied, the volume of air required, and the detention time in the air-saturation tank. A detention time, of the order of 120 minutes, is required at the floatation units. Suggested solid loadings for different types of sludges are given in the Table 5.1.

TABLE 5.1. Loading for Thickener Units.

Type of Sludge	Solid loading rate, Gravity Thickener	Kg/m²/day Floatation Thickener
Activated (unsettled)	—	25—75
Activated (settled)	20—40	50—100
Trickling Filter	40—50	—
Primary	100—150	200—275
Primary + Activated (combined)	40—80	100—200
Primary + Trickling Filter (combined)	60—100	—

5.2.3 Anaerobic Digestion of Sludge

Anaerobic Digestion is employed not only for the treatment of organic sludges, but also for different types of concentrated organic wastes. Two basic processes are involved in the anaerobic digestion—liquifaction and gasification. Two different groups of microorganisms are responsible for these two processes.

The first group of microorganisms hydrolyzes and ferments the complex organic substances to simple end products, which are primarily volatile organic acids, and alcohols. This process is known as liquifaction. These microorganisms include both facultative and anaerobic types, and are collectively known as "acid fermenters".

The second group of bacteria convert the produced organic acids and alcohols to carbon dioxide and methane gas. This process is known as gasification. The microorganisms responsible for this process are strictly anaerobes, and are known as "methane formers".

Acetic acid and propionic acid are the most common end products in the liquifaction stage. The particular strain of microorganism which converts these acids to gases in the gasification stage is the most important microorganism in the digestion process. Its rate of metabolism is very slow, and it is this rate that dictates the rate of digestion of sludge. The growth yield coefficient, Y, of the methane formers ranges between 0.04 and 0.054.

Due to the very low microbial growth rate, the production of biological sludge in the anaerobic processes is very low. Moreover, a major portion of the organic waste is converted into gases. Therefore, effluent of an anaerobic process is much more stabilized than that from an aerobic process. The nutrient requirement in the anaerobic process is also low due to the low production of microorganisms. A BOD: N: P ratio equal to 100 : 2.5: 0.5 may be maintained in this process. An anaerobic reactor operates more efficiently under an environment suitable for the rate limiting methane formers. For this, the reactor should be free from dissolved oxygen, and the contents should be, sufficiently alkaline for the proper functioning of the methane formers. It may be pointed out that, inspite of the formation of acid in the liquifaction stage, the proper level of pH is automatically maintained in a continuously operated reactor, due to the production of ammonia and other gases at the initial phase of gasification.

The temperature has got a tremendous effect in the functioning of a digester. This is graphically shown in Fig. 5.3. It has been established that two types of microorganisms, mesophilic and thermophilic, are responsible for biodegradation at two different temperature ranges. Once an operating temperature is fixed, any gross deviation from that temperature may result in an

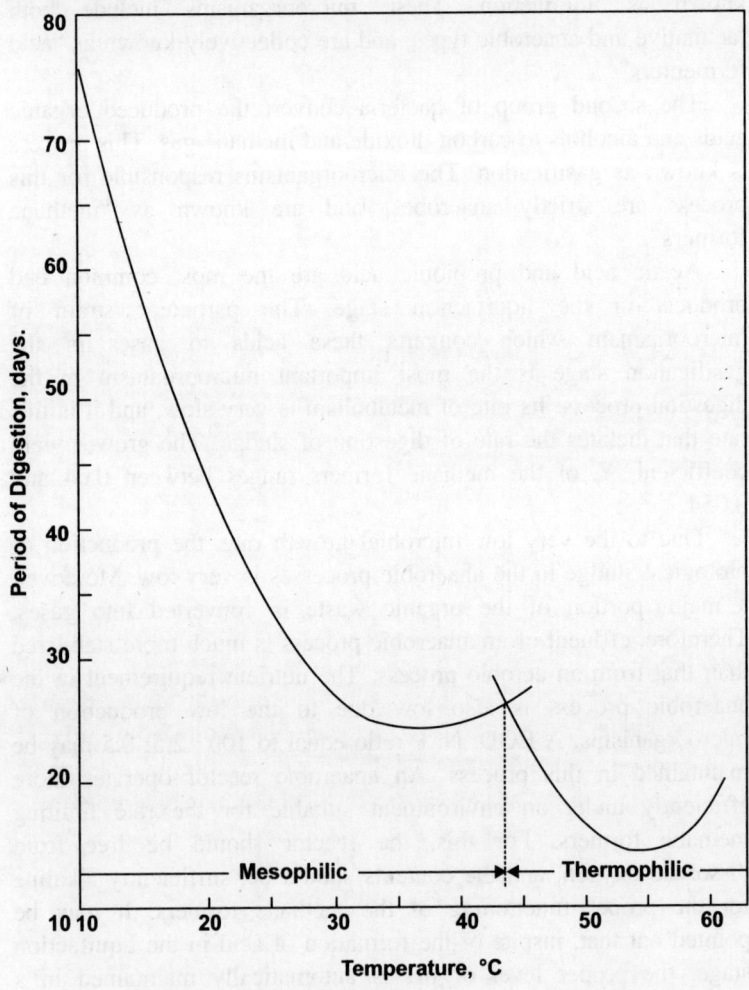

Figure 5.3: Effect of Temperature on Digestion.

unsatisfactory performance of the reactor. Therefore, the temperature is to be considered as an important factor in the design of digesters.

5.2.3.1 Process Description

The conventional sludge digestion is carried out in a single stage process, where the digestion, sludge thickening and supernatant formation take place simultaneously in one tank.

In the conventional single stage digesters, about 50% of their capacities remain unutilized due to the stratification and lack of intimate mixing. A longer detention time is also required for the complete stabilisation of the waste, which results in the larger capacity requirement of the digesters. In view of these drawbacks, in most of the recent installations, preference is given for a two stage digestion whenever the flow exceeds 4.5 MLD. In the two stage process, two tanks are provided. The first tank is meant for the digestion, and the contents are mixed continuously; the second tank is used for storage, thickening and supernatant formation. Frequently the two tanks are made identical in capacity.

The conventional two-stage digestion process differs from the so called "high rate digestion" in respect of the loading rate. In the high rate digestion the sludge is mixed and open heated to get the optimum digestion rates, in the first tank; the second tank may be a digester or just a holding tank depending upon the subsequent disposal method.

5.2.3.2 Rational method of design

The first unit in a two-stage digestion is a continuous complete mix biological reactor without recycle. Therefore, the rational design of this type of digester should be based on the knowledge of the kinetics of biological treatment or of microbiological growth.

The mass of biological solids synthesized daily, dX/dt can be given by the following equation:

$$\frac{dX}{dt} = \frac{dF}{dt} - k_d X \qquad \qquad \text{...(3.5)}$$

Now, in a complete-mix reactor, without recycle.

$$X = (dX/dt)/\theta_c \qquad \text{... (5.2)}$$

$\therefore$ Using equation 5.2, equation 3.5 may be reduced to the following form:

$$\frac{dX}{dt} = \frac{Y(dF/dt)}{1 + k_d\theta_c} \qquad \text{... (5.3)}$$

The volume of the reactor, V, can be directly computed from the following relationship:

$$\frac{V}{Q} = \theta_c \qquad \text{... (5.4)}$$

in which Q = sludge flow rate.

The rate of substrate utilisation, (dF/dt), in the equation 5.3 is estimated from the following relationship:

$$\frac{dF}{dt} = eF \qquad \text{... (5.3)}$$

in which e = fractional efficiency of waste utilisation, which ranges from 0.60 to 0.90 under satisfactory operating conditions,

and   F = Ultimate first stage BOD supplied to the reactor, kg/day.

Now, as eF indicates the rate of substrate utilisation, and 1.42 (dX/dt) usually gives the rate of substrate removal from the reactor in the form of biological mass, the difference $\{eF - 1.42(dX/dt)\}$ gives the rate of substrate stabilisation in the reactor.

$\therefore$ Stabilisation in the reactor is given by the following relationship:

$$\text{percent stabilisation} = \frac{eF - 1.42(dX/dt)}{F} \times 10$$

The end products of the anaerobic digestion are mainly methane and carbon dioxide. It can be shown that the quantity of theoretical production of methane gas is 350 litres per kg of ultimate BOD stabilised at 0°C and 1 atm pressure.

Therefore volume of methane gas produced per day can be estimated from the following relationship:

$$C = 350 \left(eF - 1.42 \frac{dX}{dt} \right) \qquad \dots \text{(5.7)}$$

in which C = volume of methane gas in litres/day,

F = BOD_L added, kg per day,

and $\dfrac{dX}{dt}$ = net growth of sludge, kg/day.

Suggested values of θ_c for use in the design of anaerobic digesters, and the usual values of Y and k_d for various types of wastes are given in Table 5.3 and 5.4 respectively.

TABLE 5.3. Suggested mean cell residence time for different operating temperatures.

Temp, °C	24	30	35
θ_c, days	20	14	10

TABLE 5.4. Growth yield coefficients and microorganism decay coefficients for anaerobic fermentation.

Waste composition	Coefficient basis	Temperature °C	Y	k_d, per day
Municipal sludge	—	—	—	—
Acetic Acid	Acetic Acid	25°C	0.054	0.011
Propionic Acid	”	25°	0.041	0.040
Dextrose Tryptone Beef extract	COD	38°	0.104	0.020
Carbohydrates	—	—	0.24	0.033

5.2.3.3 Process Parameters and Empirical Methods of Design

Tanks are usually circular and diameters range from 6 to 45 m. The depth of water ranges between 7.5 and 14 m. The minimum slope of the tank floor is 1 vertical to 4 horizontal. Several methods are available for computation of capacities of the tanks. Three empirical methods along with the suggested magnitudes of the process variables are briefly discussed.

1. Capacity based on loading rates:

For a single stage conventional digester, the capacity may be determined from the suggested solid loadings of 0.5 to 1.6 kg of volatile solids per m³ per day. Detention times of 30 to 90 days are usually provided in this type of design.

In high rate complete mix digester the loading rate varies with hydraulic retention time and the sludge concentration. Based on the assumptions that 75% of the total solids are volatile, sludge concentration is 5% and hydraulic retention time is 10 days, the volatile solid loading rate will be 3.8 kg per cubic metre per day.

2. The required volume of the conventional digester may be computed using the volume reduction formula:

$$V = [V_f - \tfrac{2}{3}(V_f - V_d)]\, t \qquad \text{.... (5.1)}$$

in which V = volume of digester, m³,

V_f = volume of fresh sludge, m³/day,

V_d = volume of digested sludge, m³/day,

and t = digestion time, days.

In the calculation of volume of sludges, it is often assumed that the moisture content of fresh sludge and digested sludge is 95% and 92% respectively, and the specific gravities are 1.02 and 1.04 respectively. It is also assumed that 60% of the volatile matter is destroyed during digestion. The suggested digestion periods at different temperatures are given in Table 5.2.

TABLE 5.2. Suggested design periods at different operation temperatures.

Temperature, °C	10	16	21	27	32	38
Digestion period, days	75	56	42	30	25	24

3. The capacities of the conventional digestion tanks may be determined on the per capita basis. Based on 35 to 45 days of detention, following is the per capita volume requirement in different types of plants.

(a) Only primary settling tank – 0.037 – 0.048 m³/capita

(b) Primary and Trickling filter – 0.071 – 0.09 m³/capita

(c) Primary and activated sludge – 0.1 – 0.128 m³/capita

The capacities computed as above give the basic space requirement. A factor of safety, usually less than 2, is applied to provide the allowance for scum, gas, sludge liquor etc.

Example 5.1:

Design a conventional single stage digestion tank for the sewage treatment plant employing activated sludge process. The following data is available:

Average flow = 250 l/capita/day
Population = 40,000
S.S. concentration in raw sewage = 100 gms/capita/day
Volatile matter = 70%
Suspended solid reduction in primary = 60%
Sp. gravity of primary wet sludge = 1.01
Sp. gravity of activated wet sludge = 1.004
Sp. gravity of digested wet sludge = 1.048
Moisture content of digested wet sludge =87%
Moisture content of primary wet sludge =90%
Moisture content of excess secondary sludge = 99%
Average temperature = 32°C

Solution:

Total flow = 10 MLD

$$\text{S.S. concentration} = \frac{100 \times 1000 \times 40,000}{10 \times 10^6}$$

$$= 400 \text{ mg/l}$$

Total SS $= 4000$ kg/day

∴ SS in primary sludge $= 0.6 \times 4000$

$$= 2400 \text{ kg/day}$$

∴ Volume of fresh primary sludge $= \dfrac{2400}{100-90} \times \dfrac{100}{1.01}$

$$= 23.8 \times 10^3 \text{ litres/day}$$

Volatile matter in primary sludge $= 0.7 \times 2400$

$$= 1680 \text{ kg/day}$$

∴ Fixed solid in fresh primary sludge $= 720$ kg/day

Solids in the activated sludge influent

$$= 4000{-}2400 = 1600 \text{ kg/day}$$

Given that 70% of the above is volatile, wt of volatile SS
$$= 1120 \text{ kg/day}$$
$\therefore$ Wt. of fixed solid = 480 kg/day

Assumed that 7.5% of the volatile SS get destroyed during activation and that 87.5% of the rest is captured in the excess sludge.

$\therefore$ Wt. of captured volatile solid = 0.875 (1–0.075) 1120
$$= .875 \times .925 \times 1120 = 906 \text{ kg/day,}$$
& wt of captured fixed solid = .875 × 480 = 420 kg/day
$\therefore$ Wt of solid in fresh secondary excess sludge
$$= 906 + 420 = 1326 \text{ kg/day}$$

$\therefore$ Volume of fresh secondary sludge $= \dfrac{1326 \times 100}{100 - 99} \times \dfrac{1}{1.004}$

$$= 132.5 \times 10^3 \text{ litres/day}$$

Now total volatile solid in fresh sludge (combined)
$$= 1680 + 906 = 2586 \text{ kg/day}$$
& total fixed solid in fresh sludge (combined)
$$= 720 + 420 = 1140 \text{ kg/day}$$

Assumed 70% of the volatile solid will be destroyed by digestion.

$\therefore$ Volatile solid after digestion = .3 × 2586 = 776 kg/day
$\therefore$ Total solid after digestion = 776 + 1140 = 1916 kg/day

$\therefore$ Wt. of digested sludge $= \dfrac{1916 \times 100}{100 - 87} = 14738.5 \text{ kg/day}$

$\therefore$ Volume of digested sludge $= \dfrac{14738.5}{1.048}$

$$= 14063.4 \text{ litres/day}$$
$$= 14.06 \times 10^3 \text{ 1/day}$$
$\therefore$ Volume of fresh combined sludge
$$= (23.8 + 132.5) \ 10^3 \text{ litres/day}$$
$$= 156.3 \times 10^3 \text{ litres/day}$$
Volume of digested sludge = 14.06 × 10³ litres/day

Now corresponding to the given temperature of 32°C the period of digestion = 25 days.

$\therefore$ Using volume reduction formula,
The capacity of the tank, required.

$$= [156.3 - \frac{2}{3} (156.3 - 14.06)] \times 25 = 1536.8 \text{ m}^3$$

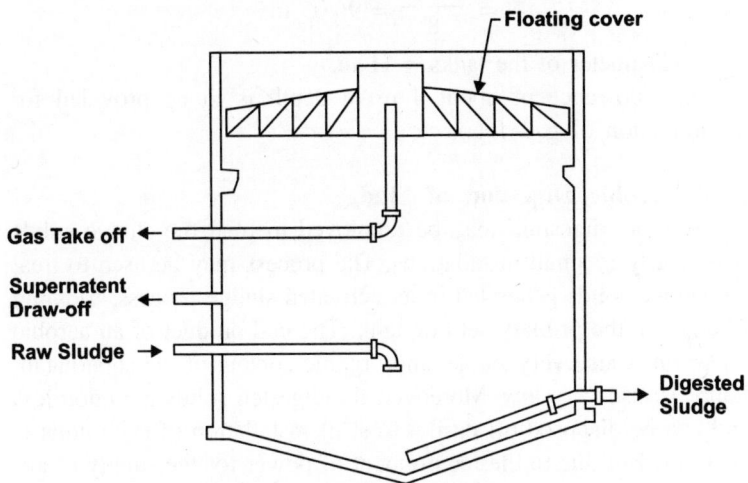

Figure 5.4: Anaerobic Digester.

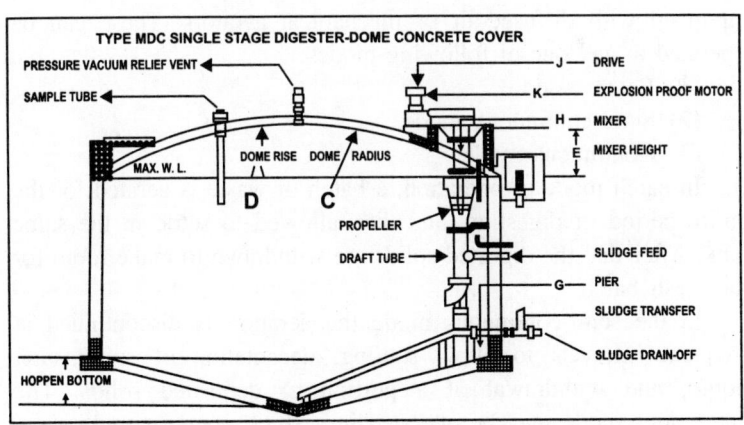

Figure 5.4(a): Dorr-Oliver Draft Tube Digester.

Providing two tanks in parallel, the capacity for each

$$= \frac{1536.8}{2} = 768.4 \text{ m}^3$$

Providing a water depth of 8 m, the surface area required

$$= \frac{768.4}{8} = 96.05 \text{ m}^2$$

$\therefore$ Diameter of the tanks = 11 m.

An allowance of about 4 m in depth is to be provided for accumulation of gas, foam etc.

5.1.4 Aerobic Digestion of Sludge

Aerobic digestion may be employed to treat the sludge solids particularly in small installations. The process may be used to treat the sludge solids generated in an activated sludge process, trickling filter, or in the primary settling tank. The end product of an aerobic digestion is also very stable, and organic content of the supernatant liquor is also very low. Moreover, the digested solids are odourless and can be disposed off easily. Cost of installation of such units is also low. But due to the requirement of power for the supply of air, the running cost may be more compared to that in anaerobic digester.

Aerobic digestion is usually accomplished in open tanks equipped with diffused-air or mechanical aerators. These can be operated in any one of following modes:

(1) Batch mode,

(2) Semi-continuous mode,

(3) Continuous mode.

In batch mode of operation, a batch of waste is aerated for the entire period of digestion, and then allowed to settle in the same tank. After that the digested solids are withdrawn to make room for the fresh batch.

In the semi-continuous mode the aeration is discontinued at frequent intervals to allow settling, decantation of supernatant liquid, and withdrawal of a portion of deposited solids. The continuous mode may be operated in a single-stage or multi-stage configuration.

The aerobic digesters are operated in such a way that the microbial activities are restricted to the endogenous respiration phase of their metabolism, i.e. no further cell synthesis is allowed to occur within the digester, and the microorganisms are forced to metabolise their own protoplasm due to the lack of supply of fresh food. It should be noted that a portion of the cell material is highly resistant to oxidation; therefore only about 75 to 80 per cent of the cell material can be oxidized by this process.

The kinetics of biological process in an aerobic digester may be given by the following relationship:

$$\frac{dX}{dt} = - k_d.X \qquad \qquad ...\ (5.8)$$

in which $\quad \dfrac{dX}{dt}$ = rate of reduction of biological sludge, mg/1/day,

$\qquad \qquad X$ = concentration of biological sludge, mg/1,

and $\qquad k_d$ = microorganism decay coefficient, per day.

It may be noted that the Eqn. 5.8 is a modified form of Eqn. 3.5, in which the first term in the RHS ($Y\ dF/dt$) is dropped due to the absence of any new cell synthesis.

For a batch mode of operation, the integration of Eqn. 5.8 yields:

$$\frac{X_1}{X_0} = e^{-k_d\,\theta_c} \qquad \qquad ...\ (5.9)$$

in which X_0 = concentration of biological sludge, or, that of biodegradable volatile suspended solids, in the digester influent, mg/1,

$\qquad X_1$ = concentration of biodegradable volatile suspended solids in the digester effluent, mg/1,

and θ_c = mean cell residence time in the digester, days
$\qquad \quad$ = hydraulic residence time in the digester, days.

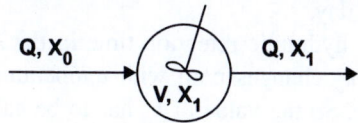

Figure. 5.5: Single Stage Aerobic Digester.

Now for a continuous mode of operation in a single stage unit, as shown in Fig. 5.5, the material balance across the digester can be given by the following relationship:

$$\begin{pmatrix} \text{net rate of change} \\ \text{of bio mass} \end{pmatrix} = \begin{pmatrix} \text{rate of decay} \\ \text{of bio mass} \end{pmatrix} - \begin{pmatrix} \text{rate of} \\ \text{outflow} \end{pmatrix} + \begin{pmatrix} \text{rate of} \\ \text{inflow} \end{pmatrix}$$

or $$V \frac{dX}{dt} = -k_d X_1 V - QX_1 + QX_0$$

Now as in a steady state, $dX/dt = 0$, the above equation may be reduced to

$$\frac{X_1}{X_0} = \frac{1}{1 + k_d \theta_c} \qquad \ldots (5.10)$$

in which V = volume of the digester = $Q\,\theta_c$,
and   Q = rate of flow through the digester.

For a continuous multi-stage system, containing n numbers of reactors, as shown in Fig. 5.6.

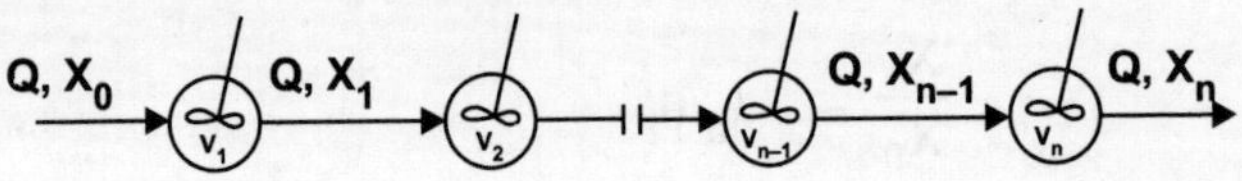

Figure 5.6:   Multi-stage Aerobic Digesters.

$$\frac{X_n}{X_o} = \mathop{\pi}_{i=1}^{n} \left[\frac{1}{1 + k_{di}(V/Q)_i} \right]^* \qquad \ldots (5.11)$$

in which k_{di} = microorganism decay coefficient in i^{th} unit, per day,
and $(V/Q)_i$ = hydraulic retention time in the ith unit, day.

The value of k_d changes both with temperature and the mean cell residence time. So the value of k_d has to be calculated properly for use in Eqn. 5.9, 5.10, and particularly in Eqn. 5.11. Ritch suggested the following relationship for k_d, upto a minimum temperature of 20°C.

---

* Product of the terms between $i = 1$ and $i = $ n.

$$k_d = \frac{0.48\,(1.05)^{T-20}}{t_3^{0.415}} \qquad \text{... (5.12)}$$

in which T = temperature of operation, °C,

t_3 = total mean cell residence time

$\quad = \theta_{cq} + \theta_c$,

θ_{cg} = mean cell residence time in the system where the olids are generated, days,

and θ_c = mean cell residence time in the digester, days.

The calculation for k_{di} in Eqn. 5.11, should be based on the corresponding location of the solids in the system.

The relationships given above are directly applicable for the treatment of solids generated in an activated sludge process. The condition in which the solids are generated in a fixed-film process (trickling filter) is quite different from that prevailing in an aeration tank. The fixed film solids are usually expected to produce a different phase of metabolism, in the initial period of digestion. The microorganism, indigenous to the digester, may utilise them as substrate, resulting in the cell synthesis or growth of microorganism. But this initial growth phase is so short, compared to the total period of digestion, that it may be neglected. The same assumption can also be made for the biodegradable solids from the primary settling tank although they do not constitute the bio-mass.

The required mean cell residence time for any desired degree of stabilisation may be obtained using the relevant expression (either Eqn. 5.9, 5.10 or 5.11) depending on the mode of operation. The mean cell residence time establishes the reactor volume and the solid loading rate. The decay coefficient in the range of 0.05 to 0.07 per day is suggested by Metcalf and Eddy Inc for the solids generated in the activated sludge process.

Oxygen is required not only for the oxidation of cell materials, but also for the oxidation of biodegradable solids from the primary, if there is any, and also for the oxidation of ammonia, which is produced due to the oxidation of cell materials. It has been found that the oxygen requirement is about 2 kg per kg of cell oxidised, and about 1.7 to 1.9 kg per kg of 5 day BOD from primary.

Operating experience suggests that a dissolved oxygen

concentration of about 1 to 2 mg/l in the digester produces a digested sludge which dewaters well.

The temperature of operation has got a profound effect in the process when the retention time is less than 15 days particularly in cold climates. After 15 days, the effect of temperature is not that important, at temperatures about 20°C. A temperature coefficient in the range of 1.08 to 1.10 is often suggested for the retention time for temperatures below 20°C and the retention times of the order of 15 days.

Depending on the origin of sludge solids, the retention time varies in the range of 12 to 22 days, the lower values being taken for the excess activated sludge.

The suggested ranges for different design parameters are given in Table 5.5:

TABLE 5.5. Suggested Design Parameters for Aerobic Digester.

Parameters	value
A. Hydraulic retention period, at 20°C, day:	
(i) Activated sludge only	12-16 days only
(ii) Primary and Activated sludge	18-22 days
(or Trickling filter sludge)	
B. Solid loading rate, kg volatile	1.6-3.2
solid/m³/day	
C. Oxygen requirement, kg/kg destroyed, for:	
(i) Cell materials	2
(ii) 5 day BOD from primary	1.7-1.9
D. D.O. level, mg/l	1-2

5.1.5 Sand Drying Beds

Sand drying beds are shallow beds of about 250 mm of sand over about equally thick well-graded gravel layer, underlain by perforated drainage lines spaced 2.5 to 6 m apart. The bed should slope towards the discharge end at a rate of 1 in 200.

For flexibility in operation the bed area may be subdivided by partitions, each approximately 6 m wide and 6 to 30 m long.

The digested sludge is applied on to the beds in 200 mm to 300 mm thick layers.

The drying time depends on weather conditions and may vary from 10 days to several weeks. The sludge cake is removed by shovel at the end of the drying period when the moisture content reduces to a satisfactory level.

The design of sand drying beds involves only the computation of the bed area. This may be done on per capita basis, or on solid loading rate basis. When data conforming to the local environmental conditions is not available, the following data may be used:

(a) Solid-loading rate: 50 to 150 kg of dry solids per square meter per year.

(b) Per capita requirement of sand bed area:

   (*i*) for digested primary sludge only = 0.9 to 1.4 m^2.

   (*ii*) for digested primary sludge and humus

$$= 1.1 \text{ to } 1.6 \text{ m}^2.$$

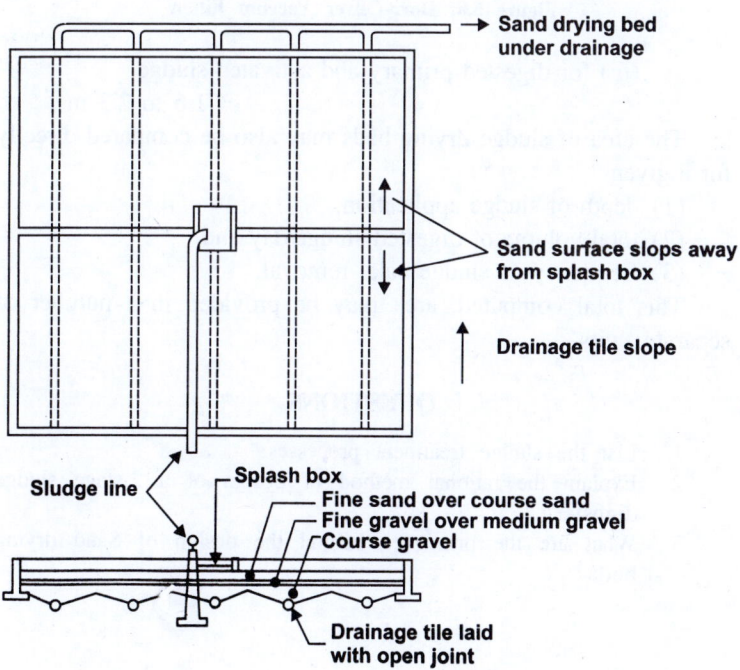

Figure 5.7: Sand Drying Bed.

Figure 5.8: Dorr-Oliver Vaccum Filter.

(iii) for digested primary and activated sludge
$$= 1.6 \text{ to } 2.3 \text{ m}^2.$$

The area of sludge drying beds may also be compared directly for a given

(1) depth of sludge application,
(2) total volume of digested sludge/day and
(3) frequency of sludge cake removal.

This total computed, area may be provided in a number of separate units.

QUESTIONS

1. List the sludge treatment processes.
2. Explain the rational method of design of a 2-stage sludge digester
3. What are the principles behind the design of Sand drying beds?

6 | Low Cost Waste Treatment Systems and their Design

6.0 INTRODUCTION

The conventional biological treatment methods employing trickling filters or activated sludge process, are very expensive to install as well as to run. The past experience with such treatment plants shows that many of them are either performing poorly due to the lack of skilled operational supervision or became idle because of mechanical break-down and lack of maintenance. A few of them again were not designed properly. On the other hand, the so called low cost treatment systems, which include Aerated Lagoon, Stabilisation Ponds, Oxidation Ditches, Anaerobic Lagoons etc., are simple to construct, have the least amount of mechanization and require little or no skilled supervision for operation. The design criteria for such low cost treatment systems have been evolved in our country. An extensive study conducted in our country revealed that, under certain conditions, the degree of treatment that can be achieved is as good as that of the conventional systems, if these low cost treatment systems are adopted.

In this chapter a brief account of the available low cost treatment systems are presented along with their design criteria.

6.1 AERATED LAGOONS

For communities producing a total waste flow of less than 4.5 MLD, the aerated lagoons may be recommended for the complete aerobic biological treatment of raw-waste water. Aerated lagoons are simple holding basins usually 2 to 4 m deep with a continuous supply of oxygen, and are seeded initially by the same type of microorganisms which are responsible for biodegradation in the activated sludge process. The oxygen is supplied to the basin by means of mechanical surface aeration units; this will also keep the contents of the basin in suspension.

Depending upon the extent of mixing, the lagoons may be classified as either complete mix or partially mixed. These lagoon types are also often identified as aerobic- and facultative aerated lagoons respectively. Power requirements for the aerators in a complete mixed aerated lagoon is about 6 watts/m^3 while that in a partially mixed system is about 1 to 2 watts/m^3.

The complete mix aerated lagoons essentially consist of two units. In the first the mechanical surface aerators are so designed that the solids do not settle to the bottom of the tank, the second unit is used as a settling tank for the removal of the solids. As such, this process is almost similar to that in the activated sludge process without any recirculation.

In partially mixed system, a large portion of the incoming solids and the biological solids produced within the lagoon settle to the bottom of the tank and soon the condition in the bottom layer of the lagoon induces an anaerobic decomposition of the settled solids. Therefore effluent from this type of tanks is more stable.

Both in complete mix and partially mixed systems, the lagoons must be followed by settling basins. The effluent may be discharged after sedimentation.

Research work conducted in India indicates that a partially mixed aerated lagoon followed by a settling basin is expected to produce a good quality effluent at low cost and low power requirement. These lagoons require cleaning once in five or six years. The depth of the lagoons may be taken around 2.5 m. To facilitated sludge removal, the lagoons may be constructed as multicell units, the various cells being operated in parallel.

6.1.1 Process Design Considerations for a Complete Mix Aerated Lagoon

A complete mix aerated lagoon may be designed as a complete mix biological reactor without cell recycle, using microorganism growth kinetics. But in places, where the variation of temperature is too much, due to the lack of information in regard to the variation of kinetic growth coefficients with temperature, the design is usually based on simultaneous and judicious application of both BOD removal and growth kinetics.

Normally in the aerated lagoons, it is assumed that BOD removal can be described by the first order removal function. So, from a steady state mass balance of the organic materials across the complete mix single lagoon, the following relationship may be written:

$$\frac{S_1}{S_0} = \frac{1}{1 + kX_1(V/Q)} \qquad \dots (3.13a)$$

in which, S_1 = effluent soluble BOD_5, mg/1,

S_0 = influent soluble BOD_5, mg/l,

X_1 = microorganism mass concentration in the reactor mg/1,

V = volume of the reactor, 1,

Q = rate of flow of waste, 1/day,

and k = specific soluble BOD_5 removal rate coefficient, 1/mg/day,

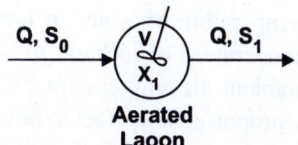

Aerated Laoon

Figure 6.1: Aerated Lagoon.

The microorganism concentration in the effluent, X_1, in a complete mix system, can be predicted using the following relationship, developed from the mass balance for the mass of microorganisms across the system in a steady state:

$$X_1 = \frac{Y(S_0 - S_1)}{1 + k_d(V/Q)} \qquad \dots (3.23)$$

in which Y = growth yield coefficient,

and k_d = specific decay coefficient, day^{-1}

The values of k, Y and k_d can be determined experimentally. For Indian conditions, where temperature rarely falls below 15°C, the following values may be assumed in the absence of any actual test results:

$$k = 0.05 \text{ 1/mg/day at } 20°C$$
$$Y = 0.5$$
$$k_d = 0.05 \text{ per day.}$$

The values of k is generally reported for a temperature of 20°C. Values at other temperatures may be estimated using the following expression:

$$k_T = k_{20} \, \theta^{T-20} \qquad \qquad ... (3.4)$$

in which θ = temperature coefficient, which may be assumed to be equal to 1.060, if no other data is available; 1.097 as the magnitude of θ is also found, in the literature, for cold climates.

The temperature of the aerated lagoon may be estimated using the following relationship (for complete mix system):

$$T_w = \frac{10^6 QT_i + AfT_a}{Af + 10^6 Q} \qquad \qquad ... (6.1)$$

in which T_w = temperature of water in lagoon, °C,
              T_i = temperature of influent to lagoon, °C,
              T_a = ambient air temperature, °C,
              f = a proportionality factor (also referred to as surface heat exchang coefficient). cal/day/m^2 °C,
              A = lagoon surface area, m^2,
and Q = flow rate through the lagoon, m^3/day.

For the coldest period in a year, the T may be estimated assuming a "f" value around 0.7×10^6 cal/day/m^2/°C.

All the relationships given above are interrelated and cannot be solved for a particular quantity independently. So the process variables are fixed in the design using an iterative method of solution, starting from the given process parameters.

As a first approximation, the microorganism mass concentration, X_1, in the reactor, and also in the effluent of a complete mix lagoon is taken as follows:

$$X_1 = Y (S_0 - S_1) \qquad \qquad \text{... (6.2)}$$

The approximate hydraulic retention time (V/Q) is then obtained from Eqn. 3.13a:

$$\frac{V}{Q} = \frac{S_0 - S_1}{S_1 k X_1} \qquad \qquad \text{... (6.3)}$$

On the basis of this hydraulic retention time, the capacity and surface area of the lagoon is determined. The temperature of water in the lagoon is then determined using Eqn. 6.1.

Actual value of X_1 is then estimated, and using the corrected value of k in Eqn. 6.3, the actual detention time is calculated. The design of the lagoon is revised, keeping the surface area unchanged.

6.1.1.1 Oxygen requirement

Oxygen requirement may be estimated using the method as outlined for the activated sludge process in chapter 4.

Usually mechanical surface aerators are used to cater to the requirement of oxygen. The surface aerators are usually rated to supply certain amount of oxygen per unit of power supply under certain standard conditions. Depending upon the horse power or wattage of each aerator, and the rated capacity, the number of such aerators is fixed. In the design, the expected capacity of the aerators under field conditions is usually estimated using the following relationship:

$$N = N_0 \left[\frac{C_s - C_0}{C_w} (1.024)^{T-20}\alpha \right] \qquad \text{... (6.3)}$$

in which, $N =$ Kg of O_2/Watt/hr transferred under field conditions,

$N_0 =$ rating of the aerator used, i.e. kg of O_2/Watt/hr transferred in water at 20°C and zero D.O,

$C_s =$ saturation oxygen concentration for waste at operating temperature,

$C_0 =$ operating oxygen concentration,

$C_w =$ Saturation oxygen concentration for water at 20°C = 9.17 mg/l,

and
$$T = \text{temperature of lagoon in } °C,$$
$$\alpha = \text{salinity-surface tension correction factor, usually}$$
$$0.8 \text{ to } 0.85.$$

6.1.2 Process Design Considerations in Partially Mixed Aerated Lagoon

In a partially mixed lagoon, the value of X_1 is a function of aeration power unit to the system. As such, it is very difficult to estimate the magnitude of X_1 in this case. However, field observations suggest that a power input of 1 Watt/m^3 may result in a biomass concentration of 20 mg/l or less. So hydraulic retention time and then the capacity of the lagoon is estimated with an assumed value of X_1 using Eqn 3.14. This value of X_1 may however be checked by Eqn 3.23.

The oxygen requirement in a partially mixed lagoon depends upon the extent of mixing. This may be computed using the following relationship.

$$O_2 \text{ requirement/day} = \beta \ Q(S_0 - S_1) \qquad \dots \ (6.4)$$

in which β = a constant which varies from 0.8 mg O_2/mg BOD_5 removed in winter to 1.5 mg O_2/mg BOD_5 removed in summer.

6.1.3 Process Design Considerations in Dual Aerated Lagoon

Research work conducted in USA suggested that, instead of a single aerated lagoon of either type, a combination of a complete mix lagoon followed by a partially mixed lagoon may be adopted to arrive at a very low value of effluent BOD, at a lower cost.

In a dual aerated lagoon system, with a complete mix lagoon followed by a partially mixed lagoon, as indicated by Rich, the effluent soluble BOD_5 of the system will be minimum when:

$$S_1 = \frac{kd}{Yk} + \sqrt{\frac{X_2 S_2}{Y}} \qquad \dots \ (6.5)$$

In which S_1 = effluent soluble BOD_5 of complete mix lagoon, mg/1

S_2 = effluent soluble BOD_5 of partially mixed lagoon, mg/1,

$$X_2 = \text{biomass concentration in the effluent}$$
from partially mixed lagoon, mg/1,

$$k_d = \text{microorganism decay coefficient per day,}$$

and k = specific soluble BOD_5 removal coefficient, 1/mg/day

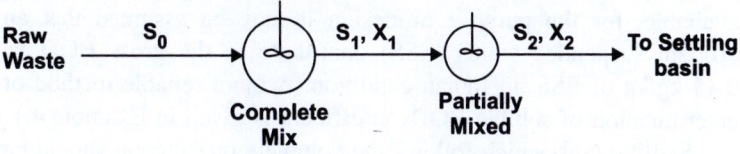

Figure 6.2: Dual Aerated Lagoon.

Once the value of S_1 is fixed for the system as shown in Fig. 6.2, the hydraulic retention time of the partially mixed lagoon, $(V/Q)_2$ may be estimated using the following relationship:

$$\left[\frac{V}{Q}\right]_2 = \frac{S_1 - S_2}{X_2 k S_2} \qquad \dots (6.6)$$

From Eqn. 3.14 and Eqn 3.23, the hydraulic retention time of complete mix lagoon, $(V/Q)_1$, is given by:

$$\left[\frac{V}{Q}\right]_1 = \frac{1}{Y k S_1 - k_d} \qquad \dots (6.7)$$

The minimum hydraulic retention time $(V/Q)_1$ in the complete mix lagoon, to prevent any washout of biomass is given by:

$$\left[\frac{V}{Q}\right]_{1C} = \frac{1}{Y k - k_d} \qquad \dots (6.8)$$

The hydraulic retention time in the second lagoon in the dual or two-cell system is normally kept less than 3 days to prohibit the growth of algae. These lagoons may be cleaned at an interval of two years. Therefore the designs are to be done on the basis of present flow condition with only a marginal capacity for future increases in flow rates.

For a dual aerated lagoon system described above, the order of magnitude of $(V/Q)_1$ and $(V/Q)_2$ are respectively 3 days and 2 days. For most waste waters the minimum total hydraulic retention time for optimal sedimentation is around 5 days.

All the kinetic equations given above are applicable for the soluble part of the BOD of the waste. If no other information is available, for the purpose of design it may be assumed that an effluent suspended solids (ESS) contribute to the gross BOD @ 0.45 kg/kg of ESS, in Indian condition. A more reliable method of determination of soluble BOD_5 at effluent is given in Example 4.1.

Settling tank which follows the complete mix lagoon should be designed in the same way as was done in the case of Activated sludge plant.

Example 6.1

Design a complete mix aerated lagoon system to treat a domestic sewage flow of 2 MLD. Following conditions and requirements may be assumed:

1. Influent suspended solids = 200 mg/l
2. Effluent suspended solids after settling $\not> 20$ mg/l
3. Influent BOD_5 = 200 mg/l
4. Effluent BOD_5 $\not> 20$ mg/l
5. Volatile suspended solids = 80% of total solids produced.
6. Summer temperature = 35°C
7. Winter temperature = 15°C
8. Waste water temperature = 20°C
9. Oxygen concentration to be maintained in the lagoon = 1.5 mg/l
10. Lagoon depth = 2.5 m
11. Specific substrate removal coefficient, k = 0.05 1/mg/day at 20 °C
    Growth yield coefficient, Y = 0.5
    Decay coefficient, k_d = 0.05 per day
12. Surface heat exchange coefficient, $f = 0.7 \times 10^6$ cal/day/m²/°C
13. Oxygen transfer coefficient of the aerator α = 0.8
14. Temperature coefficient θ = 1.065.

Solution:

Flow = 2 MLD = 2×10^3 m³/day

S_o = 200 mg/l
Now the total effluent BOD_5 = 20 mg/l
and effluent SS after settling = 20 mg/l
∴ In Eqn. 6.4 C_1 = 20 mg/l
X = 20 × .8 = 16 mg/l
(assuming VSS = 80% of total SS)
∴ S_1 = soluble BOD_5 in the effluent = 20 – 0.54 × 16
= 11.36 mg/l (assuming ESS BOD_5 @ 0.54 kg/kg)
As first approximation X_1 = (200 – 11.36) 0.5
= 94 mg/l
Now neglecting for the time being the variation of k value with temperature, from Eqn. 6.3,

$$\frac{V}{Q} = \frac{S_0 - S_1}{S_1 k X_1} = \frac{200 - 11.36}{11.36 \times 0.05} \times 94 = 3.5 \text{ days.}$$

∴ Providing a retention time of 3.5 days, the volume required

$$= 3.5 \times 2 \times 10^3 \text{ m}^3$$
$$= 7 \times 10^3 \text{ m}^3$$

∴ Surface area required $= \dfrac{7 \times 10^3}{2.5} = 2.8 \times 10^3 \text{ m}^2$

(assuming a depth = 2.5 m)
Summer temperature of the water in the lagoon, (from Eqn. 6.1):

$$T_w = \frac{10^6 \times 2 \times 10^3 \times 20°C + 2.8 \times 10^3 \times 0.7 \times 10^6 \times 35°C}{2.8 \times 10^3 \times 0.7 \times 10^6 + 10^6 \times 2 \times 10^3}$$

= 27.4°C
Value of k at temp 27.4°C
= 0.05 × $(1.065)^{7.4}$ = 0.05 × 1.593
= 0.0796 1/mg/day

Now k_d = 0.05 per day

∴ $X_1 = \dfrac{Y(S_o - S)}{1 + k_d (V/Q)} = \dfrac{0.5(200 - 11.36)}{1 + 3.5 \times 0.05}$

= 81 mg/l

Using Eqn 6.3, again,

$$\frac{V}{Q} = \frac{200 - 11.36}{11.36 \times .0796 \times 81} = 2.57 \text{ days}$$

Now modifying our design, after using $V/Q = 3$ days,
Volume required $= 3 \times 2 \times 10^3 \text{ m}^3 = 6 \times 10^3 \text{ m}^3$
Now keeping the surface area same as in the 1st trial i.e.
$2.8 \times 10^3 \text{ m}^2$ the depth required

$$= \frac{6 \times 10^3}{2.8 \times 10^3} = 2.15 \text{ m}$$

Provide a depth of the aerated lagoon: 2.2 m.

The remaining part of the design, i.e. the dimensioning and calculation for oxygen requirement and power requirement is left as an exercise to the readers.

6.2. STABILIZATION POND (OXIDATION POND)

A Stabilisation Pond is simply a shallow body of water contained in an earthen basin, open to sun and air. The ponds are classified, according to the nature of the biological activity which takes places within the pond, as aerobic, facultative, and anaerobic.

In aerobic stabilisation pond the oxygen is supplied by natural surface aeration and by algal photosynthesis. Except for the algal population, the microbiological population present in the ponds are similar to that in activated sludge system. These aerobic population releases carbon dioxide which is taken up by the algae for their growth; algae in turn release oxygen which help in maintaining the aerobic condition of the pond. Very shallow depth of aerobic pond (of depth 0.15 m to 0.45 m,) is used for the treatment of irrigation return water or any other industrial waste, where the aim is the removal of nitrogen by algal growth. For the treatment of domestic waste water, pond depth between 1 m and 1.2 m may be used. For better results, the contents of the aerobic pond must be mixed periodically. The hydraulic regime within the pond is assumed to be in between complete mix and plug flow.

Most of the stabilization ponds in India are facultative in nature. Three zones exist in this type of ponds. The top zone is an aerobic zone in which the algal photosynthesis and aerobic biodegradation takes place. In the bottom zone the organics of the municipal waste water settle and undergo anaerobic decomposition. In the intermediate zone which is partly aerobic and partly anaerobic, the decomposition of organic wastes are done by the facultative bacteria. The nuisance associated with the anaerobic reactions are eliminated due to the presence of top aerobic zone; the foul smelling end products of anaerobic decomposition are carried to the top zone by mixing currents and are oxidized or utilised by the aerobic population there. As such, the maintenance of an aerobic layer in the facultative pond is important; the depth of this layer is, however a function of solar radiation, waste characteristics, BOD loading, and temperature.

In anaerobic pond the entire depth is in anaerobic condition except an extremely shallow top layer. Normally these tanks are used in series after facultative pond for complete treatment of a waste.

6.2.1 Process Design Considerations

Design parameters for the stabilisation ponds are not well defined. The mathematical modelling of the process of stabilisation in the ponds is very difficult due to the simultaneous involvement of the operations like sedimentation, oxidation, digestion, gas exchange, photosynthesis, evaporation, seepage, etc. Inspite of this, numerous methods of design have been proposed in the literature. Out of them, two methods, the first applicable for aerobic pond only, and the second applicable both for aerobic and facultative ponds, will be described here.

In the first method, which is applicable for aerobic stabilisation pond, the oxygen resources of the pond are equated to the applied organic loading. The principal source of the oxygen is photosynthesis, and that too is dependent on solar energy.

The solar energy again is related to geographical, meteorological and astronomical phenomena, and varies principally with time in year and the latitude of the place. Based on the studies conducted by the National Enviromental Engineering Research Institute, Nagpur, the following values of the yield of photo-synthetic oxygen in different latitudes are suggested, for India (Siddiqi):

Latitude	Yield of photosynthetic O_2
8°N	325 kg/hectare/day
12°N	300 „ „
16°N	275 „ „
20°N	250 „ „
24°N	225 „ „
28°N	200 „ „
32°N	175 „ „
36°N	150 „ „

Yield of photosynthetic oxygen may be calculated directly if the amount of solar energy in cal/m^2/day and the efficiency of the conversion of light energy to fixed energy in the form of algal cells are known.

As flow × detention time = depth × surface area, the organic surface loading in kg of BOD per hectare per day can be estimated using the following relationship.

$$L_0 = 10 \left(\frac{d}{t} \right) BOD_L \qquad ... (6.9)$$

in which L_0 = organic loading in kg/hectare/day,

d = depth of pond in m,

t = detention time in days,

and BOD_L = ultimate soluble BOD in mg/1

The Eqn. 6.9 also gives the oxygen requirement for the aerobic decomposition of the waste, in Kg/hectare/day, in aerobic ponds.

The oxygen requirement thus found, when equated to the yield of oxygen, gives the magnitude of (d/t). By providing a suitable depth, the magnitude of t can be fixed; t should be atleast 6.5 days. Now from the rate of flow of the waste, the required capacity of the pond, and thus the surface area can be determined.

The above method may be applied with slight modification in the design of facultative ponds, taking into consideration of the BOD stabilization of solids in the bottom zone by anaerobic reaction. In this case the oxygen requirement is corrected to take into account the non-settleable portion of BOD which undergoes the aerobic decomposition in the aerobic zone. Normally a correction factor of 0.5 is taken. A greater value of depth is taken for this type of ponds; however a deeper pond may not increase the efficiency of the treatment.

As basically the stabilization ponds are biological treatment methods, in the second method, the principle of biological treatment kinetics are applied. Assuming a retarded first order reaction kinetics for the removal of BOD in the pond, the Eqn. 3.13 gives the relationship between the efficiency of the treatment in BOD removal, 1st order BOD removal rate constant, and the detention time, in a complete mix system. But it should be recognized that in practice the stabilisation ponds are never the complete mix type. The hydraulic regime within the tank assumes an intermediate condition in between plug flow and complete mix system; so these may be considered as plug flow systems with certain amount of axial dispersion of the materials. Wehner and Wilhelm have developed an equation correlating the efficiency, axial dispersion, detention time, first order BOD removal rate constant etc for this type of hydraulic condition (Eqn. 3.14). To facilitate the use of that equation, Thirumurthi developed design curves, as shown in Fig. 3.7 which gives the product of 1st order BOD removal rate constant (K) and detention time (t) corresponding to per cent BOD remaining (S_1/S_0) for different values of dispersion factors. It may be noted that the dispersion factor for idealized complete mix system is ∞, that for idealized plug flow is zero, and for most stabilization ponds it varies between 0.1 and 1.0.

In stabilization ponds, K varies from 0.05 to 1.0 per day. Under Indian conditions, in the temperature range of 20-35°C, this may be taken as 0.22 per day at 20°C, for primary ponds ; for secondary and tertiary ponds this value reduces to 0.1 and 0.06 per day at 20°C. Values of K for other temperatures may be calculated using Eqn. 3.4 with suitable temperature coefficient (θ) values.

Pond dispersion factor in aerobic ponds may be taken as 1.0, while that in facultative ponds varies in between 0.3 and 1.0 (a typical value of 0.5 may be assumed in design).

As most ponds are bound to receive some amount of settleable organics, and as the thick layer of accumulated sludge will turn the bottom layer in the. pond anaerobic, it is better to design the ponds as facultative ponds.

General Considerations

Accumulation of sludge and grit in the pond may be assumed to be 0.060 to 0.090 m³/capita/year. The ponds need cleaning once in five or six years.

Inlet pipes with the bell mouth at its end discharge near the centre of the pond. The effluent usually overflows near a corner on the wind war side. The overflow arrangements are box structures with multiple valved draw-off lines to permit operations even under seasonal variations in depth.

The height of the dikes enclosing the pond should be adjusted to provide with a free board of 1 m in the pond depth.

If the surface area of the tank is too large, it is advisable to provide two or more numbers of ponds.

If the soil is too pervious, the bottom and dikes should be sealed to prohibit seepage; a commonly used sealing agent is bentonite.

Example 6.2

Design a facultative stabilisation pond to treat a domestic sewage flow of 2 MLD, at a place, the latitude of which is 20°N. The 5 day 20°C BOD of the sewage is 200 mg/l. Suitable other data may be assumed apropos to Indian conditions.

Solution: Method I

5 day 20°C BOD = 200 mg/l

Assuming 1st stage BOD removal rate constant $K = 0.23$ per day in a BOD bottle test, the ultimate BOD

$$\text{of the sewage} = \frac{200}{1 - e^{-5 \times 0.23}} = \frac{200}{1 - 0.316}$$

$$= 293 \text{ mg/l}$$

At 20° N, the yield of photosynthetic oxygen

$$= 250 \text{ kg/hectare/day.}$$

Now from Eqn. 6.9 organic loading L_0

$$= 10 \times \frac{d}{t} \times 293 \text{ kg/hectare/day}$$

Assuming that (*i*) 50% of this load is non settleable, and (*ii*) it undergoes aerobic decomposition in the top layer,

Oxygen requirement $= 10\dfrac{d}{t}$ 293 × 0.5 kg/hectare/day

$\therefore$ $\quad 10\dfrac{d}{t}$ 293 × 0.5 = 250

$\therefore$ Solving $\dfrac{d}{t}$ = 0.171

Now providing d = 1.5 m

$$t = \frac{1.5}{0.171} = 8.77 \text{ days}$$

As flow × detention time
$\qquad$ = depth × surface area,

The required surface area $= \dfrac{2 \times 10^3 \times 8.77}{1.5}$ = 11.70 × 10^3 m^2

Providing two tanks in parallel, area of each = 0.585 × 10^4 m^2
Assuming length/width ratio = 2, the size of each tank
$\qquad\qquad\qquad\qquad$ = 54 m × 108 m

Method 2

Assuming effluent BOD_5 desired = 30 mg/l, and 50% of the BOD is stabilized by anaerobic decomposition alone,

$$\% \ BOD_5 \text{ remaining} = \frac{S}{S_0} = \frac{30}{200/2} = 30\%$$

Assuming a dispersion factor in the facultative pond = 0.5, from Fig. 3.7,

$$Kt = 1.3$$

Now assuming winter temperature in the pond = 15°C, and the value of K in the pond = 0.22 per day at 20°C,

$K_{15°c} = 0.22(1.06)^{15-20}$ = 0.1645 per day

$$\therefore \quad t = \frac{1.3}{0.1645} = 7.9 \text{ days}$$

Design flow = 2 × 10^3 m^3/day
$\therefore$ $\quad$ Capacity of the tank = 2 × 10^3 × 7.9
$\qquad\qquad\qquad\qquad$ = 15.8 × 10^3 m^3

150

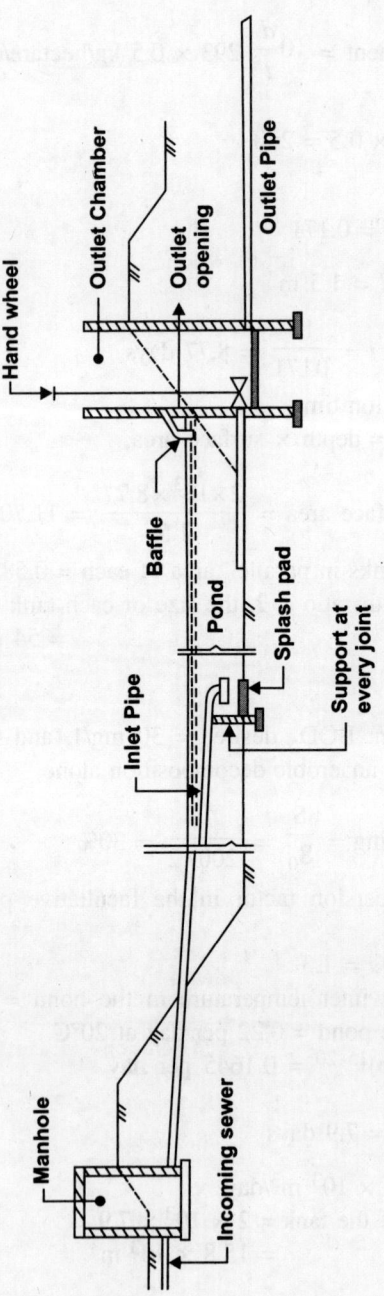

Figure 6.3: Waste Stabilisation Pond.

Providing a depth = 1.5 m,

$$\text{Surface area} = \frac{15.8 \times 10^2}{1.5} = 10.38 \times 10^3 \ m^2$$

Providing two tanks, area of each = $5.19 \times 10^3 \ m^2$

∴   Let us provide two tanks of 51 m × 102 m size.

As the maintenance of an aerobic layer in the top at all conditions is essential, the probable minimum depth of the top aerobic zone in the worst condition should be estimated before a final recommendation of the design. In India, a BOD loading in the range of 330 to 560 kg/hectare/day, will probably produce an aerobic layer in the top at all times.

6.3 EXTENDED AERATION PROCESS

For a small community, producing 4.5 MLD of waste or less, when sufficient land area is not available, an extended aeration process may be recommended for waste treatment instead of a stabilization pond or aerated lagoon. In fact, an extended aeration process is only a modification of the activated sludge process, and consists of an aeration tank, followed by a clarifier-thickener, and a recycle line. But in this process, a longer retention time and a lower food-

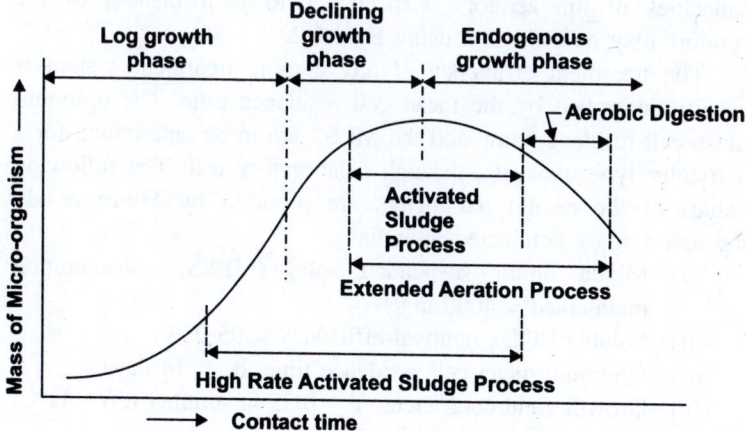

Figure 6.4: Sludge Growth Relationships for Various Aerobic Biological Reactors.

to-microorganisms ratio are provided to capitalize the microbial activities in their endogenous growth phase, as shown in Fig. 6.4. The biological mass which is synthesized at the declining growth phase is oxidized by the micro-organism itself at the endogenous growth phase. However, the biomass, synthesized at the declining growth phase, contains a small fraction which is not biodegradable, and accumulates as sludge in the system. The excess sludge thus produced is highly stable, and can directly be dried over a sand drying bed. Thus the process eliminates the sludge digestion arrangement. The primary settling tank is also very often eliminated in the treatment process, to simplify the sludge treatment and disposal. The aeration and mixing is accomplished by either surface mechanical aerators or diffused air system.

6.3.1 Process Design Considerations

The hydraulic regime in an extended aeration tank is the complete mix type with cellular recycle. Therefore the reactors can be designed employing usual microbiological growth kinetics, but employing proper kinetic growth coefficient for extended aeration process.

The surface aerators, if provided, are to be designed to provide the system with the oxygen requirement, on the basis of the rated capacities of the aerators. Expected field performances of the aerators may be estimated using Eqn. 6.2.

The treatment efficiency of any aerobic treatment system is usually controlled by the mean cell residence time. The optimum mean cell residence time and the MLSS are to be determined for a particular type of waste, through a laboratory test. The following values of the design parameters are reported by Mudri et al., through a study conducted in India.

(i) Mixed liquor suspended solid (MLSS) concentration maintained = 4000 mg/1

(ii) Soluble BOD_5 removal efficiency = 95-98%,

(iii) Optimum mean cell residence time, θ_c = 14 days,

(iv) Growth yield coefficient, Y = 0.55 in summer (26°- 32°C) to 0.66 in winter (19°–23°C)

(v) microorganism decay coefficient, k_d = 0.017/day in winter to

0.03/day in summer,

(*vi*) specific gross BOD removal rate constant, $k = 0.042$ in winter to 0.053 in summer,

(*vii*) Effluent suspended solids (ESS) contribute to the overall BOD of the effluent @ 0.45 kg/kg of ESS,

(*viii*) Volatile suspended solids in MLSS = 60-85%.

6.3.1.1 General requirements

The size and shape of an aeration tank is important for better mixing and aeration. The dimensioning of an extended aeration tank may be done in the same way as in conventional activated sludge processes. With diffused air system, the depth of the liquid should not be more than 3 to 5 m. When surface aerators are used the depth may vary from 1.2 m to 3.6 m.

6.3.1.2 Oxygen requirement

The expression for the oxygen requirement, derived in chapter 3, indicates the amount which is required for the stabilization of the soluble BOD only. Additional oxygen will be required for the oxidation of ammonia, produced during the oxidation of biomass in the endogenous phase. Again when a raw municipal sewage is directly admitted into the process, some of the suspended solids are converted to soluble components by microbiological enzymatic hydrolysis, and are utilised for cell synthesis. So additional oxygen is to be supplied to satisfy the demand exerted by this portion of the waste. In conservative design, the oxygen requirement is assumed to be equal to the total ultimate BOD removed or is calculated @ 2 kg of O_2/kg of total BOD_5 removed.

6.4 OXIDATION DITCHES (PASVEER DITCHES)

Oxidation ditches are simple artifical ditches, dug in the ground for the purification of sewage using a very simplified technique. The basic form of the plant is a ring shaped circuit or ditch. Across the length of this ditch, an aeration rotor is mounted, which provides for oxygenation as well as circulation of the sewage.

The raw sewage when it enters at A, gets diluted with large amounts of purified sewage which is present in the ditch. The intensive aeration, which follows, not only produces a clean and

154

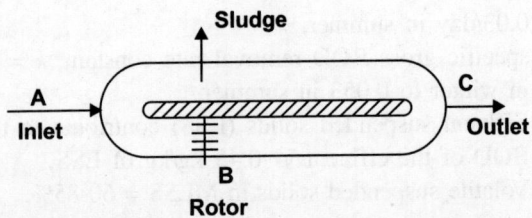

Figure 6.5: Oxidation Ditch Intermittent Flow Type.

purified sewage, but the sludge formed in the purification process is mineralized to such an extent that it can be dried without causing any objectionable odours.

Oxidation ditches are constructed in two types—(*i*) continuous flow type and (*ii*) intermittent flow type. In a continuous flow type ditch, the mixed liquor is allowed to settle in a separate settling tank and a part of the settled sludge is returned to the ditch.

In an intermittent flow type ditch, no separate settling tank is used; the flow in the ditch remains suspended during a predetermined period and the ditch itself is used for settling. The clarified supernatant is withdrawn through the outlet; the removal of surplus sludge is effected with the aid of a "sludge trap", shown

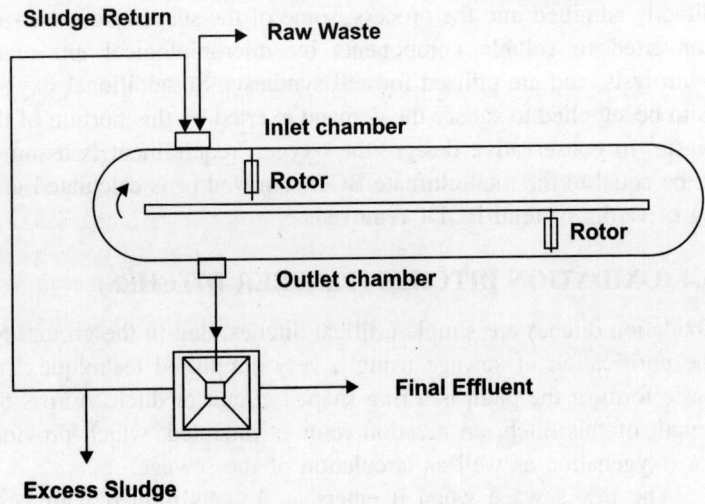

Figure 6.6: Oxidation Ditch Continuous Flow Type.

in Fig. 6.5 (Sludge traps are small metal or concrete sedimentation funnel which is placed in or partly beside the circuit, and whose edge projects above the water; small inlet and outlet apertures are incorporated in the edge. From the amount of circuit liquid flowing regularly through the tap, a certain portion of sludge is deposited. Accumulated portions are removed by pump).

6.4.1 Design Considerations

Oxidation ditch is, in fact, a particular form of Extended Aeration tank. As such, the oxidation ditches can be designed using the same principle, as elaborated in the previous section.

In regard to the general requirements, the depth of the ditch is usually kept in between 1.0 to 1.5 m. The top width of the ditches are selected to suit the length of the aeration rotor. A trapezoidal section, with a side slope of 1: 1, may be more effective as it has a greater exposed area.

6.4.1.1 Aeration rotor design

The success of any aerobic biological process depends to a great extent on the selection of an aeration system and its operating conditions. Therefore the design of the aerator rotors in the oxidation ditch should get a careful attention, as their performances vary widely under different operating conditions. Usually two types of aerator rotors are in use:

(*i*) cage rotor, as shown in Fig. 6.6, consisting of two circular discs at two ends of the rotor connected by a hollow shaft and a number of equally spaced angle sections, on which rectangular strips are mounted, the positions of the strips being staggered.

(*ii*) Angle-iron aerator rotor (Kessener Brush type) which consists of hollow shaft, on which a number of equally spaced iron angles sections are mounted in a number of rows, the positions of the iron angles being staggered on alternate rows.

The oxygenation capacity of the aerators vary with the shape and size of both the aerator and the aeration tank. The capacity increases with the submergence of the rotor in the waste, upto a certain critical value, and then decreases with the depth of submergence. It also increases with the increased speed of the rotor, but the power requirement for increased RPM increases at a higher rate. An optimum revolution of 72 RPM and a submergence of 135 mm is often recommended for Indian conditions. The information regarding the capacity of oxygenation at different speed and depth of submergence with different types and sizes of aerator rotors can be obtained through the performance characteristic curves. Usually these curves are prepared for a standard condition of temperature and pressure. Pasveer prepared such characteristic curves for two particular types and sizes of rotors for a standard temperature of 10°C and pressure of 760 mm of mercury. Characteristic curves for other types and sizes of rotors may be obtained through model studies.

The oxygenation capacity obtained from the characteristic curves, are usually applicable for either tap water or distilled water at a standard condition. The expected oxygenation capacity for the waste water under the field conditions may be estimated using the following relationship:

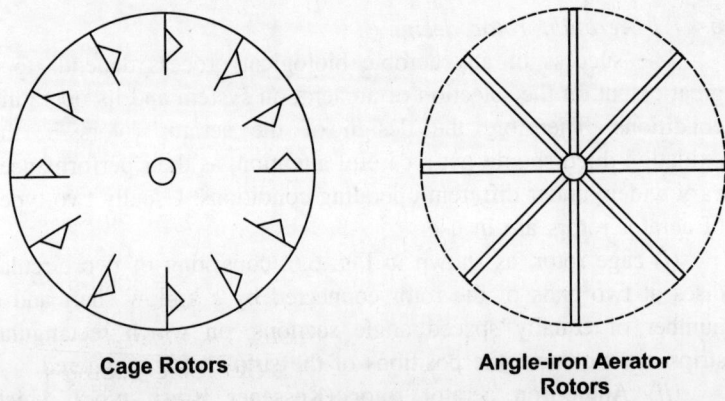

Cage Rotors   **Angle-iron Aerator Rotors**

Figure 6.7: Aerator-Rotor for Oxidation Ditches.

$$OC_{ST} = \alpha. \; OC_{20} \left[\frac{\beta C_{SST} - C_{SOT}}{C_{WS}} (1.016)^{T-20} \right] \quad ... \; (6.10)$$

in which, OC_{ST} = oxygenation capacity with waste water at the operating temperature of T°C, kg of O_2/hr/meter length,

OC_{2o} = rated oxygenation capacity with water at 20°C,

kg of O_2/hr/metre length of rotor,

α = Oxygen transfer correction factor for waste water, usually 0.8-0.85,

β = salinity-surface tension correction factor, usually 1,

C_{SST} = saturation dissolved oxygen concentration of, waste water at operating temperature,

C_{SOT} = operating dissolved oxygen concentration of the sewage,

and Cws = saturation dissolved oxygen concentration of water at 20°C.

The oxygenation capacity of the rotor under the field conditions is now to be used, along with the oxygen requirement for the treatment, to find out the total length of the rotors required. The length is then to be divided into a number of rotors, equally spaced along the length of the ditch.

Example 6.3:

Design a continuous flow type oxidation ditch to treat a domestic sewage flow of 2 MLD. Following values of the parameters and conditions may be assumed.

Total 5 day 20°C BOD of the raw sewage	= 150 mg/1
Desired total 5 day 20°C BOD at the effluent	= 20 mg/1
Suspended solids in the raw sewage	= 230 mg/1
Desired suspended solids at the effluent	= 30 mg/1
Mixed liquor suspended solids	= 5000 mg/1
Growth yield coefficient	= 0.55
Microorganism decay coefficient	= 0.03
Mean cell residence time	=15 days.

Solution

1. Kinetic equations derived in chapter 3, are applicable for any soluble waste, or for the soluble part of any waste. As such, it is necessary to modify the given data before its application in the said equations. Moreover, in any system treating soluble waste, the mixed liquor volatile suspended solid are taken as the mass of the active microorganism in the reactor. This may not be true for a system treating waste like the unsettled domestic sewage which consists a large portion of inactive volatile suspended solids in it. So

necessary corrections are also to be made in the determination of microbiological mass.

Assuming usual association of BOD with suspended solids in the municipal sewages @ 0.25 kg/kg of SS, in Indian conditions, the influent soluble BOD_5, $S_0 = 150\text{-}230 \times 0.25$

$$= 92 \text{ mg/l}$$

Assuming that the effluent suspended solids (ESS) contribute to the overal BOD of the effluent @ 0.45 kg/kg of ESS, the soluble BOD_5 at the effluent, $S_1 = 20\text{-}0.45 \times 30$

$$= 6.5 \text{ mg/l}$$

Figure 6.8: Dorr-Oliver Surface Aerator.

Now assuming the microbial mass concentration to be 60 % of the MLSS (to take care of the inactive volatile suspended solids in the raw sewage).

The microbial mass concentration $X = 0.6 \times 5000 = 3000$ mg/l

2. Now $X = \dfrac{Q}{V} \theta_c \dfrac{Y(S_0 - S_1)}{1 + k_d \theta_c}$

∴ Volume of the reactor,

$$V = \frac{2 \times 10^6 \times 15 \times 0.55(92 - 6.5)}{3000(1 + 0.03 \times 15)} = 324 \text{ m}^3$$

3. Again, $\dfrac{dX}{dt} V = \dfrac{XV}{\theta_c}$

∴ Microbial mass in the excess sludge/day, $\dfrac{dX}{dt} V$

$$= \frac{3000 \times 324 \times 10^3}{15 \times 10^6} \text{ Kg} = 64.8 \text{ kg/day}$$

∴ Excess total solids per day $= \dfrac{64.8}{0.6} = 108$ kg/day.

4. If wasting of excess sludge is accomplished from the tank itself, the rate of wasting $= \dfrac{324}{15} = 21.6$ m³/day.

5. Now assuming a 50% recirculation, and same percentage of active microorganisms in the return sludge, the desired concentration in the underflow of the clarifier—thickener, X'_r, is given by:

$$5000 (Q + 0.5 Q) = X'_r (0.5 Q).$$

∴ Solving, $X'_r = 15000$ mg/1.

6. Oxygen requirement = total ultimate BOD removed
$(150-20)^2 \times 1.47 = 382$ kg/day.

7. Checks:

(i) Hydraulic retention time provided $= \dfrac{V}{Q}$

$$= \frac{324 \times 10^3}{2 \times 10^6} \text{ days}$$

$$= 3.89 \text{ hrs.}$$

(ii) Process loading $= \dfrac{(150-20)2}{5000 \times 324 \times 10^{-3}}$

$= 0.160$ kg total BOD/kg of MLSS.

8. Ditch dimensions:

Assuming a trapezoidal cross section for the ditch with 1:1 side wall slope, 1.2 m liquid depth and 1.0 m bottom width,

cross sectional area $= \dfrac{1+3.4}{2} \times 1.2$

$= 2.64$ m^2

$\therefore$ Total length of ditch required $= \dfrac{324}{2.64}$ m $= 122.7$ m

Now providing a partition within the ditch with 16 cm top width, radius of curvature of the centre line of the curved portion of the ditch

$= 0.08 + \dfrac{3.4}{2} = 1.78$ m.

$\therefore$ Length of the curved portions at two ends
$= 2\pi \ 1.78 = 11.18$ m

$\therefore$ Straight length of the ditch

$= \dfrac{122.7 - 11.18}{2} = 55.75$ m

9. Assuming oxygenation capacity of a 70 cm dia cage rotor $=$ 2.5 kg of O_2/hr/meter length, at 72 RPM and 135 mm of depth of submergence, at the field operating condition,

total length of rotors required $= \dfrac{382}{2.5 \times 24} = 6.35$ m.

If length of each rotor is 2.25 m, number of rotors required = 3 nos.

QUESTIONS

1. Compare the process design considerations in aerated lagoons and oxidation ditches.

2. Design a facultative stabilization pond to treat a domestic sewage flow of 3 mld, at a place with a latitude of 32 degrees north. The 5 day BOD of the sewage is 230 mg/l. Assume suitable data for Indian conditions.

7 | Experimental Studies in Biological Waste Treatment Design

7.0 INTRODUCTION

The objectives in any biological waste treatment studies in the laboratory are usually two fold:

(*i*) to examine whether the waste is amenable to biological treatment or not, and

(*ii*) to determine the kinetic growth coefficients, that may be used in the design and operation of pilot—or prototype plants.

The experimental studies may be carried out either in a batch reactor or in a continuous flow reactor. However, a continuous flow reactor should be employed wherever possible, as it simulates a prototype closely.

Depending on the characteristics of the raw waste, and the expected response to the particular type of biological treatment, the tests are to be conducted to assess the treatability under either aerobic or anaerobic conditions. Tests are to be conducted at a constant room temperature. The feed is to be pretreated for the removal of heavy suspended solids and floating oil or grease. The pH of the feed is to be adjusted for the optimum metabolism of the microorganisms.

7.1 AEROBIC TREATMENT STUDIES IN CONTINUOUS REACTOR

A continuous treatment unit as shown in Fig. 7.1 constructed by

162

plexiglass and both aeration chamber and settling chamber being combined in one unit may be used for the studies. The aeration and settling chambers are separated by an adjustable baffle.

The waste is continuously fed from a feed bottle employing an air-pressure flow system. It is necessary to connect the air pressure regulator with the feed bottle by a capillary tube.

The mixed liquor of the aeration chamber may be aerated by compressed air or by using porous difusers. Alternatively aquarium aerators may also be employed for this purpose; but aquarium aerators may become choked at higher MLSS concentration.

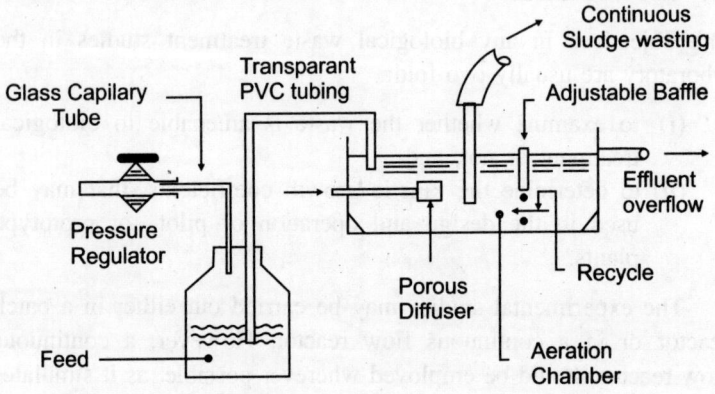

Figure 7.1 Laboratory Scale Activated Sludge Unit.

Continuous wasting of sludge may be accomplished from the aeration chamber using diapham pump or vacuum.

The effluent may be drawn off by pump or vacuum, or through a gravity overflow system, as shown in Fig. 7.1.

To start with the experiment the unit is filled with the waste along with seed sludge if required, and aerated for certain period (say one day). After that the aerated waste is allowed to settle and the supernatant is decanted, and a fresh batch of waste is added. The entire cycle is repeated for several days till the MLSS builds up to a certain level.

For some industrial waste the seed sludge may have to be acclimatized in the tank or in a separate reactor. Acclimatization involves the exposure of the microorganism in an aerated chamber in a gradually increasing dose.

When MLSS concentration reaches a desired level (say 4000 mg/1) with or without acclimatization, the continuous feeding operation is started, with proper adjustment of air pressure regulator, to have a desired hydraulic retention time. The sludge is withdrawn continuously at a rate consistent with the desired mean cell residence time. The mixed liquor is aerated continuously at such a rate that the operation of settling is not disturbed in the settling chamber.

The above operations are continued till a steady state condition is reached, indicated by a constant effluent substrate concentration. Now in the treatment system, flow Q, and influent substrate concentration S_0 are fixed, and mean cell residence time θ_c is adjusted by fixing the rate of withdrawal of sludge, $\Delta X/\Delta t$. The following other parameters are determined in the steady state condition:

(i) effluent substrate concentration, S_1,

(ii) Food-to-microorganism ratio, $\Delta F/\Delta t$, using the following relationship (on finite time and mass basis)

$$\frac{\Delta F}{\Delta t} = \frac{Q(S_0 - S_1)}{V} \qquad \ldots (7.1)$$

in which V = volume of the aeration chamber.

(iii) Mixed liquor volatile suspended solids, X

The test is repeated for different values of θ_c and corresponding sets of values of Q, S_0, S_1, $\Delta X/\Delta t$, $\Delta F/\Delta t$, and X are recorded. For each set of reading the percentage removal of the substrate is calculated.

If condition demands sets of such tests are to be carried out under different temperatures of operation, and mixed liquor suspended solids concentrations.

The treatability and the optimum value for θ_c is determined from a "percent substrate removal Vs. mean cell residence time" plot, as shown in Fig. 7.2(a).

Now the specific utilization, on a finite time and mass basis is given by the following equation:

$$\frac{\Delta F / \Delta t}{X} = \frac{k_m S}{K_3 + S} \qquad \ldots (7.2)$$

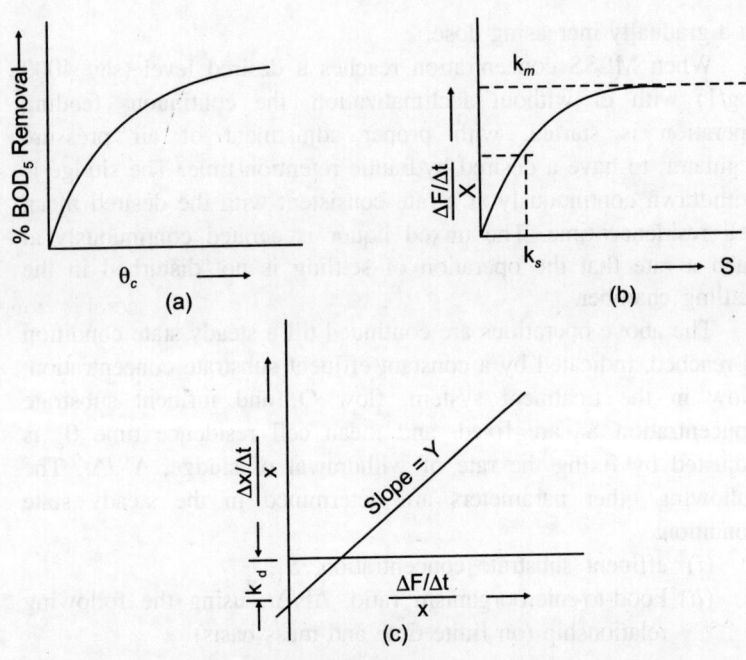

Figure 7.2: (a) Studies for Treatability, (b) Determination of k_m and k_s, (c) Determination of Y and k_d.

So, from the values determined in each test, a "$\Delta F/\Delta t/X$ Vs S_1" plot will determine the values of k_m and K_s as shown in Fig. 7.2(b).

The values of Y and k_d may be determined from a plot of "$(\Delta X/\Delta t)/X$ Vs. $(\Delta F/\Delta t)/X$". The slope of the best fit straight line obtained will give the value of Y, while the intercept is equal to k_d, as shown in Fig. 7.2(c).

7.2. AEROBIC TREATMENT STUDIES IN SEMI-CONTINUOUS REACTORS

Semi-continuous reactors, similar to that shown in Fig. 7.1 may be used in the aerobic treatment studies, if continuous feeding and sludge withdrawal appear to be difficult.

In the semi-continuous mode the seed microorganisms are also to be acclimatized if necessary and the MLSS concentration is to be built up to a predetermined level. The content of the reactor is aerated continuously. When the MLSS builds up to a desired level the mixed liquor is wasted at a fixed time interval; prior to the wasting the rate of aeration is to be increased to allow a thorough mixing of the contents in the reactor. The amount withdrawn is immediately to be replaced by a fresh batch of feed.

The rate of wasting of sludge establishes the mean cell residence time. Other measurements include the influent and effluent substrate concentration (in this case BOD_5), volatile suspended solid concentration at different time, and the DO depletion rate using a DO meter and DO probe.

The oxygen requirement in any aerated biological reactor may be given by the following equation:

$$O_2 \text{ required/day } = Q(S_0 - S_1)(1.47 - 1.42Y) + 1.42\ k_d\ XV$$

$$.... \quad (4.6)$$

Now for a semicontinuous reactor as described above, Eqn. 4.6 may be modified as below:

$$\frac{\Delta O_2 mg/1/min}{X} = \frac{\Delta S\ mg/1}{\Delta t.X} \cdot A + B \qquad ... \quad (7.3)$$

in which, $A = 1.47 - 1.42Y$,

$\qquad\qquad B = 1.42\ k_d$,

$\qquad\quad \Delta O_2 = $ mean rate of depletion of oxygen in the time interval of Δt, mg/l/min,

$\qquad\qquad \Delta t = $ time interval between two sampling, minutes,

$\qquad\qquad \Delta S = $ removal of BOD_5 in time Δt,

and $\qquad X = $ volatile suspended solid, mg/1.

The Eqn. 7.3 provides the basis for determining Y and k_d from the BOD_5 removal and oxygen depletion rate data.

To start with the test, prior to the daily wasting of sludge, two or three mixed liquor samples are withdrawn and the rate of oxygen utilization per minute is determined using a D.O. meter. The depletion rate divided by the value of X represents the constant B in Eqn. 7.3. Immediately after the above determination, the sample is replaced in the reactor.

Just after the addition of fresh feed, initial BOD_5, S_0, is determined. Then at short intervals, over the next two or three hours, both BOD_5 remaining, S_t, and the D.O. depletion rate are determined.

Now, $ln\ (S_t/S_0)$ is plotted against time. The slope of the straight line thus obtained (Fig. 7.3a) gives the value of K; the value of k is obtained by dividing K by X.

Now from the observed oxygen depletion rate at the begining and at the end of a time interval the mean oxygen depletion rate is obtained and divided by X. The corresponding average rate of BOD_5 removal during this interval is also calculated, and divided by X. Now the values of $(\Delta O_2/X)$ are plotted against the corresponding values of $(\Delta S/\Delta t)/X$, in a graph as shown in Fig. 7.3(b). The intercept B is determined first from the observations taken prior to the wasting and feed step. The slope of the best fit straight line represents the constant A in Eqn.7.3.

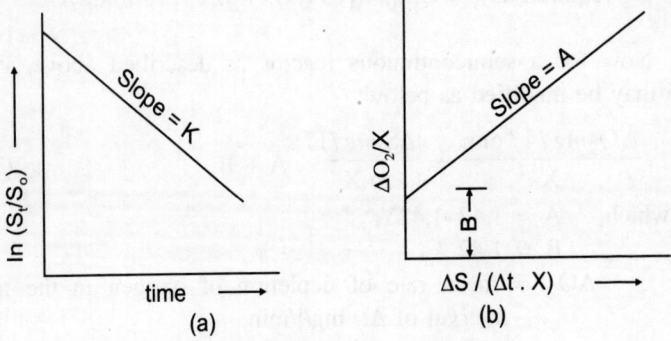

Figure 7.3 (a): Determination of K, (b) Determination of Constants A and B, and hence Y and k_d.

The values of A and B thus found may be used now to calculate the values of Y and k_d.

7.4. ANAEROBIC TREATMENT STUDIES IN BATCH REACTORS

In some types of the anaerobic digesters, feeding of raw sludge, withdrawal of supernatant liquor, and wasting of sludge are done on an intermittent pumping schedule. On the other hand a continuous feeding and withdrawal of mixed liquor is preferred in two-stage digestion processes. Depending on the preference the laboratory tests may be carried out on bench scale models of either batch or continuous-flow reactors.

A simple batch reactor, as shown in Fig. 7.4, may be used for the purpose of assessing treatability.

The tests are to be conducted under a constant room temperature.

The test assembly includes (*i*) one stoppered digester bottle equiped with sludge wasting and feed arrangements, (*ii*) a graduated bottle for the collection of gas, and (*iii*) a brine collector, which collects the brine solution, displaced by the gas collector bottle.

A digester being a complete-mix-no-recycle system, the θ_c is equal to θ. And the θ_c value may be established by fixing the wasting rate. Now as the volume of liquid removed as wasting must equal the volume of feed, the raw sludge must be diluted accordingly with distilled water.

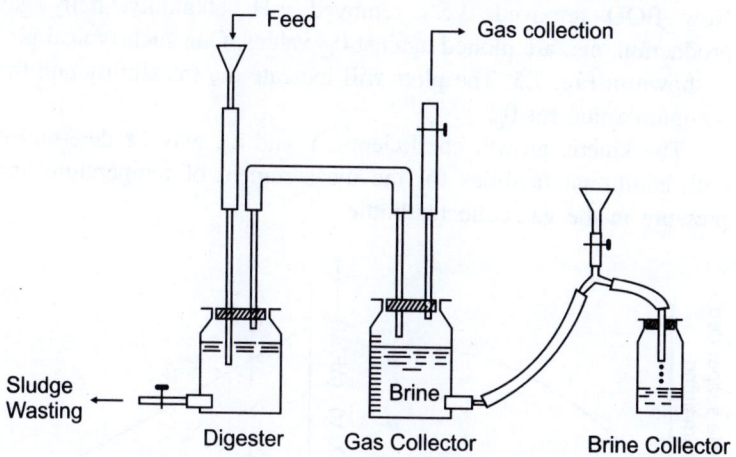

Figure 7.4. **Laboratory Scale Anaerobic Reactor.**

The experimentation is done as follows. The seed sludge is kept in the sealed digester bottle for one day. This results in the removal of dissolved oxygen if any, from the waste by the facultative bacteria. A predetermined amount of sludge is then wasted, the amount being dictated by the value of θ_c. An equal volume of diluted raw sludge is simultaneously fed into the digester bottle. It may be noted that the digester bottle must be thoroughly shaken several times in a day and positively before the wasting.

At the start of the test the gas, if there is any, should be expelled from the gas collector bottle. Accumulation of gas is indicated by the fall of brine layer in the bottle.

The cycle of wasting and feeding is to be continued till a steady state condition is reached. The steady state condition is indicated by the accumulation of equal volumes of gas in each day.

At steady state, daily gas production, BOD of settled effluent and the room temperature are recorded. The removal efficiency in terms of BOD, and percent stabilization based on methane gas production are calculated. For the later, of course, the analysis of the gas is required to determine the fraction of methane in it. Temperature, pH, volatile solids concentration, alkalinity etc. of the withdrawn slurry may also be noted.

The entire experiment is repeated for different values of θ_c. Now BOD removed, VSS removed, pH, alkalinity, daily gas production, etc. are plotted against θ_C values. One such typical plot is shown in Fig. 7.5. The plots will indicate the treatability and the optimum value for θ_c.

The kinetic growth coefficients, Y and k_d, may be determined with additional facilities for the measurement of temperature and pressure in the gas collector bottle.

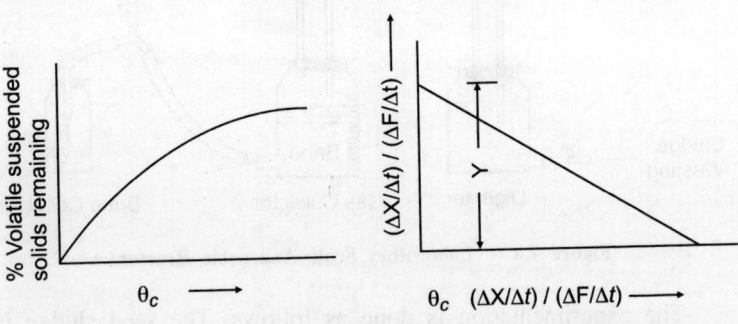

Figure 7.5: Studies for Treatability by Anaerobic Digestion.   Figure 7.6: Determination of Kinetic Coefficients in anaerobic Digestion.

Recalling Eqn. 5.3 and modifying that for a finite time and mass basis, we get:

$$\frac{\Delta X}{\Delta t} = \frac{Y(\Delta F / \Delta t)}{1 + k_d \theta_c} \qquad \text{... (7.4)}$$

Now again from. Eqn. 5.7 on a finite time and mass basis,

$$C = 350 \ (eF - 1.42 \ \Delta X / \Delta t) \qquad \text{... (7.5)}$$

in which C is the litres of methane gas produced/day at standard condition (0°C and 760 mm of mercury).

The volume of methane gas produced at the operating condition can be reduced to that at standard condition using the following relationship.

$$C = 0.37 \frac{(p_1 - p_w)C_1}{273.1 + T_1} \qquad \text{... (7.6)}$$

in which, p_1 = pressure under which the volume of gas is measured, in mm of mercury,

T_1 = temperature under which the volume of gas is measured, °C,

p_w = vapour pressure of water at T_1°C, in mm of mercury,

and C_1 = volume of methane, in litres, in the collected gas.

So using Eqns. 7.5 and 7.6 the value of eF can be estimated since $\Delta X/\Delta t$ is known.

But, on a finite time and mass basis, from Eqn. 5.4.

$$\frac{\Delta F}{\Delta t} = eF \qquad \text{... (7.7)}$$

Therefore in Eqn. 7.4, all the terms excepting the kinetic growth coefficients are known. To facilitate plotting the said equation is modified as below:

$$\frac{(\Delta X / \Delta t)}{(\Delta F / \Delta t)} = Y - k_d \left[\frac{(\Delta X / \Delta t)}{(\Delta F / \Delta t)} \ \theta_c \right] \qquad \text{... (7.8)}$$

Now from the observed and calculated values of $(\Delta X/\Delta t)$ and $(\Delta F/\Delta t)$, the ratio $(\Delta X/\Delta t) / (\Delta F/\Delta t)$ is calculated for different values of θ_c. The ratio $(\Delta X/\Delta t) / (\Delta F/\Delta t)$ is then plotted against the product $\theta_c . (\Delta X/\Delta t) / (\Delta F/\Delta t)$. as shown in Fig. 7.6.

The intercept of the straight line thus obtained gives the value of Y, while the slope of the line indicates the value of k_d.

QUESTIONS

1.   Write an explanatory note on Treatability Studies.

8 | New Concepts in Biological Waste Treatment

8.0 INTRODUCTION

The wide spread interest regarding the treatment and handling of waste water because of their growing deleterious effects on the receiving streams, lead to many new developments in the field of waste water treatment. These developments include improvements of the conventional treatment processes for more effective removal of the pollutants, and new treatment processes capable of removing the pollutants not ordinarily removed by conventional treatment. Few of the new methods, although conceptually similar to the conventional treatment processes, are radically different from the later in regard to the operation and performances.

In this chapter, some of the promising latest developments in the field of biological waste water treatment processes will be described briefly.

8.1 NITROGEN REMOVAL BY BIOLOGICAL NITRIFICATION AND DENITRIFICATION

Most of the existing waste water treatment plants are not designed to remove nitrogen and phosphorous. Removal of either or both of these nutrients may be required to prevent the degradation of the receiving water courses, caused (1) by the usual problem of eutrophication associated with the discharge of these nutrients, and also (2) by the oxygen depletion, resulting out of the ammonia oxidation by nitrifying bacteria.

Nitrogen can be removed from the waste water by the

progressive biological oxidation of nitrogen compounds to nitrites and nitrates followed by conversion to nitrogen gas. In fact, the nitrogen removal by biological means consists of the following two distinct steps: (i) Nitrification, where the ammonia is oxidised to nitrites and then to nitrates by aerobic nitrifying autotrophic bacteria, and (ii) Denitrification, where nitrates are reduced to nitrogen gas by either autotrophic or heterotrophic anaerobic bacteria.

8.1.1 Nitrification

Nitrification can be achieved in an activated sludge process along with the usual carbonaceous oxidation, with some modifications in the design. But in regard to the necessary conditions for the 100% nitrification in the activated sludge process, there are many contradictions and discrepancies in the available literature on the subject. Nitrifying bacteria have got a very slow growth rate, compared to heterotrophic bacteria. Generally a mean cell residence time of 3 to 4 days at 20°C which corresponds to a loading factor of 0.3 to 0.4 kg of BOD_5 per day per kg of MLSS, at a MLSS concentration around 2000 mg/l, is found adequate for the desired degree of nitrification.

The kinetic growth equations, used in Chapters 3 and 4 are applicable for the analysis and design of the reactors for biological nitrification. However, in most of the studies the microorganism decay coefficient, k_d, is neglected. The kinetic coefficients for biological nitrification in some typical wastes are given in Table 8.1.

TABLE 8.1: Kinetic Coefficients for Biological Nitrifications

Waste type	Y	k_d	k_m	K_s	μ_m	$[\theta_c]_c$	Reference
Synthetic river water	0.29	0.05	1.8	3.6	—	2.1	Stration and McCarty (at 20°C)
Activated sludge	0.05	—	6.6	1.0	0.33	3.0	Downing et al (at 20°C)
Synthetic waste	—	—	—	8.33	1.28	0.86*	Siddiqi et al

* Computed from the given data of influent ammonia-N concentration of 80–85 mg/l.

It may be noted that the process of nitrification in a mineral solution is much faster than that in a municipal waste water, If an activated sludge process is designed with proper attention to nitrification, it is expected that both carbonaceous oxidation and nitrification will be achieved simultaneously in the same reactor.

The trickling filters can also achieve the desired degree of nitrification along with the stabilisation of organic carbonaceous substances. As explained in section 4.2.1 in chapter 4, the performance of the trickling filter, in nitrification also, is profoundly influenced by specific surface area of the supporting medium, and the influent substrate concentration. The filters may be designed by the method as described in section 4.2.1 of chapter 4.

The kinetic coefficients Y, k_m and K_s, and the product $k_m \times Xd$ (in the case of trickling filter only), should be determined, before the design is carried out, in a pilot plant study, using the actual waste and the type of media used.

Though the nitrification can be achieved in the conventional aerobic biological treatment processes, some times it is recommended to separate the two processes, *viz.* the carbonaceous oxidation and nitrification in two stages using separate reactors having separate sludge systems at each stage. (Fig. 8.1). In the first stage, about 80% of the carbonaceous material is oxidised in the activated sludge process. The first stage effluent with residual carbon and almost entire amount of ammonia is then passed on to the second stage reactor, for nitrification, enriched with nitrifying autotrophic bacteria. This method will offer greater flexibility of operation, and each stage can be designed for their optimum performances.

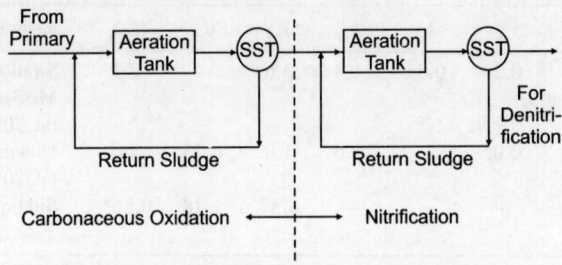

Figure 8.1: Two Stage Biological Nitrification.

8.1.2 Denitrification

In the process of denitrification, the nitrates present in the waste are reduced to nitrogen gas by the anaerobes, while utilising them as hydrogen-acceptors in the process of biological oxidation of the supplied substrate. Therefore, a denitrification process essentially requires a fully nitrified influent, an anaerobic condition and a supply of proper substrate. While oxygen is supplied in a balanced amount to any conventional aerobic biological treatment process for the stabilization of the substrate, in the process of biological denitrification, a balanced amount of substrate is supplied to reduce the nitrate in the process of stabilisation of the supplied substrate under anaerobic conditions.

Most of the denitrifiers are anaerobic heterotrophic microorganisms. As such the process of denitrification needs to be supplemented by the organic carbon. The carbon source may be a part of the organic carbonaceous waste, or an organic chemical like methanol, or methyl alcohol. However, methanol is considered as the most economical source of supplemental carbon.

Denitrification can alternatively be accomplished by employing autotrophic nitrifiers with inorganic sulphur supplementation in the form of thio-sulphate and sulphide. However additional sources of nutrients for cell synthesis *viz.* inorganic carbon, ammonia, and trace elements are also to be added into the reactor.

Both complete mix fluid bed biological reactors Fig. 8.2, and fixed film biological reactors are tried for the denitrification. The anaerobic upflow filters are found to be more effective in the denitrification of secondary effluents. Pilot plant studies indicate that a mean cell residence time of 3 to 4 days is required for 90% reduction of nitrate in complete mix fluid bed reactors. On the other hand a little more than 2 hours of detention time is adequate for similar efficiency in an upflow anaerobic filters—probably due to a very high mean cell residence time established in those filters.

Keeping in mind that the nitrates are reduced in the process of oxidation of the supplied substrate, the kinetic growth equations of the activated sludge process may be adapted for the analysis and design of the denitrification process. The oxygen concentration terms are to be replaced by oxygen-equivalent concentration of the nitrates. The rate of nitrate reduction therefore can be given by

174

suitably modifying the Eqn. 3.37 in the following form:

$$OE_0 = (S_0 - S_1) (1-1.42\ Y) + 1.42\ k_d\ X \frac{V}{Q}\quad .... \quad (8.1)$$

in which OE_0 = oxygen equivalent (OE) concentration of the nitrate plus residual dissolved oxygen, if there is any, in the influent,

S_0, S_1 = ultimate BOD of the substrate in the influent and effluent respectively,

Y = growth yield coefficient, on BOD_{ult} basis,

k_d = microorganism decay coefficient,

V = volume of the reactor,

and Q = rate of flow.

The contribution of second term in Eqn. 8.1, i.e. that of endogenous respiration, to the rate of nitrate reduction is small compared to the rate of nitrate reduction, and hence may be neglected for all practical purposes.

TABLE 8.2: Kinetic Coefficients for Biological Denitrification.

Substrate	Coefficient basis	Y	k_d	K_3	k_m	reference
Milk solids	BOD_5	0.61	0.071*	150	2.5	Johnson
Methonol	COD	0.18	0.04	9.1	..	Stencel
Glucose	Glucose	0.25	—	—	—	Hadjpetrou (as quoted by Johnson)
$S_2O_3=$	Nitrate	0.703+	0**	—	—	Bisogni
$S=$	Nitrate	0.704+	0**	—	—	,,

* computed value

** Endogenous decay is assumed to be negligible for autrotrophic microorganisms

+ Thermodynamically predicted values; observed yield varied from 0.4 to 1.4 for the thiosulphate-substrate system. The values are equivalent to about 0.246, calculated on BOD_{ult} basis.

The other kinetic equations, as described in chapters 3 and 4 in connection with the design of both fluid bed and fixed film biological reactors, will be applicable in the process of biological denitrification without any modification. Kinetic coefficients for biological denitrification are given in Table 8.2. In a fluid bed reactor, a mixed liquor suspended solids concentration of around 2000 mg/1 and mean cell residence time in the range of 1 to 5 days are found to be adequate.

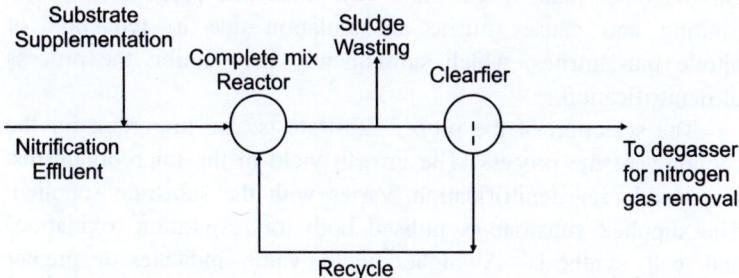

Figure 8.2. Complete Mix Fluid Bed Reactor for Denitrification.

As pointed out earlier, in the process of denitrification the substrate or carbon source is to be supplied, in a specific proportion to the nitrate present in the waste. This proportion is often defined by the "substrate-nitrate ratio" by weight, C_F. The substrate portion of this ratio is usually expressed in terms of either the actual weight of the substance used, or COD, or BOD. However, it will be shown later that it is convenient to express this ratio in terms of oxygen equivalent of both the substrate and the nitrate.

$$C_F = \frac{(\text{weight of the substrate added})}{(\text{weight of the nitrate present})} \qquad \text{.... (8.2)}$$

or

$$C_F = \frac{[\text{BOD}_5 \ (\text{or COD}) \ \text{of the substrate added}]}{[\text{weight of the nitrate present}]} \qquad \text{.... (8.3)}$$

or

$$C_F = \frac{[\text{Oxygen equivalent of the substrate added}]}{[\text{Oxygen equivalent of the nitrate present}]} \qquad \text{....(8.4)}$$

In most of the denitrification studies, employing heterotrophic organisms, the methanol is used as a carbon source. And, as such, the substrate-nitrate ratio (by weight) is usually expressed as the weight of the methanol added to the weight of the nitrate present.

The substrate-nitrate ratio plays an important role in the process of denitrification. A higher ratio will make the process nitrate-limiting and causes higher BOD in the effluent. On the other hand a low ratio will make the process substrate-limiting and causes nitrite accumulation due to reduction of nitrate into nitrites, which subsequently may inhibit the process of denitrification.

The selection of the proper substrate is also important for the economy of the process. The growth yield of the microorganisms, responsible for denitrification, varies with the substrate supplied. The supplied substrate is utilised both for respiration (oxidation) and cell synthesis. A higher yield value indicates a greater expenditure of the substrate for cell synthesis, which in fact has no direct relation with the nitrate reduction. On the other hand, a lower yield indicates more efficient use of the substrate in respiration, and thus in the reduction of nitrate. As such a substrate must be used, which has got the least growth yield coefficient. Lower the yield, greater is the economy.

A term "consumptive ratio" is often used to compare the relative economy of different substrates in their use in the denitrification. Consumptive ratio, C_R, is usually defined as the ratio of the total quantity of substrate utilized during denitrification, to the stoichiometric requirement of the substrate for denitrification.

$$C_R = \frac{\text{grams of substrate consumed}}{\text{Stoichiometric requirement of the substrate in grams}}$$

$$\text{... (8.5)}$$

The consumptive ratio, C_R, of any substrate is directly related to the growth yield coefficient Y, of the microorganism in the same substrate and practically represents the inverse of the quantity $(1 - 1.42\ Y)$. Higher the yield, greater will be the consumptive ratio of the substrate. The above expression for C_R may also be modified in terms of oxygen equivalent of both substrate and nitrate as follows:

$$C_R = \frac{\text{oxygen equivalent of substrate consumed}}{\text{Oxygen equivalent of nitrate reduced}}$$

$$\text{.... (8.6)}$$

Therefore, when both the substrate-nitrate ratio, C_F, and consumptive ratio, C_R, are expressed in terms of oxygen equivalent of substrate and nitrate by Eqns. 8.4 and 8.6 respectively, their relative values can effectively be used to maintain a nitrate limiting growth in the reactor. For a specific substrate, to prohibit the nitrite accumulation, the C_F value should not be less than the value of C_R. Again a higher value to C_F compared to C_R is undesirable for economical reasons.

It has been reported that the consumptive ratio, C_R, of both sulphide and thiosulphate in autotrophic denitrification is 1.35, while that of methanol in heterotrophic denitrification is found to be about 1.33.

8.1.2.1 Substrate requirement in denitrification

The substrate-nitrate ratio, or more specifically, the amount of substrate required to be added in the reactor may be computed either using the above concept of consumptive ratio and the oxygen equivalent, or directly by using Eqn. 8.1. The following conversions may be used in the calculation:

$$\text{OE of nitrate} = \frac{48}{14} \times \frac{5}{6} \text{ or } 2.86 \times \text{nitrate}$$

OE of waste $(BOD_{ult}) = 1.5 \times BOD_5$

OE of methanol $= 1.5 \times$ Methanol.

And also neglecting endogenous respiration, as pointed out earlier, it may be assumed that:

$$C_R = \frac{1}{1-1.42Y}, \quad \quad \text{.... (8.7)}$$

in which Y is computed on BOD_{ult} basis.

Using Eqns. 8.1 and 8.7 and neglecting endogenous respiration, on a finite mass and time basis we get:

(Kg of nitrate) 2.86 = (Ultimate BOD of substrate consumed) × (Consumptive ratio of
the substrate). ... (8.8)

However, the estimate will be more accurate when endogenous respiration is considered using Eqn. 8.1.

Methanol has got a consumptive ratio of about 1.33. This value can also be verified by Eqn. 8.7 using the observed values of Y (on

BOD_{ult} basis). So when methanol is used for substrate supplementation, Eqn. 8.8 reduces to the following form:

(Kg of Nitrate) 2.86 = (ultimate BOD or OE of methanol consumed) × 1.33.

Therefore, Kg of OE of methanol or Kg of ultimate BOD required

= (2.86 × 1.33) (Kg of Nitrate)

= 3.8 (Kg of Nitrate)

Now if dissolved oxygen is present in the reactor influent, additional quantity of BOD_{ult} or OE will be required to reduce the residual DO. Therefore, Kg of OE of methanol required

= 3.8 (Kg of Nitrate) + (Kg of DO)

Assuming that the entire quantity of residual BOD_{ult} will be utilized by the microoganisms as carbon source, in the nitrate reduction, the methanol requirement has to be reduced by the equivalent amount of ultimate BOD available in the waste itself. Therefore, Kg of OE of methanol required to be added

= 3.8 (Kg of nitrate) + (Kg of DO)

— (Kg of residual BOD_{ult} in the reactor influent) .... (8.9)

The Kg of oxygen equivalent of methanol, thus computed, is to be divided by 1.5 to get the Kg of methanol required for substrate supplementation.

The substrate requirement, or in general, the substrate-nitrate ratio not only varies with the substrate used because of the different yield of the microorganisms in different substrates, but also varies with the type of reactor employed for denitrification. The reported values of the required methanol-nitrate ratio (by weight) varies from 2.48 to 5, due to its dependence on the type of reactor employed and the different composition of the waste used.

Methanol is a costly material. So excess addition will not only produce a system effluent high in BOD, but also makes the process uneconomical. Studies are also to be conducted to search out alternative and cheaper chemical additives for biological denitrifications.

8.2 PHOSPHATE REMOVAL FROM THE ACTIVATED SLUDGE PROCESS

The presence of phosphate in the raw waste in a greater proportior to that required for bacterial growth (BOD: N: P = 100 : 5: 1) may result in an effluent phosphate concentration, sufficient to cause eutrophication in the receiving stream. The low phosphate requirement of the microorganisms compared to that of carbon and nitrogen makes the phosphate removal by conventional activated sludge process impossible. Though some times an excess biological uptake of phosphate occurs in the aeration tank, release of phosphate, in the form of orthophosphate, to the solution occurs during the extended period of aeration, and in the secondary clarifier. The phosphates which are precipitated by some chemical changes in the aeration tank, and further removed after being entrapped in the biological floc matrix, go again into the solution either in the secondary clarifier, or in the supernatant liquor of the digester.

The phosphate removal from the aeration tanks can be accomplished by chemical precipitation only. This can be done by adding sodium aluminate to the tank. One part of sodium aluminate is required for one part of phosphate in the waste. Alternatively it can be done by providing excess air in the tank, which results in the purging of carbon dioxide from the mixed liquor, followed by precipitation of phosphate by carbon dioxide. However to prohibit the phosphate precipitates from going back to solution, the sludge solids in the clarifier need quick separation.

Consequently the digester supernant becomes a concentrated source of phosphates. It therefore requires tertiary treatment by lime or alum. Two parts of alum is required for precipitation of one part of phosphate, and alum treatment is preferable for low phosphate concentration. For higher phosphate concentrations, lime treatment is economical and is preferred. Lime mud, however, needs a careful disposal. It may be thickened, dewatered, reclaimed, slaked and reused.

8.3 ANAEROBIC FILTERS:

Anaerobic filters are upflow fixed film biological reactors, originally developed to treat high strength industrial wastes. The filters are usually 0.8 to 1.2 m deep; media size varies in the range of 2 to 2.5 cm. The anaerobic condition is provided by the complete submergence of the filter bed in the waste. The anaerobic microorganisms adhere to the surface of the media and are not sloughed off the filter. As such very long mean cell residence time can be achieved, even at a very short hydraulic retention time, which is essential for efficient treatment.

Though the anaerobic growth rate is characteristically a very slow process, due to the accumulation of biological solids, and entrapment of residual solids of the influent flow, the anaerobic filters need occasional desludging by flushing the bed from the top. But sometimes the media may be required to be removed from the filter bed and washed thoroughly. A continuous operation for a period of one to two years is not expected to affect the overall performance of the filter.

Head loss through the filter usually varies from 2.5 to 15 cms during normal functioning of the filter bed.

Though originally developed for the treatment of high BOD wastes, the filters may be employed to treat low BOD wastes from small communities. In the secondary treatment of the Septik Tank effluents, the anaerobic upflow filters can be used very effectively. In a particular series of study, the anaerobic filters, of depth 60 to 75 cms, are found to reduce the BOD by 70 to 78%, while influent BOD varied from 170 to 250 mg/l (Raman & Khan). However, most of the anaerobic filters are found to improve their overall performance, at a much higher substrate concentration, say beyond COD of 1000 mg/l. The anaerobic filters are also found to be very effective in the process of denitrification, as mentioned in the section 8.1.2. A BOD reduction of 70 to 75% is obtained in a pilot plant sewage treatment study, at a flow rate of 160 l/m^3/hr.

The gas production in the anaerobic filters is almost similar to that in Anaerobic Digesters. Methane constitutes about 65 to 70% of the total gas produced.

Depending upon the nature of the waste, all the filters take 30 to 40 days to attain a steady state condition.

The anaerobic filters are essentially plug flow type fixed film reactor. As such both Eqn 4.13 and Eqn 4.15 may be used conveniently in the design of an anaerobic filter:

$$-\frac{\Delta S}{\Delta t} = \frac{k_m S.(AXd)}{K_s + S} = \frac{(k_m Xd)S}{K_s + S}A \qquad .... \ (8.10)$$

$$(k_m Xd) \left(\frac{AH}{Q} \right) D = (S_0 - S_1) + K_s \ln (S_0/S_1) \qquad .... \quad (8.11)$$

The reported experimental observations, though scanty, reveal that the anaerobic filters behave similar to trickling filters. At similar conditions of influent substrate concentration, the mass flux of the substrate increases at higher hydraulic loading. Though the response to the higher loading is similar in both the anaerobic filter and the trickling filter, the reasons cannot be the same. The response of the trickling filter to higher hydraulic loadings was discussed in section 4.2.1. The mass flux in an anaerobic filter increases with flow rate probably due to the greater rate of contact of the substrate with active bio-film.

As in trickling filter, at a specific hydraulic loading, the mass flux in the anaerobic filters increases with an increase in the influent substrate concentration.

The experimental observations indicate that in anaerobic filters also, the rate of reduction of substrate concentration per unit depth decreases with flow rate, as in trickling filters Fig. 4.5.

Therefore curves similar to Figs 4.5 and 4.6, in chapter 4 may be plotted from the experimental observations in a pilot plant study with anaerobic filters. For a specific type of media, the "specific surface area", A is fixed. The choice of the particular value of Q and the dimensions of the filter, H and D, can be made judiciously, based on those sets of curves, considering the relative overall performances in alternative designs. It may be noted that the continuous function relating the substrate concentration and the filter depth is not linear in lower hydraulic and organic loadings. As such, in all the above curves, the rates of flow should be expressed in terms of flow per unit wetted perimeter.

The experimental results (Khan and Siddiqi) obtained on a synthetic waste having the substrate concentration range of 890 to 2470 mg/1 of COD, have been used to examine the effect of substrate concentration on the mass flux of the substrate at different flow rates. A plot (similar to Fig. 4.5) with this data indicates that $K_s \gg S_0$. Under this condition, the Eqns 4.13 and 4.15 may be conveniently modified in the following form:

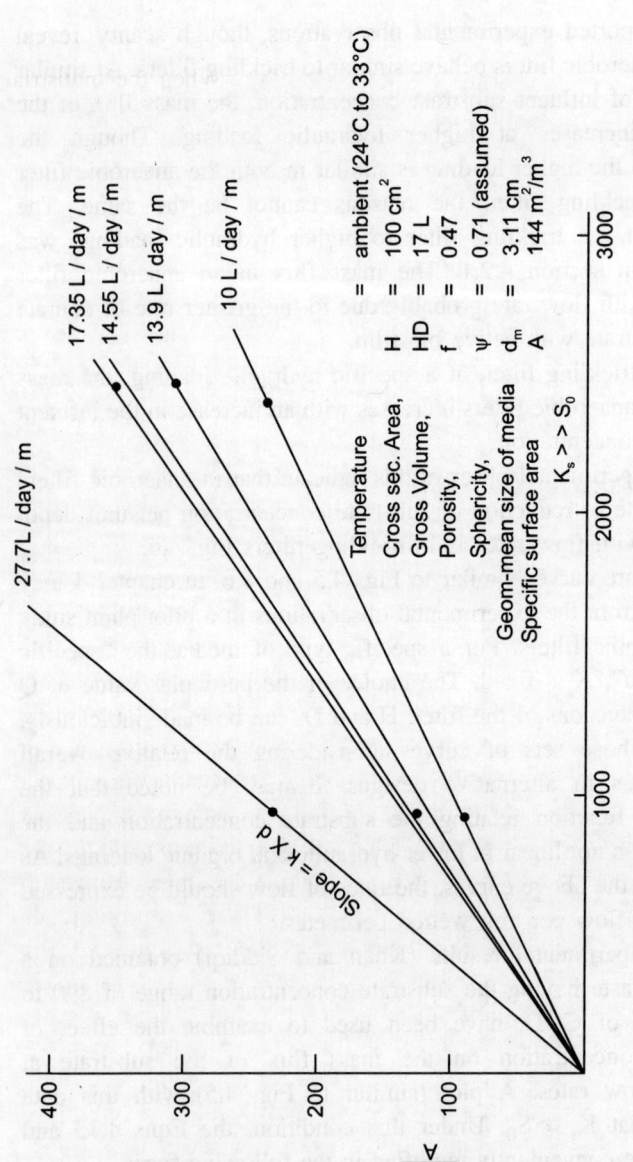

Figure 8.3. Effect of Applied Substrate Concentration on the Mass Flux, at Different Flow Rates. [Based on Data Presented by Khan and Siddiqi (1976)].

$$\frac{\Delta S}{\Delta t} \simeq \frac{k_m}{K_s} S(AXd)$$

$$= (kXd)\ S.A \qquad\qquad .... (8.10)$$

in which k = specific substrate removal coefficient, 1/mg/day,

$$= k_m/K_s,$$

and $$S_1 = S_0 \exp\left[-(kXd)\left(\frac{AH}{Q}\right)D\right] \qquad .... (8.11)$$

The magnitudes of the quantity kXd for various flow rates are given by the slopes of the curves, as shown in Fig. 8.3. A set of substrate concentration vs filter depth curves, like Fig. 4.6, is to be drawn using Eqn 8.11. The selection of a particular flow rate and filter depth will then complete the design.

8.4 ROTATING DISC BIOLOGICAL CONTACTOR

The rotating disc biological contactor (RBC) is one of the newest methods of biological treatment of wastes. The treatment unit consists of a cylindrical bottomed horizontal flow tank on which a partially submerged light weight rotor unit is placed, along the direction of flow. The rotor unit consists of a series of closely spaced biologically inactive discs mounted on the rotating shaft. The microbial film grows and covers the entire available surface of the discs. This film adsorbes the organic pollutants during the submerged period of rotation cycle, and the major amount of oxygen transfer occurs when the bio-film is exposed to the atmosphere during the other half of the rotating cycle. However, the substrate utilisation within the microbial film is a continuous process. The process of 'sloughing' is also continuous thus maintaining a constant thickness of the microbial film on the discs. Therefore, a RBC is a continuous flow aerated fixed film biological reactor. While the discs offer the necessary surface area for biological growth, the aeration is provided by its rotation.

The discs can be made of PVC, polystyrene, polythelene, galvanized steel, asbestos cement or any other light weight material. The diameters usually vary in between 1 m and 3 m but may go up to 5.9 m. The reported thickness of the discs varies from 100 mm to 127 mm but may be as thin as 4 mm. The clear spacing between the discs varies from 3 to 5 cms. The discs are operated normally with 40 to 60% submergence ratio, and the speed varies usually from 2 RPM to 6 RPM, but may be as high as 12 RPM.

Due to the operational limitations of the numbers of discs on one shaft, i.e. in one unit, the process may consist of a number of stages in series. The number of stages or modules may vary from 2 to 6 depending on the characteristics of the waste and the effluent characteristics required.

The rotating biological disc contactor should be followed by a secondary settling tank to allow the removal of biological solids. No recycling of sludge is, however, required.

Pilot plant studies with RBC indicates that about 90% reduction in BOD can be achieved with hydraulic retention times of 1.25 hrs and 1 hr respectively in disc unit and settling tank.

The rotating disc biological contactors are essentially plug flow reactors, without recycle. The substrate removal occurs due to microbial metabolism both in film phase and liquid phase. However, the mixed liquor suspended solids concentration in the liquid phase of the reactor is always found to be very low, and the hydraulic retention time (as well as mean cell residence time in the liquid phase) is also very short. As such the substrate removal in the liquid phase is comparatively insignificant. Assuming the substrate removal occuring only in the rotating biological films, the Eqn 4.13 may be modified for RBC as follows:

$$-\frac{\Delta S}{\Delta t} = \frac{k_m 2_\pi (r_0^2 - r_n^2) \, dXNS}{K_s + S} \cdot \frac{1}{V} \quad .... \ (8.12)$$

Therefore, for plug flow continuous model of RBC, Eqn 4.15 reduces to:

$$(S_0 - S_1) + K_s \ln(S_0/S_1) = (k_m Xd) \frac{2\pi N(r_0^2 - r_u^2)}{Q} \quad ... \ (8.13)$$

in which $S_0/S_1 =$ substrate concentration in the influent and effluent respectively,

$$N = \text{numbers of discs in the stage (or module),}$$

and $\quad V = \text{volume of the reactor,}$

$r_0, r_u = \text{total radius of the discs and the radius of the unsubmerged portion of the disc, respectively,}$

$d = \text{effective thickness of the microbial film,}$

$X = \text{unit mass of biological film on dry weight basis,}$

$k_m = \text{maximum rate of specific substrate utilisation,}$

$K_s = \text{substrate concentration at the utilisation rate of } k_m/2,$

and $\quad Q = \text{rate of flow.}$

In the Eqn. 8.13, the term $k_m X d$ is the maximum mass flux, and the inverse of the term $2\pi N(r_0^2 - r_u^2)/Q$ indicates the flow per unit surface area.

The performance variables in the RBC are number of discs, submergence, hydraulic loading, influent substrate concentration, and the RPM of the discs. Studies on the effect of submergence have shown that the overall performance remains constant beyond a submergence ratio of 0.5. The effect of the submergence may be taken as insignificant under the normal operating conditions of 40 to 60% submergence. Increased rate of revolution of the discs is believed to be capable of supplying more oxygen to the system. However the effect of RPM is also found to be negligible under the normal range of operating conditions. The number of discs and their radii (total and unsubmerged) remain only to be adjusted to provide with the required surface area for biological growth.

While designing a RBC, it is required to conduct a pilot plant study to determine the values of $k_m X d$ and K_s corresponding to different hydraulic loadings, expressed in terms of flow per unit surface area. As flow rate is fixed for a particular system, the only parameter which remains to be selected is the surface area, consistent with the operational and structural limitations.

Depending on the characteristics of the wastes, the reported hydraulic loadings range from 110 litres/day/m^2, in the case of a primary effluent, to 3800 litres/day/m^2 for an anaerobic lagoon effluent. However, in the latter case, it is necessary to use three stages in the system.

The RBC is also found to be very effective in the process of nitrification. The loss of head through one stage is very low (less

186

than 25 cms) compared to trickling filters. It also occupies much less floor area compared to that in trickling filter, under comparable overall performance. However the operation of the RBC depends on the supply of power. The power requirement depends on the RPM of the discs. The rate of power consumption increase rapidly beyond 5 RPM. As such the RPM must be selected, also through pilot plant study, for an optimum condition. The excessive oxygenation caused by higher rate of revolution is not desirable for economical reasons. The power consumption may be reduced by reducing the RPM in the lower stages in the multistage units. The power requirement is estimated to be around 4 to 9 KW per million litres per day.

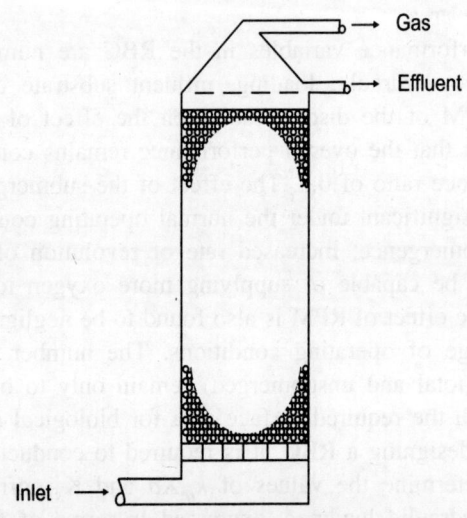

Figure 8.4: Anaerobic Filter.

The only reported experimental results with low strength synthetic waste, under Indian conditions, indicate that $K_s \gg S_o$. (vide Fig. 8.5). As such Eqn 8.13 may be simplified, like Eqn 8.11, as follows:

$$S_1 = S_o \exp\left[-kXd\left\{\frac{2\pi N(r_o^2 - r_u^2)}{Q}\right\}\right] \quad \text{.... (8.14)}$$

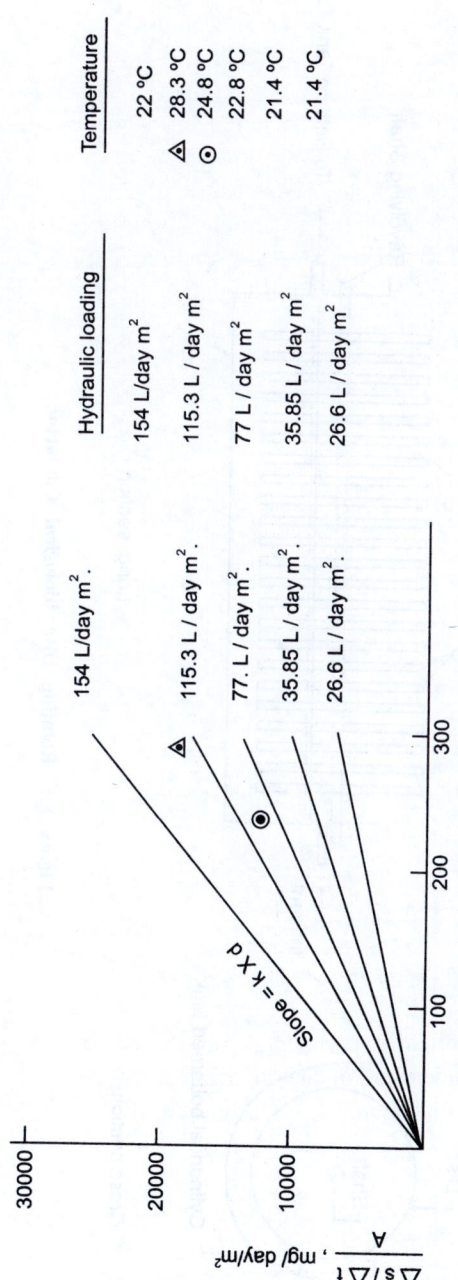

Figure 8.5: Effect of Applied Substrate Concentration on the Mass Flux, at Different Flow Rates. [Based on data Presented by Khan and Siddiqi (1972)].

188

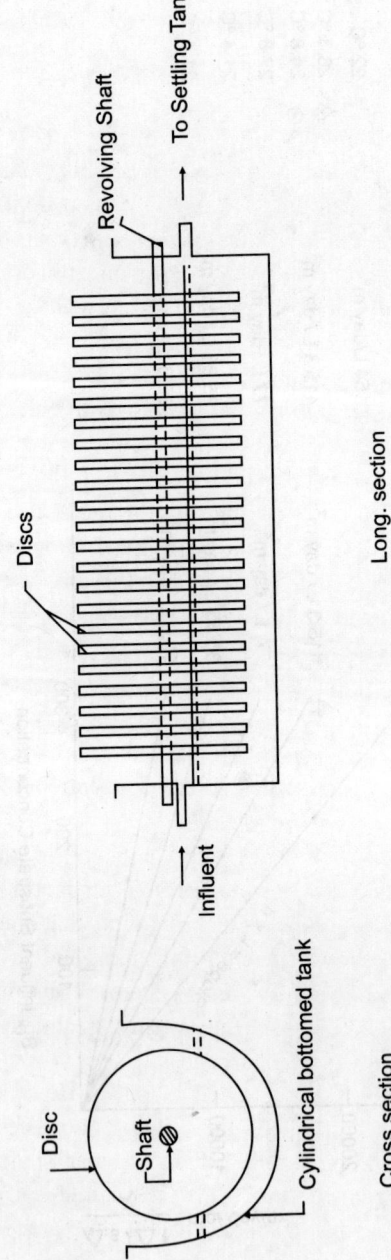

Figure 8.6. Rotating Disc Biological Contactor.

However, pilot plant studies are to be conducted before any design to determine the response of the RBC, under different process variables, with actual type of wastes to be treated.

8.5 U-TUBE AERATION SYSTEM

U-tube aeration tank, which derived its name from the configuration of flow of liquid through it, is a deep hole, sunk in the ground, and is essentially a high efficiency oxygen transfer system. The configurations include: (*i*) deep split trench with a baffle suspended from the top in the middle, but a little distance off the bottom of the trench, (*ii*) deep circular shaft with a smaller diameter "downcomer" tube suspended centrally and freely from the top, but a little distance off the bottom of the shaft. (Fig 8.7)

The liquid is circulated through the U-tube, and the air is injected somewhere in the downcomer tube. A high degree of aeration is then achieved, at the cost of much less energy, taking the advantage of (1) higher oxygen saturation concentration at naturally occuring higher hydrostatic pressure within the U-tube, resulting in the higher "dissolved oxygen-deficit-driving force" for oxygen transfer, (2) higher bubble contact time, in the order of 3 to 5 minutes, compared to 15 sec in diffused air units, and (3) the high degree of turbulence. Depending upon the depth of the units, i.e. the pressure available within the tube, an oxygen transfer rate, many times higher than that with the conventional aerators, can be achieved.

The application of the U-tube concept in the waste water treatment, though a new development, is in fact a modification of the activated sludge process. It was found that much acceleration of treatment, reduction of energy utilisation and saving of valuable space can be achieved by this system. Thus it results in an economical installation of the waste treatment facility, with minimum environmental disturbances.

The sizes of U-tube aeration tanks, used in waste water treatment, vary according to the quality of waste water, and the degree of treatment required. When circular shafts are used, the diameter varies from 0.5 to 2.0 m. The diameter of the downcomer tube is adjusted to have approximately the same cross-section as the annular space between the two pipes. A 130 m deep and almost 2 m diameter shaft has been installed at one place in USA to treat domestic waste mixed with a strong industrial effluent. In order to ensure a smooth flow pattern, massive bell mouthed T-pieces may be mounted at the top of the shaft, as shown in Fig 8.7(b). This enlarged portion of the shaft at the top also acts as a gas disengagement tank.

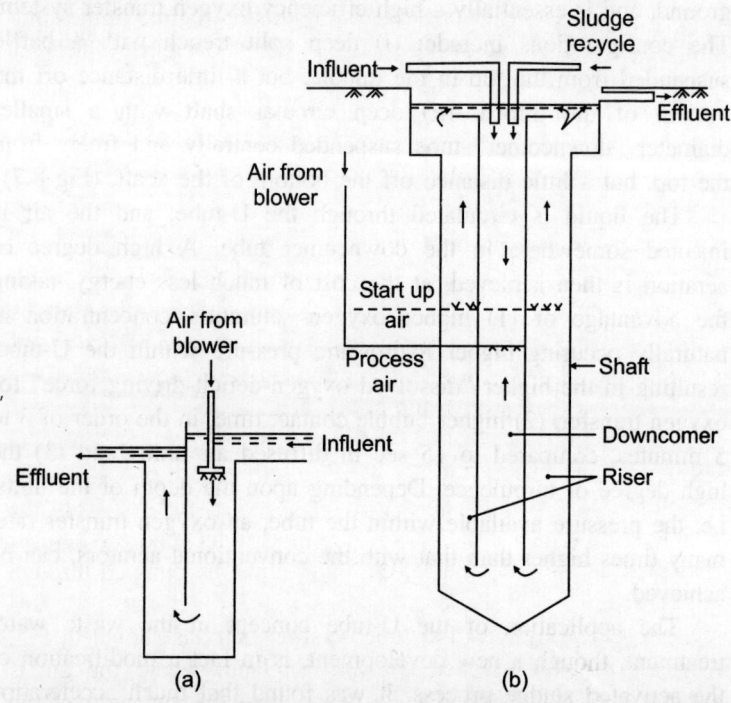

Figure 8.7: U-tube Aeration Systems: (a) Split Trench, (b) Deep Shaft, with Internal Circulation of Waste.

The influent waste passes down the central downcomer pipe and returns up through the annular riser space between the wall of the shaft and the outside of the tube. A greater extent of internal circulation can be achieved with a design shown in Fig. 8.7(b). The effluent is then clarified in the secondary settling tank and the settled sludge is recycled if necessary, after necessary wasting. A high degree of aeration can be achieved due to sufficient mixing caused by the internal circulation of the waste, and the turbulence within the tube. The biological solids produced in a U-tube aeration system have got good settling and dewatering characteristics. However, the rate of production of biological solids in the system is very low.

The U-tube aeration system for biological treatment may be designed like any other biological reactors. However, sufficient information in regard to the kinetic coefficients or any other design criteria is not available at present. The hydraulic regime in the split-trench U-tube as shown in Fig. 8.7(a) is plug flow type, while that in the deep shaft with provision for internal circulation is complete mix type, and the unit resembles a horizontal flow oxidation ditch.

U-tube aeration system, however, demands more attention in the hydraulic design of the aeration unit, to take the maximum advantage of this new concept.

The head losses through the aeration unit include frictional losses, velocity head, that due to geometry, and that due to air injection in the system. Head loss due to air injection depends on the air-water ratio, i.e. on the volumetric flow rate of air in the aerator tube. The system should be provided with sufficient driving force for circulation against the above head losses.

In the configuration of the U-tube aeration tank, as shown in the Fig. 8.7(a), where the air is injected near the top of the downcomer section, the required driving force is derived from the available water head, as shown, at the inlet end. Wherever sufficient head is not available, the liquid must be pumped through the U-tube, using a propeller pump.

In the configuration of U-tube aeration tank, as shown in Fig. 8.7(b), originally developed by ICI, England, where the air is injected sufficiently below the top surface, in the downcomer, the driving force for circulation is derived from the difference in densities between the nonaerated downcomer liquid and the aerated, low density liquid beyond the point of air injection. So in this process, the air is injected in the upflow section first and then gradually it is switched over to downflow section. As such, the air supply alone can bring about the circulation of liquid, with a proper design. The velocity of flow through the aeration tube should be sufficient to exceed the buoyant velocity of the bubbles. A higher velocity through the tube, lowers the size of the unit; but such increase of velocity becomes prohibitive due to the increased velocity head and frictional losses. The velocity of flow plays another important role in the design of U-tube aeration tanks. The aeration unit should be supplied with sufficient amount of air. But the maximum air-water ratio in the aeration unit practicable is a function of the velocity. Higher velocity is needed for a desired higher air-water ratio. The air requirement in U-tube aeration system is much smaller than that in the conventional aerobic biological reactors, because of the greater solubility and hence greater oxygen transfer rate at the very high hydrostatic pressure. It has been found that a maximum of 20% air-water ratio can be achieved at a velocity of 1.2 m/sec. Higher velocity in the range of 1.8 m/sec to 2.4 m/sec is required for attaining an air-water ratio of around 25%. It may be noted that surging occurs in the aeration tube, when maximum practicable air water ratio is exceeded.

The velocity of flow within the aeration tube usually ranges from 1 to 2 m/sec, and an oxygen transfer rate upto 3 kg/hr/m^3 can be achieved.

The power requirement depends on the oxygen requirement, oxygen transfer efficiency and the depth of the air-injector. Higher the depth, higher will be the power consumption. But higher depths are required to generate higher density difference between the downflow and upflow sections, for circulation. The air injector depth ranges in between 20 m and 40 m. The resulting power economy ranges from 4.5 to 3.0 Kg of oxygen per KWH.

The inherent efficiency of the U-tube aeration system and the associated economy in land and power consumption in its use, will soon establish its position in the field of biological treatment of domestic and even strong industrial wastes. As the U-tube aeration units are capable of handling a very large volume of waste, occupying comparatively very small area, the system may be employed where there is no room for installation of a conventional treatment system, particularly for the treatment of wastes at the sea out-fall.

QUESTIONS

1. Explain the design and operation of Anaerobic filters.
2. Draw a neat sketch of a Rotating Disc Biological contactor and explain its function.
3. Explain the working of an U-tube Aeration system.

9 | Industrial Waste Treatment

9.0 INTRODUCTION

While a huge amount of water is required for different industrial processes, only a small fraction of the same is incorporated in their products and lost by evaporation; the rest finds its way into the water courses as waste water. Thus the industries join the municipalities to contribute to the "pollution" of the natural bodies of water.

The industrial wastes either join the streams or other natural water bodies directly, or are emptied into the municipal sewers. Thus these wastes affect in some way or other, the normal life of a stream or the normal functioning of sewerage and sewage treatment plants. Much attention is now given in India on the treatment of industrial wastes, due to its growing pollution potential arising out of the rapid industrialization of the country. Streams can assimilate certain amount of wastes before they are "polluted", and the municipal sewage treatment plants can be designed to handle any kind of industrial wastes. Thus we have three alternatives for the disposal of the industrial wastes *viz* (*i*) The direct disposal of waste into the streams without any treatment, (*ii*) discharge of the wastes into the municipal sewers for combined treatment, and (*iii*) separate treatment of the industrial wastes before discharging the same into the water bodies. The selection of a particular process depends on various factors like the following:

(*i*) Self purification capacity of the streams,

(*ii*) Permissible limits of the pollutants in the water bodies, established by law, (Tolerance limits for the inland surface waters, subjected to pollution, as given by ISI are shown

194

in Table 1.3)

(*iii*) Economic interests of both the municipalities and the industries,

(*iv*) Technical advantages, if any, in mixing the industrial wastes with domestic sewage.

After a thorough economic and technical appraisal, if it is decided to treat the industrial waste either independently or along with the domestic sewage, the treatment plants are to be designed after (*i*) a thorough investigation of the characteristics of the wastes, and (*ii*) a cost study for the final choice of the particular method of treatment.

9.1 CHARACTERISTICS OF THE INDUSTRIAL WASTES

Unlike the domestic sewage, the industrial wastes are very difficult to generalize. The characteristics of the industrial wastes not only vary with the type of the industry, but also from plant to plant producing same type of end products. Different types of liquid wastes originate front various types of industrial processes. The pollutants include the raw materials, process chemicals, final products, process intermediates, process by-products, and impurities in raw materials and process chemicals. Broadly, these pollutants can be classified as follows:

(*a*) Organic substances that deplete the oxygen content of the receiving streams and impose a great load on the biological units of the sewage treatment plant.

(*b*) Inorganic substances like carbonates, chlorides, nitrogen etc. that render the water body unfit for further use and some times encourage the growth of some undesirable micro-plants in the body of water.

(*c*) Acids or alkalis which make the receiving stream unsuitable for the growth of fish and other aquatic life there, and cause serious difficulties in the operation of sewage treatment plants.

(*d*) Toxic substances like cyanides, sulphides, acetylene, alcohol, petrol etc. which cause damage to the flora and fauna of the receiving streams, affect the municipal treatment processes and some times endanger the safety of the workmen. A list of the toxic substances from some selected industries is given in Table 9.1.

(*e*) Colour-producing substances like dyes, which though not toxic, are aesthetically objectionable when present in the water supplies.

(*f*) Oil and other floating substances, which not only render the streams unsightly, but interfere with the self-purification of the same, and the operations of the sewage treatment plants.

Whenever an industrial waste is decided to be discharged into a sewage treatment plant, special attention must always be given to toxic components of the waste, if there are any.

Characteristics of the wastes from some selected Indian Industries "are given in Table 9.2.

9.2 TREATMENT OF INDUSTRIAL WASTES

The treatment of industrial waste water may be accomplished in part or as a whole either by the biological processes, as done in the case of sanitary sewage, or by processes very special for the industrial waste water only. The important factors, which affect the planning for a industrial waste water treatment plants are (1) the discontinuous and some times seasonally discharged wastes, (2) high concentration of the waste, and (3) Non biodegradability and toxicity of some wastes. Depending upon the mode of discharge of the waste, and the nature of the constituents present in it, the treatment may consist of any one or more of the following processes.

(*i*) Equalization
(*ii*) Neutralization
(*iii*) Physical treatment
(*iv*) Chemical treatment
(*v*) Biological treatment.

The above processes may be carried out partly or entirely in a municipal sewage treatment plant along with the domestic sewage, or in a separate treatment plant.

When the characteristics of the waste vary in a day and also the discharge rate is not uniform or continuous, the waste may require Equalization before it is subjected to the treatment. Equalization consists of holding the waste for some pre-determined time in a continuously mixed basin, which produces an effluent of fairly uniform characteristics.

When the waste contains excessive amount of acid or alkali (particularly acid), the waste requires Neutralization in the neutralization tank. Neutralization may be carried out in the Equalization Tank, when the conditions permit.

When the industrial waste is treated along with the municipal sewage or discharged into a stream, the waste may be subjected to another prior unit operation, known as proportioning. Proportioning consists of the control of the discharge of the waste into the receiving sewer or stream, in a fixed proportion to the flow of domestic sewage or the stream. This helps not only in protecting the treatment device from the shock load but also in improving the sanitary quality of the treated effluent.

Before an industrial waste is subjected to a chemical or biological treatment, or both, it may be required to separate the suspended matter by physical operations like Sedimentation, and Floatation. Sedimentation tanks are to be provided only when the waste contains high percentage of settleable solids. Floatation is employed to separate fine particles with very low settling characteristics. Floatation consists of creation of fine air bubbles in the waste body by the introduction of air to the system. The rising air bubbles attach themselves to the suspended particles and thereby increase the buoyancy of the particles. The particles thus lifted to the liquid surface are removed by skimming.

Some of the industrial wastes, amenable to biological treatment, may require prior chemical treatment; some requires only chemical treatment without any biological treatment. Some of the chemical and physico-chemical processes, employed in the Industrial wastes treatment, for the removal of dissolved inorganic materials are:

(i)   Reverse osmosis (Hyper filtration),

(ii) Electrodialysis,

(iii) Chemical oxidation,

(iv) Chemical precipitation,

(v) Adsorption,

(vi) Ion Exchange,

(vii) Thermal reduction,

(viii) Air stripping.

When a permeable membrane separates a dilute and a concentrated solution, the osmotic pressure drives the water molecules from the dilute solution, through the membrane to establish an equilibrium. This natural response is reversed in the reverse osmosis process, where the waste water containing dissolved salts are filtered through a semipermeable membrane such as cellulose acetate at a pressure higher than the osmotic pressure. In this type of treatment, pre-treatment of the waste such as Activated carbon adsorption or chemical precipitation followed by some kind of filtration is necessary.

In electrodialysis, the flow of ionic substances is initiated by providing an electrical protential between two electrodes and the substances are filtered by using an ion-selective membrane ahead of the electrodes. The separation of substances in this process requires sets of alternate cation and anion permeable membranes in between the electrodes as shown in the following schematic diagram Fig. 9.1.

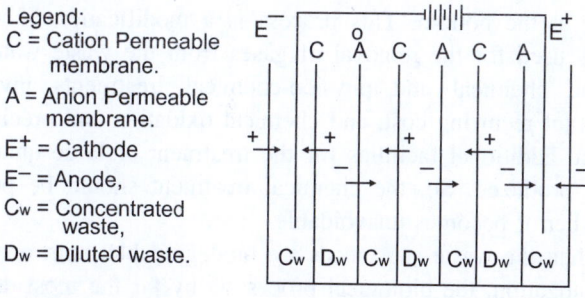

Legend:
C = Cation Permeable membrane.
A = Anion Permeable membrane.
E^+ = Cathode
E^- = Anode,
C_w = Concentrated waste,
D_w = Diluted waste.

Figure 9.1. Schem. Diagram of Electrodialysis Process.

When the waste water is passed through such cells, alternate cells of concentrated and dilute wastes are formed. Electrodialysis also requires some pretreatment as in Reverse Osmosis.

The chemical oxidation consists of addition of chemicals like chlorine and ozone to reduce the BOD loading on the subsequent biological process, or to reduce substances like ammonia, cyanide etc.

Chemical precipitation consists of coagulation either by alum or ferric salts as well as treatment by lime.

Adorption involves the passage of pre-treated waste water through fixed bed Activated Carbon column, and is used only as a tertiary treatment unit, to remove non-biogradable organics like synthetic detergents, colour and odour. The regeneration of the exhausted bed is accomplished by oxidizing and thus removing the adsorbed organics in a furnace at a temperature of about 925°C. Other commercially available materials like clay, fly ash etc. are also recommended as potential adsorbents.

Ion exchange involves similar passage of waste water through a fixed bed synthetic ion exchange resin bed, where some of the undesirable cations or anions of the waste are exchanged for Sodium or Hydrogen ions of the resin. Ion exchange bed requires regeneration, and special care should be taken for the treatment of the waste produced due to regeneration.

Thermal reduction involves the burning and thereby oxidation of some refractory and toxic substances (like organic cyanide).

Air stripping involves the passage (down flow) of a liquid waste through a packed tower, which is equipped with an air blower at the bottom. This process is a modification of aeration process used for the removal of gases from the waste water.

The chemical and physico-chemical treatments involve a significant recurring cost, and chemical oxidation and precipitation required additional facilities for the treatment of large quantity of sludge produced. So, the chemical treatment should be provided only when it becomes unavoidable.

When the waste substances are biodegradable, with or without acclimatization, the biological process is by far the most desirable treatment process. The treatability of an industrial waste may be assessed by conducting laboratory tests on BOD/ COD ratio. If the ratio is greater than 0.6, the wastes are biologically treatable without acclimatisation; if the ratio ranges from 0.3 to 0.6, the waste needs acclimatisation for biological treatment; if the ratio is less than 0.3, other methods are suggested for the treatment. The acclimatisation involves the gradual exposure of the waste in increasing concentration to the seed or initial microbiological population, under a controlled condition.

The design criteria for the conventional biological treatment processes may be different for different types of industrial wastes.

The system parameters for particular type of industrial wastes may be determined by laboratory experiments. In the absence of any actual test result, the performance data of similar type of waste may be used for design.

In some cases, while the commonly available microbiological population fails to achieve the biological oxidation, some special type of microorganisms do well. So, for the effective metabolism of some complex organic substances, development of suitable microbial culture containing specific group of organisms is necessary. (As for example, some micro-organisms which are phenolytic in action, and often found in well manured soil, have been identified, and are employed in the activated sludge treatment of coke oven effluents).

Most of the industrial wastes do not contain sufficient amount of nutrients for good microbial growth. For effective biological treatment of this type of wastes nutrients are added to the reactors in the form of Urea, Superphosphate or any other compound containing Nitrogen and Phosphorus. For a balanced growth of microorganisms in a reactor, the BOD: N: P ratios of 100: 5: 1 in aerobic systems and 100: 2.5: 0.5 in anaerobic systems are to be maintained.

Special care should be taken in regard to the toxic wastes. Toxicity may be of acute or chronic type, and may be to humans, plants, animals or to microorganisms responsible for aerobic or anaerobic biological treatment. Some of the toxic wastes like phenols, cyanides, formaldehydes etc. yield to acclimatised growths of normal or special type of bacteria. Some other toxic metal ions like Copper, Zinc, Chromium etc. interfere with the biological oxidation by tying up the enzymes essentially required for microbial growth. As such these must be pretreated chemically before the waste is subjected to biological treatment.

So far, only the principles of industrial waste treatment are described; the selection of the particular process depends on the effluent requirements and the characteristics of the waste. The methods of treatment of wastes from some typical industries, along with their characteristics will be dealt with in the subsequent chapters. It must be borne in mind that before a treatment policy is fixed up for a particular industry the scope of recycling and

Figure. 9.2: Dorr Clariflocculator.

reclamation (recovery) of the wastes must be considered, for a better management of Industrial waste waters. Seggregation of strong wastes from the weak wastes some times reduces the problem.

TABLE 9.1. Toxic Chemicals from some Selected Industries.

Industry	Toxic Pollutants
1. Fertilizers	Ammonia, Arsenic
2. Coke Ovens	Phenols, Cyanide, Thiocyanate, Ammonia.
3. Metullurgicals	Heavy metals, e.g. Copper, Cadmium, Zinc.
4. Electro plating	Hexavalent chromium, Cadmium, Copper, Zinc.
5. Synthetic Wool	Acrylonitrile, Acetonitrile, HCN
6. Petrochemicals	Phenol, Heavy Metals, Cyanides.

TABLE 9.2. Pollution Characteristics of Different Industries.

Industry	Pollution Characteristics	Suggested Treatment Methods
Paper and pulp	Strong Colour High Bod, High COD/BOD ratio, Highly alkaline, High sodium content.	Chemicals recovery, Lime Treatment for colour, Biological treatment
Tannery	Strong colour, High Salt content, High BOD, High dissolved solids, Presence of sulphides, Lime and Chromium.	Chemical treatment, Biological treatment
Textile (cotton)	Highly alkaline, High BOD, High suspended solids	Chemical and Biological Treatment
Distillery and Brewery	Strong colour, High chloride, High sulphate, Very high BOD.	Biological Treatment
Petrochemicals	Oil, High BOD & COD, High total solids.	Chemical Treatment, Biological Treatment

Industry	Pollution Characteristics	Suggested Treatment Methods
Parmaceuticals	High total solids, High COD, High COD/BOD ratio, Either acidic or alkaline.	Chemical Treatment, Biological Treatment
Coke oven	High ammonia Content, High phenol Content, High BOD, Low suspended solids, High cyanide.	Chemical Treatment Biological Treatment
Oil refineries	Free and emulsified oil.	Oil separation, Chemical Treatment, Biological Treatment
Fertilizer	High Nitrogen Content.	Biological Treatment
Dairy	High dissolved solids, High suspended solids, High BO D, Presence of oil and Grease.	Biological Treatment
Suger	High BOD, High Volatile Solids, Low pH	Biological Treatment

QUESTIONS

1. What are the toxic pollutants in wastes coming from Fertilizer plants, Petrochemicals and Electroplating industries?

10 | Pulp and Paper Mill Wastes

10.0 INTRODUCTION

The pulp and paper industry is one of the oldest industries in our country. But there has been a tremendous expansion in this industry during the last twentyfive years. Varieties of papers and similar products are now manufactured in different mills throughout the country. The paper industry, as it stands now, is one of the largest industries in our country. And also this is one of the major industries which contributes a lot towards the pollution of our water environment.

The paper mills use the 'pulp' as the raw material, which is again produced utilizing different cellulosic materials like wood, bamboo etc in the pulp mills. However, most of the paper producing units in India are integrated pulp and paper mills. Few mills in our country produce pulp only, using the same manufacturing process, for the production of Rayon fabrics.

The pulp and paper mill wastes characteristically contain very high COD and colour; the presence of Lignin in the waste, which is derived from the raw cellulosic materials and is not easily biodegradable, makes the COD/BOD ratio of the waste very high. It may be noted that, the pollution potential of the paper mills are negligible compared to that of the pulp mills. As such, it is the pulp making process which is responsible for the pollution problems associated with the integrated pulp and paper mills.

The peculiar pollution potential of the industry, their location mostly on the banks of the small rivers, and the general awareness for the conservation of the water resources has led to a considerable research work in search of an economical solution of the pollution abatement problems of the industry. In this chapter, an attempt has been made to enlighten these problems and to discuss the suitable solutions of the same.

10.1 MANUFACTURING PROCESS AND THE SOURCES OF THE WASTES

The volume and the characteristics of the wastes depends on the type of manufacturing process adopted and the extent of reuse of water employed in the plant. The process of manufacturing of paper may be divided into two phases—pulp making, and then making of final product of paper.

In the pulp making phase, the chipped cellulosic raw materials are digested with different chemicals in one tank under high temperature and pressure. The process thus loosens the cellulose fibres and dissolves the lignin, resin and other non-cellulosic material in the raw material.

The Kraft process (or sulphate process) of pulp making uses sodium sulphate, sodium hydroxide and sodium sulphide as the above mentioned digester chemicals; another process of pulp making, known as Sulphite process uses magnesium or calcium bisulphite, and sulphurous acid as the digester chemicals; sodium hydroxide or lime is used in Alkali process of pulp making.

The spent liquor produced by the above process of digestion is known as "Black liquor". This is not only very rich in lignin content, but also contains a large amount of unutilised chemicals. Therefore, this black liquor is treated separately for the recovery of chemicals. In Sulphite process, however, the black liquor is not treated for the chemical recovery. So while the entire quantity of the liquor in Sulphite process makes the coloured waste from this section, in the Kraft process, the same is produced due to the leakages, spillages or overflows only from the digester.

The cellulosic fibre after being separated from the black liquor is washed and then partially dewatered in a cylindrical screen called 'Decker'. A concentrated wash water may be sent for chemical recovery (in the Kraft process); the dilute wash water forms the waste water. This volume of waste water is known as 'Brown stock wash' or 'Unbleached Decker Waste'.

The washed cellulosic fibres are then sent for the bleaching in three stages, where chlorine, caustic and hypochlorite are used in successive stages. Waste waters from the first and the last stages are light yellow in colour, while that from the caustic extraction stage is highly coloured.

The bleached pulp is then sent for the paper mill. In the paper mill, the pulp is disintegrated, and mixed with various filler materials like alum, talc etc, and dyes, in an oblong shaped specially made tank known as "Beater". After beating, the pulp is refined in a machine known as "Jordan".

The refined pulp is then diluted to proper consistency for paper making and passed through a screen to remove lumps or knotes. Now this pulp is carried by a travelling belt of fine screen to a series of "Rolls", where the final product, the paper, is produced. The drained water, often called as "White water" forms the waste water from the paper mill section; this waste water contains fine fibres, alum, talc etc. Usually the fibres are recovered from this waste, and the treated liquid is reused for the wet chipping process.

The black liquor of the Kraft process is concentrated by evaporation, and then incinerated with the addition of sodium sulphate. The organics like lignin, resin etc are burnt out, and the smelt is dissolved in water. The resulting liquid is known as "Green Liquor". Lime is now added to this liquor, resulting in the formation of "White Liquor", and "Lime mud", containing chiefly calcium carbonate. White liquor contains desired cooking (digesting) chemicals and is sent for use in digester. The lime mud is calcined (by burning) to form calcium oxide, which is reused to recaustise other green liquors into white liquors.

Besides those mentioned above, a small volume of waste water is also produced when the bark is removed from the raw wood and the later is reduced to chips by wet process.

Some very toxic waste material are also generated during the process of chemical recovery from black liquor. Toxic materials like Dimethyle Sulphide, Methyl Mercaptan etc also comes out with digester relief gases, and forms a colour-less waste water after condensation.

A simplified flow diagram of a Kraft pulp and paper mill is given in Fig. 10.1.

206

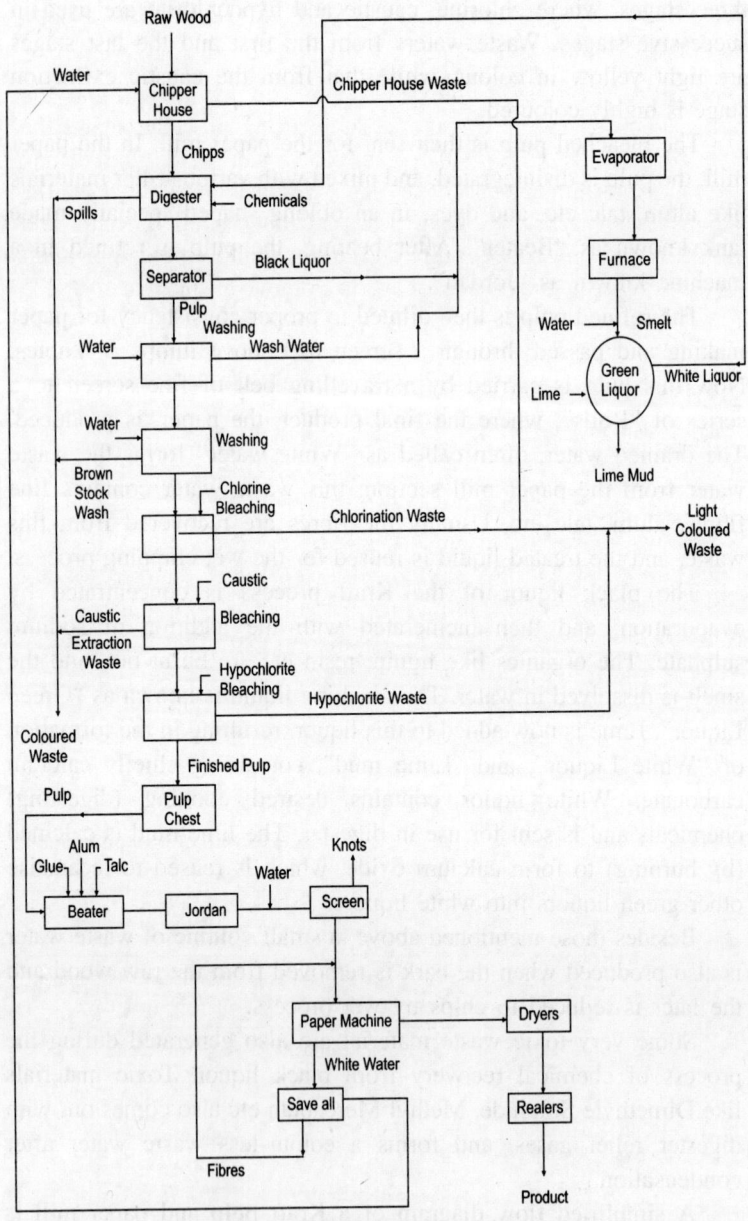

Figure 10.1: Simplified Flow Diagram of a Kraft Pulp and Paper Mill.

10.2 CHARACTERISTICS OF PULP AND PAPER MILL WASTES

In regard to the waste volume, it must be mentioned that the volume depends mainly on the manufacturing procedure, and the water economy adopted in the plant. The mode of discharge may also be of intermittent type if the black liquor is not treated for the chemical recovery. It has been observed that a well operated and well managed integrated pulp and paper mill, employing Kraft process for pulping, produces a waste volume in the range of 225 to 320 m^3 per tonne of paper manufactured. The mills manufacturing special quality of paper produce larger amount of water for washing and bleaching. Incidentally, most of the modern Indian mills operate on the Kraft process.

Like the volume of waste, the chemical composition of the wastes will also depend on the size of the plant, manufacturing process, and to great extent on the material economy practised (by the way of recovery of chemicals and fibres) in the plant. In most of the small paper mills in India, the chemical recovery is not practised due to economical reasons. As such the pollution potential of the wastes of smaller mills is higher than that of the larger mills. Generally the pulp and paper mills wastes are characterised by very strong colour, high BOD, high suspended solids, and high COD/BOD ratio. Characteristics of the combined effluent of two pulp and paper mills, one small and another large, are tabulated in Table 10.1.

TABLE 10.1. Characteristics of the combined effluent of the pulp and paper mills.

Item	Small Mill (Sastry et al)	Large Mill (Subrahmanyam et al.,)
	Produces 20 tonnes of paper/day	Produces 2000 tonnes of paper/per day.
Flow per day	330 m^3/tonne	222 m^3/tonne
Colour	–	7800 units
pH	8.2–8.5	8.5–9.5
Total solids, mg/l	–	4410
Suspended solids, mg/l	900–2000	3300
C.O.D., mg/l	3400–5780	716
B.O.D., mg/l	680–1250	155
C.O.D/BOD ratio	3.9–5	4.6

TABLE 10.2. Distribution of Pollution Load from Different Sections of Pulp and Paper Mills.

Item		Digester Section	Bleaching section	Paper mill section
1. Flow %	Small	45.5	16.2	10.8
	Large	9.75	27.8	16.7
2. BOD, %	Small	66	18.4	2
	Large	32.5	32.5	1.43
3. Suspended solids, %	Small	60	14.5	7.75
	Large	3	1.35	3.4

The distribution of pollution load from the different sections of the pulp and paper mills are shown in Table 10.2.

It may be noted that too much of difference, in per cent suspended solids contribution from different sections, between the large and small mills, arises due to the following reasons. The large mill cited in the Table 10.2, produces a large amount of Lime mud, and without being calcined it is discharged as a waste; this lime mud in the large mill contributes 86.5% of the total suspended solids in the effluent. As such the per cent suspended solids from the digester section, bleaching section, paper mill section etc assumes a very low value. On the other hand, no question of lime mud arises in the small mill cited, as it does not have the chemical recovery plant. The waste volume percentage from the digester section is also higher in the case of small mill, as the entire quantity of the black liquor is wasted in such mills, while only the leakages and spillages from the digester constitute the waste flow from the digester section in a large mill.

10.3 THE EFFECTS OF WASTES ON RECEIVING WATER COURSES OR SEWERS

Crude pulp and paper mill wastes, or insufficiently treated wastes cause very serious pollution problems, when discharged into the streams. The pollution extends over a very long stretch, sometimes as long as 80 kms in the downstream, due to the presence of slowly decomposing components in the waste. The fine fibres often clog the water intake screens in the down stream side. A toxic effect may also be induced upon the flora and fauna of the stream due to sulphites and phenols in the waste. However, in a particular case, the waste is found not to be toxic upto a concentration of 35%. Again, the bottom deposite of Lignino-cellulosic materials near the point of discharge of the waste in a stream undergo slow decomposition and may lead to the dissolved oxygen depletion followed by the creation of anaerobic condition and destruction of the aquatic life.

The question of discharge of this waste into the municipal sewers does not arise due to the large amount and strong nature of the waste.

10.4 TREATMENT OF PULP AND PAPER MILL WASTES

The treatment of the waste may consist of all or a combination of some of the following processes.

(a) Recovery: The recovery of the process chemicals and the fibres reduces the pollution load to a great extent. Where the economy permits the colour bearing 'black liquor' is treated for the chemical recovery. The process of recovery is described earlier. However, in this process the lignin is destroyed. The same may also be recovered from the black liquor, by precipitation by acidulation with either carbon dioxide or sulphuric acid (Das and Tapadar). These recovered lignins have got various uses in other industries. The alkali lignins of Kraft process may be used as a dispersing agent in various suspensions. Lignins may be used as raw materials for various other substances like Dimethyl sulphoxide, which is used as spinning solvent for polyacrylonitrile fibres. Activated carbons may also be manufactured from the lignins, recovered from the black liquors.

The fibres in the white water from the paper mills are recovered either by sedimentation or by floatation using forced air in the tank.

The recovery of lime from the lime mud, by the process of calcination, is not practiced in India, due to undesirable content •of silica in it.

(b) Chemical treatment for colour removal: The chemical coagulation for the removal of colour is found to be uneconomical. Attempts have been made to remove colour from the waste using the Lime sludge; the results are not at all encouraging (Sastry et al, 1974). "Massive Lime Treatment" process, developed by the National Council for Stream Improvement in USA is said to be capable of removing 90% of colour and 40% to 60% of BOD from the waste (Subrahmanyam). In this process, entire quantity of lime, normally required for the recaustisation of green liquor into white liquor, is taken and allowed to react first with the coloured waste effluent. The colour is absorbed by the lime, and the sludge after settling is used in recaustising the green liquor. The treatment of the green liquor with coloured lime sludge results in the formation of dark brown liquor, containing both desired cooking chemicals and colour-producing component like lignin. This lignin bearing liquor is used as digester liquid, and then is destroyed along with the fresh lignins, in the subsequent operations of concentration and incineration in the process of chemical recovery.

Flow diagram for the massive lime treatment of the coloured waste is given in Fig. 10.2.

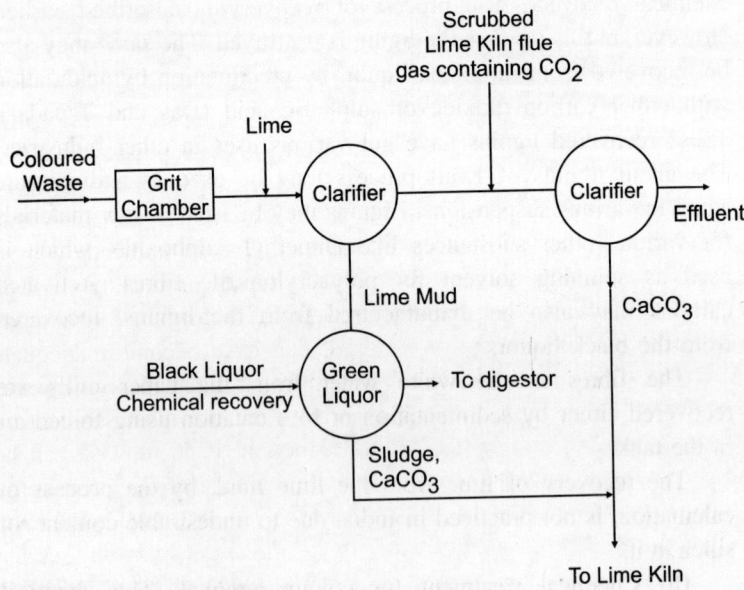

Figure 10.2: Massive Lime Treatment for Colour Removal in Pulp and Paper Mill.

(c) Activated carbon for colour removal: In a study conducted by NEERI, it has been observed that Acidic Activated Carbon can remove 94% colour from the Pulp mill waste. However, pH of the waste required to be reduced to 3.0 before this activated carbon treatment.

(d) Physical treatment for clarification: Mechanically cleaned circular clarifiers alone are found to be capable of 70-80% removal of the suspended solids from the combined mill effluent. About 95 to 99% removal of settleable solids can be accomplished in the clarifiers (Gillespie). However, the BOD reduction is comparatively small and of the order of 25-40% only. A surface loading in the order of 30 to 31 $m^3/m^2/day$ is found to be adequate for the 79% removal of suspended solids and 52% removal of COD, at a detention time of 30 minutes (Saxena et al).

The primary sludge produced in the clarifiers can be thickened to such a consistency in the clarifier itself, that can be easily dewatered mechanically.

(e) Biological treatment of the waste: Considerable reduction of BOD from the waste can be accomplished in both conventional and low cost biological treatment processes. Some are also effective in the reduction of colour from the waste.

If sufficient area is available, the waste stabilisation ponds offer the cheapest means for treatment. Depth of these ponds vary from 0.9 m to 1.5 m; the detention period may vary from 12 to 30 days. A minimum of 85% removal of BOD is found to be achievable elsewhere (Gillespie), with a loading rate upto 56 kg/hectare/day.

Aerated lagoons are the improved forms of the stabilisation ponds; they can be adopted to upgrade the performance of already present quiescent stabilisation ponds, that have become inadequate because of the increased loading or more stringent receiving water criteria. The mechanical surface aerators are the most satisfactory oxygen transfer device. The BOD reduction of 50 to 95% can be achieved in the aerated lagoons by varying nutrients feed, air supply and detention time (3 to 20 days), at the loading range of 670 to 1340 kg of BOD per hectare per day (Gillespie). In India Aerated lagoons are employed either, when the effluent BOD is moderate (where a partial chemical recovery from the black liquor is practiced), or, as a polishing device. In one particular case, a detention period of 5 days is found adequate for 90% BOD removal; the system rate constant is found to be 0.21/day (Kothandaraman). It may be noted that the pulp and paper mill waste does not contain necessary nutrients for the bacterial growth, and hence Nitrogen and Phosphorous are to be added into the lagoons in the form of Urea or Ammonia and Phospheric acid in BOD : N : P ratio of 100 : 5 : 1. The nutrient addition is found to be not necessary when a detention period of more than 10–15 days are provided (Gillespie).

Segregated strong waste or combined wastes may be well treated in Anaerobic lagoons with nutrient supplementation. A BOD loading of 0.048 kg/m^3/day and a detention time of 20 days is found to be adequate for 72.5% removal of BOD in a particular case (Sastry et al 1977). In another case (Sastry et al 1974), 77.5% removal of BOD is reported at a detention time of 6—8 days and loading of 0.017 kg of BOD/m^3/day. Aerated lagoons may be employed after the anaerobic lagoons where a high effluent quality is required. The system rate constant in the aerated lagoons, preceeded by the anaerobic lagoons may be taken as 0.15-0.17/day. A detention period of 7 days in these aerated lagoons are adequate.

Activated sludge process is the most satisfactory and sophisticated system for the effluent treatment. Instead of porous diffusers, the surface aerators are often suggested as the oxygen transfer device in the aeration tanks treating pulp and paper mill effluent. It is reported that about 80 to 90% BOD removal can be achieved with a loading rate of 0.2 to 0.3 kg of BOD per kg of MLSS at a detention time of 3 to 9 hrs, MLSS concentration of 2000-4000 mg/1, recirculation ratio of 0.3-0.5, and a nutrient supplementation at the BOD: N: P ratio of 100: 5: 1 (Subrah manyam et al).

While designing the secondary settling tank, in an activated sludge process it should be born in mind that a large portion of the fine fibres are not biodegradable and also do not settle easily.

Trickling filter has got a limited use in the treatment of the pulp and paper mill effluent, due to the greater chances of clogging of the media with fibrous material. Also the trickling filter system is incapable to provide a high degree of treatment—even with the new plastic media with greater specific surface area the BOD removal is

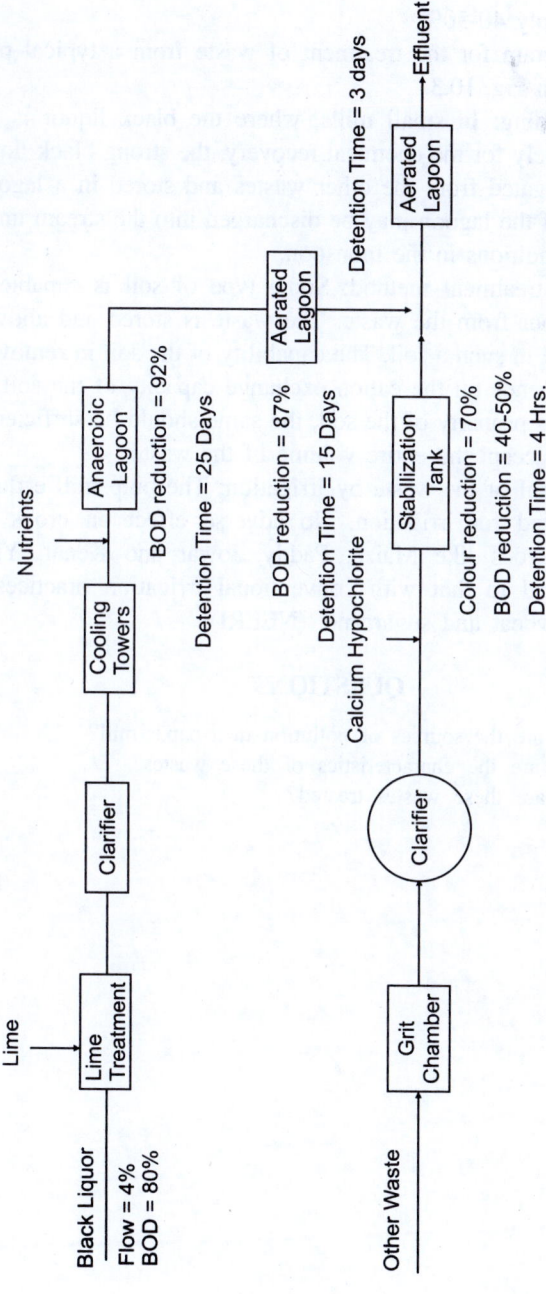

Figure 10.3: Flow Diagram for Treatment of Waste of a Typical Pulp Mill.

found to be only 40-50%.

Flow diagram for the treatment of waste from a typical pulp mill is given in Fig. 10.3.

(f) Lagooning: In small mills, where the black liquor is not treated separately for the chemical recovery, the strong black liquor must be segregated from the other wastes and stored in a lagoon. The content of the lagoon may be discharged into the stream under favourable conditions in the monsoon.

(g) Land treatment method: Some type of soil is capable of removing colour from the waste. The waste is stored and allowed to be absorbed in such a soil. The capability of the soil in removing the colour depends on the cation exchange capacity of the soil. In addition to this property of the soil, the same should be sufficiently permeable to accept the entire volume of the waste.

(h) Disposal of the waste by Irrigation: The pulp mill effluent may be utilized for irrigation. No adverse effect on crops are reported for crops like Maize, Paddy, Jowar and Kenaf. Yield almost identical to that with conventional irrigation practices is reported for wheat and sugarcane. (NEERI).

QUESTIONS

1. What are the sources of pollution in a paper mill?
2. What are the characteristics of these wastes?
3. How are these wastes treated?

11 | Breweries Wineries and Distilleries Waste

11.0 INTRODUCTION

While Breweries and Wineries produce beer and wine respectively, a large number of products of varying origin are obtained in Distilleries. The range of products from distilleries include industrial alcohols, rectified spirit, silent spirit, absolute alcohol, beverage alcohol etc. But two things are common in all the products mentioned above—

(*i*) all the above products are obtained through the biochemical process of fermentation by yeast, using carbohydrates as raw materials, and

(*ii*) all the products contain ethyl alcohol in different proportions.

In all the industries mentioned above, the final product, the fermentation medium, and the unwanted residue obtained at the time of preparation of the medium are all characteristically of high BOD and they present a formidable threat to the environment when discharged either into the water courses or onto the land without treatment. Due to their varying pollution potential and origin the wastes from all the three industries, breweries, wineries, and distilleries, will be discussed separately in the following sections:

11.1 ORIGIN AND CHARACTERISTICS OF BREWERIES WASTES

Making of beer essentially consists of two stages:
(1) preparation of malt from grains like barley and (2) brewing the barley.

In malt making, the barley grains are steeped to bleach out colour, and then made to sprout under aerobic conditions. The grain malt is then dried and stored after screening the sprouts out. The malt from the malt house is then transported to the brewing section, where the wort, the medium of fermentation, is prepared by making a mash of coarse grained malt with hot water, and by transforming the starch to sugar by boiling with hops. The wort is then filtered and cooled. The filtered wort is then innoculated with a prepared suspension of yeast, which ferments the sugar to alcohol. When the fermentation is complete, the yeast and malt residue is filtered out and finally the beer is carbonated before packing for sale. As the flavour of the product is of prime importance in all breweries the selection of raw materials and the control in the process is done accordingly to arrive at the objectives.

Brewery wastes originate in both these stages. One being the spent water from the steeping process from the malt house. The waste includes the water soluble substances of the grain that are diffused into it. Characteristically it contains a large amount of organic soluble solids indicated by a high BOD in the order of 400-800 mg/1 and low suspended solids concentration. In the brewing plant, the major potential pollutant is the fermentation residue or the spent grains. Wastes also originate in the preparation of yeast suspension (i.e. prefermentation section), from washing of containers, equipments and floors, and in the process of by-product recovery from spent grains. Large volume of almost unpolluted water also comes out as waste cooling water. The waste from the brewing plant contains a high suspended solids and also a high BOD. While malt house waste is usually alkaline in nature, the brewing plant is generally acidic. Composition of a typical malt house and combined brewery waste is given in the Table 11.1.

11.2 ORIGIN AND CHARACTERISTICS OF DISTILLERIES WASTE

The beverage alcohol industries utilize different grains, malted barley and molasses as raw materials. On the other hand the molasses (black strap type) are exclusively used as raw materials in the industrial alcohol industry. The final stages being indentical, the preparation of the fermenting, medium or mash is slightly different

TABLE 11.1. Composition of Malt House and Combined Brewery Wastes.

	Malt house waste	Brewery waste
pH	6.9–9.5	4.0–7.0
COD (permanganate value), mg/1	31–175	30–1225
BOD, mg/1	20–204	70–3000
Total solids, mg/1	428–700	272–2724
Suspended solids, mg/1	22–339	16–516
Total Nitrogen (N), mg/1	14–56	7–42

in these two industries. In beverage alcohol industry, the preparation of mash consists of (*i*) preparation of 'green' malt, (*ii*) preparation of cooked slurries of the grains, (*iii*) mixing of the above two followed by pH adjustment, and nutrient (ammonium salts and phosphates) supplementation. On the other hand in molasses distilleries, the preparation of mash consists of (*i*) dilution by water to a sugar content of about 15% (*ii*) pH adjustment to 4.0-4.5 to prohibit bacterial activities and (*iii*) nutrient addition. The yeast suspension is prepared separately in the laboratory with part of the diluted molasses and then innoculated into the mash for fermentation under controlled conditions. The fermented liquor containing alcohol is then sent to a overhead tank without separation of the solid materials. The same is then degasified, and then the alcohol is stripped leaving a "spent wash". The crude alcohol is then redistilled and stored in vats. Some of the beverage alcohols like gin attain their final form at this stage, some others like whisky require aging in charred oak wood barrels.

The "spent wash" is the major polluting component of the distilleries, and it is reported to be ten to fifteen times the final product in volume. The other pollutants include yeast sludge, which deposits at the bottom of fermentation vats. In most of the distilleries in India this yeast sludge is mixed with the spent wash and discharged. Malt house wastes also contribute towards pollution in beverage alcohol distilleries. In addition to these major BOD and solids contributing wastes, floor washes, waste cooling water, and wastes from the operations of yeast recovery or by-products recoveries processes also contribute to the volume of these wastes. The composition of some of the wastes from distilleries are given in Table 11.2.

TABLE 11.2. Characteristics of Composited Combined Waste, Spent Wash and Yeast Sludge from Different Indian Distilleries.

	Yeast sludge	Combined waste	Spent wash	Spent wash
pH	4.8	3.9–4.3	3.5–3.65	4.0–4.5
Temperature, °C	—	—	—	90–95
COD, mg/l	368000	27900–73000	118000	70000–151200
BOD, mg/l	165000	12230–40000	41380	28000–80000
Total solids, mg/l	—	16640–26000	99000	59100–114500
Suspended solids, mg/l	73000	4500–12000	350	1000
Total Nitrogen, mg/l	—	—	1135	14800–19000
Alkalinity	—	380–510	—	—
Colour	—	—	Dark brown	Dark brown
DO	—	—	NIL	—
Rate of waste flow	—	0.9–1.0 MLD	—	0.82 MLD

11.3 ORIGIN AND CHARACTERISTICS OF WINERIES WASTE

The wineries utilize the fruit juices as the raw materials. So the first operation in any winery is the pressing of fermentable juice from the fruits like grape, 'mahua', etc. The waste from this operation ressembles that from the canning industry, and includes the spent fruits or pomace, wastage of fermentable juices and floor wastes etc.

The second stage in any winery consists of fermentation of this juice employing the method described earlier. The wine attains its final form at this stage and requires only decantation, blending and bottling for sale. The waste from this stage comes from fermenting, decanting spillages, floor washes etc, and resembles that from a brewery. The composition of waste from a factory producing different brands of wine and other alcohol beverages, and some fruit juices, are given in Table 11.3.

Some form of wines, usually used as intermediates for brandy, are obtained by fermenting the crushed fruits without separating the pomace. This method apparently reduces the pomace residue problem at the first stage, but transfers the problem to the brandy plant.

In the third stage, i.e. in the brandy plant, wine of either type or the fermentation residue in the wine making is distilled to obtain brandy. Depending upon source of the brandy, the waste may have low to very high solid concentration, and resembles distillery wastes, very much.

TABLE 11.3. Characteristics of Combined Wastes from Brewery, Winery and Food Processing Units.

Temperature	20-29°C
Colour	Brownish yellow
pH	4.0
COD, mg/l	1800-3000
BOD, mg/l	1500-2000
Dissolved solids, mg/l	6800-9400
Suspended solids, mg/l	15-17

11.4 EFFECTS ON RECEIVING STREAMS/SEWERS

All the above types of wastes discussed earlier are not toxic to the aquatic life of the receiving stream. But due to their high BOD content, they deplete the dissolved oxygen of the receiving water. This results in anaerobic decomposition of this organic solids, both settled and suspended, producing a malodorous condition over a fairly long stretch of the stream. The conditions further deteriorate due to the strong growth of sewage fungi. The dark colour of the stream renders it unaesthetic.

Brewery waste, which is comparatively of lesser strength may be discharged in a fresh condition into the sewers to the extent of 3-5% of the domestic sewage. A strongly acidic or putrefied brewery waste will disrupt the normal biological activities of the waste treatment plants. For the sake of safety the brewery waste, if discharged into the sewers, must be screened and pretreated by lime.

The very high BOD content of the distillery waste makes it non amenable to the aerobic biological treatment, and as such it cannot be discharged into municipal sewerage system directly.

11.5 TREATMENT OF THE WASTES

Brewery wastes being comparatively less strong can be treated by aerobic biological treatment, after screening and neutralization. Usually, the biological treatment is accomplished by two stage process for 90-94% BOD reduction. A flow sheet of one such Brewery waste treatment plant employing high rate trickling filters is shown in Fig. 11.1. When sufficient land is available the brewery wastes may be used for broad irrigation after neutralization to utilize the fertilizing components of the waste.

The yeast sludge from the distilleries which contains very high suspended solids and BOD and is rich in proteins, carbohydrates, vitamins, may be treated separately for by-product recovery. But in practice, they are mixed and discharged along with the spent wash, thereby increasing the pollution load of the plant effluent.

The raw spent wash with low pH, high dissolved solids, high temperature, high sulphates, and high BOD is not amenable to aerobic biological treatment. Physico-chemical methods are also found to be ineffective in the treatment of spent wash or combined distillery wastes. Two stage biological method of treatment consisting of an anaerobic treatment, followed by an aerobic treatment of the waste, have been widely accepted as the only methods of treatment of the wastes from the distilleries.

Both closed anaerobic digestion and open anaerobic lagooning has been tried in India as a means of anaerobic treatment. A single stage digester is usually adopted for the anaerobic treatment when land available is limited. Different laboratory studies with different wastes suggest different loading rates of the digester. For a BOD reduction of 90.8% a BOD loading of 2.5 kg/m^3/day and a detention time of 15 days is suggested in one case (Subbarao et. al).

In another case a digester with a BOD loading of 3.01 kg/m^3/

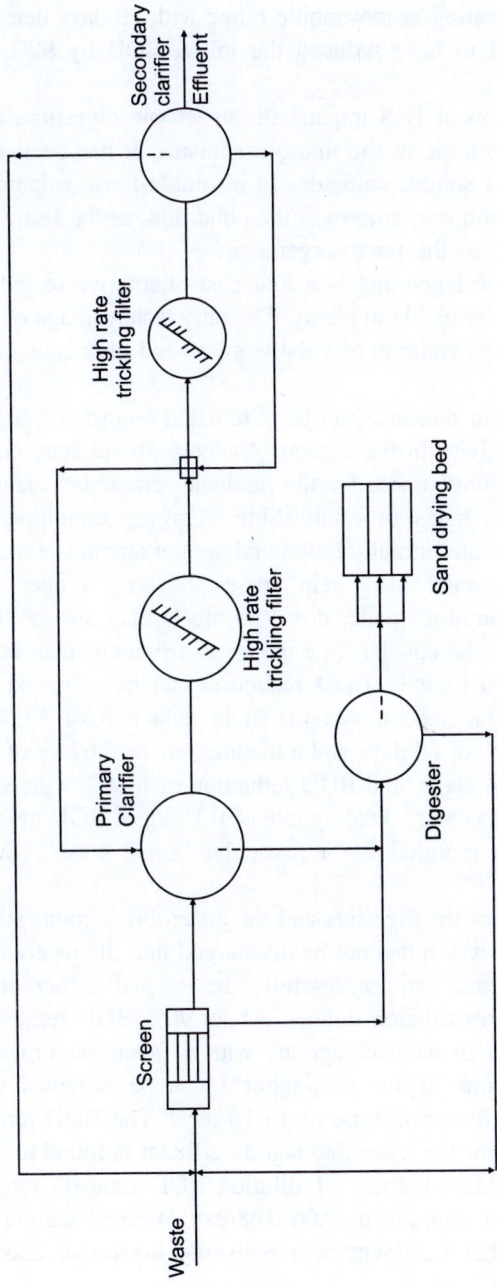

Figure 11.1: Flow Sheet for the Treatment of Brewery Waste.

day, and operating in mesophilic range with 10 days detention time was reported to have reduced the initial BOD by 90% (Sen and Bhaskaran).

Production of H_2S impairs the anaerobic digestion, as soluble sulphides are toxic to the micro-organisms. It has been found that conversion of soluble sulphides to insoluble ferric sulphides by the addition of iron salts improves the condition, as the ferric sulphides are not toxic to the micro-organisms.

Anaerobic lagooning is a low cost alternative to the digesters when land is available in plenty. The only disadvantage of anaerobic lagoons is the evolution of volatile gases and obnoxous odour from the ponds.

This odour nuisance can be eliminated by establishing a proper anaerobic activity in the lagoon. As the high sulphate content and low pH is unfavourable for the methane fermenters, neutralization of the waste helps in establishing a proper condition for their activity. A greater initial dilution and greater amount of acclimatized seed sludge may also help in establishing proper anaerobic activities. Iron at a smaller dose (50 mg/1) may also be used as an alternative to the costlier lime treatment (Swaminathan et. al).

About 90 to 95% BOD reduction can be achieved in a two stage anaerobic lagoon system. BOD reduction of 85.5% with a detention time of 60 days and a loading rate of 0.67 kg of BOD/m³/ day in the 1st stage, and BOD reduction of 65.5% with a detention time of 40 days and a loading rate of 0.15 kg of BOD/m³/day in the 2nd stage is reported for a particular "spent wash" (Subba Rao et. al).

Effluent of the digesters and the anaerobic lagoons still contain a high BOD, which can not be discharged into the receiving waters. These effluents can successfully be treated either in aerated lagoons, or in oxidation ditches. About 90% BOD removal can be accomplished in aerated lagoons with a detention time of 20-28 days. The same degree of treatment can be achieved with 1: 1 dilution at a detention time of 15-16 days. The BOD removal rate constant, K, for the anaerobic lagoon effluent is found to vary from 0.01 to 0.05/day; with 1 : 1 dilution BOD removal rate constant was found to improve to 0.06-0.68/day. Aerated lagoon effluents requires further treatment in a polishing lagoon of about 24 hrs

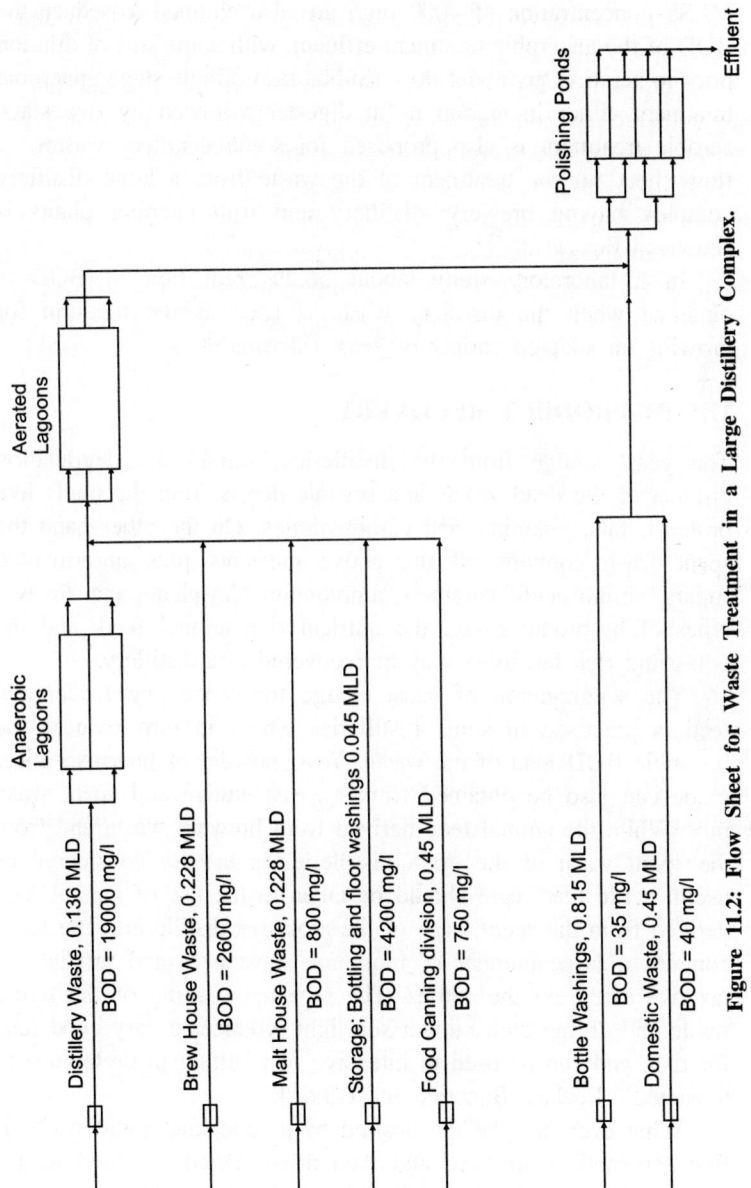

Figure 11.2: Flow Sheet for Waste Treatment in a Large Distillery Complex.

Anaerobic Lagoons

Aerated Lagoons

Polishing Ponds

Effluent

Distillery Waste, 0.136 MLD

BOD = 19000 mg/l

Brew House Waste, 0.228 MLD

BOD = 2600 mg/l

Malt House Waste, 0.228 MLD

BOD = 800 mg/l

Storage; Bottling and floor washings 0.045 MLD

BOD = 4200 mg/l

Food Canning division, 0.45 MLD

BOD = 750 mg/l

Bottle Washings, 0.815 MLD

BOD = 35 mg/l

Domestic Waste, 0.45 MLD

BOD = 40 mg/l

detention time. Oxidation ditches with 50-60 hrs detention time and MLSS concentration of 4000 mg/l are also claimed to reduce the BOD of the anaerobic treatment effluent, with some sort of dilution prior to aeration, by about 90% (Subba rao). Single stage anaerobic treatment either in lagoon or in digester followed by two stage aerobic treatment is also proposed for some distillery wastes. A flow sheet for the treatment of the waste from a large distillery complex having brewery, distillery and fruit canning plants is shown in Fig. 11.2.

In a laboratory study, about 56.2% reduction of BOD is obtained when the distillery waste is used as the medium for growing an adapted variety of yeast (Sharma etc.)

11.6 BY-PRODUCT RECOVERY

The yeast sludge from the distilleries contain the degradation product of the dead yeasts and organic debris from the malts like proteins, fats, vitamins, and carbohydrates. On the other hand the spent wash contains all the above nutrients plus unfermented sugars, amino acids, caramels, ammonium phosphates etc. So two types of by-products *viz.,* the nutrient rich animal feed, and the potassium rich fertilisers may be recovered in a distillery.

The seggregation of yeast sludge for processing for animal feed is practised in some distilleries which in turn reduces the insoluble BOD load of the waste. Yeast powder of pharmaceutical grade can also be obtained from a yeast sludge and spent wash mix. While the animal feed derived from brewery waste and from the spent wash of the grain distilleries is usually considered as useful cattle feed, care should be taken in the use of animal feed derived from the spent wash of the molasses distilleries. The latter contains a large number of inorganic substances and produces a laxative effect on the catties. The repeated soaking of the liquid waste and drying under direct sun light produces a very good feed for fish, and can be used in intensive fish culture in preference to mustated oil cakes (Banerjee and Pakrasi).

What ever may be the desired by-product the liquid waste is first screened, evaporated and then dried. Dried screened wastes are known as "dried distillers grains". The evaporation and concentrating of soluble waste is accomplished in different type of evaporators. The concentrated waste is then dried on conventional spray or drum dryers. The product is known as "dried distillers' solubles (DDS)", which is normally used as an animal feed. The DDS can further be incinerated in hearths (at temperatures not exceeding 700°C) to produce inorganic ash rich in potassium salts. In the process outlined by Chakavarty and Bhaskaran, the concentrated spent wash is directly fed into the incinerator to produce ash. The ash containing potassium salt can further be purified by a sequence of operations like leaching with water, filtration, and acidifying by sulphuric acid. It is further concentrated in vacuum evaporators, and finally crystallization of potassium chlorides and sulphates is done. It has been reported that about 3.85 tonnes of Potash, or 5.75 tonnes of potassium chloride, and 1.27 tonnes of potassium sulphate can be derived from about 320 m^3 of spent wash.

It may be noted that the condensate water arising out of the process of evaporation of spent wash still contains a high BOD and should get proper attention before its disposal. It has been reported that about 84% of original waste volume comes out as condensate water, but the COD and BOD are drastically reduced by about 90% and 85% respectively. COD: BOD ratio of the condensate is also reduced to about 1.16 to 1.2 from that of the original range of 1.7 to 1.8. This indicates that the condensate is more biodegradable. The recycle of the condensate water for such uses whenever water of this quality will be tolerated is also advocated.

QUESTIONS

1. What are the characteristics of distillery wastes?
2. Draw a flow sheet for treatment of distillery wastes.
3. How are the by-products recovered in distillery wastes?

12 | Tannery Waste

12.0 INTRODUCTION

The tanning industry is one of the oldest industries in India. Usually the tannery wastes are characterised by strong colour, high BOD, high pH and high dissolved salts. Disposal of these wastes into water courses or onto land, with or without prior sedimentation, was not a problem perhaps decades ago. But in the recent years, the concentrated growth of this industry in certain localities has shown how the waste from this industry can cause irreversible damage to the water environment in the vicinity. In view of its peculiar pollution potential, and the increasing demand for good quality of water, both for domestic and other industrial purposes, it has become essential to treat the waste to a certain degree prior to its disposal.

12.1 MANUFACTURING PROCESS

Various unit operations are involved in the processing of raw animal skins or hides into different types of leather. The tanning process consists of three basic stages: (1) preparation of the hides for tanning, (2) tanning proper, and (3) finishing.

In the first stage, the hides are washed to remove dirt and preservative salts used earlier, and then soaked in fresh water containing sodium chloride and preservative chemical like "Antimucin", for one to five days. The soaked hides are then washed again with sufficient running water. The washed hides are then "limed" with a paste of lime and sodium sulphide (8:1). Limed hides are then mechanically cleaned of hairs and fleshings in wooden vats with running fresh water. The subsequent operations

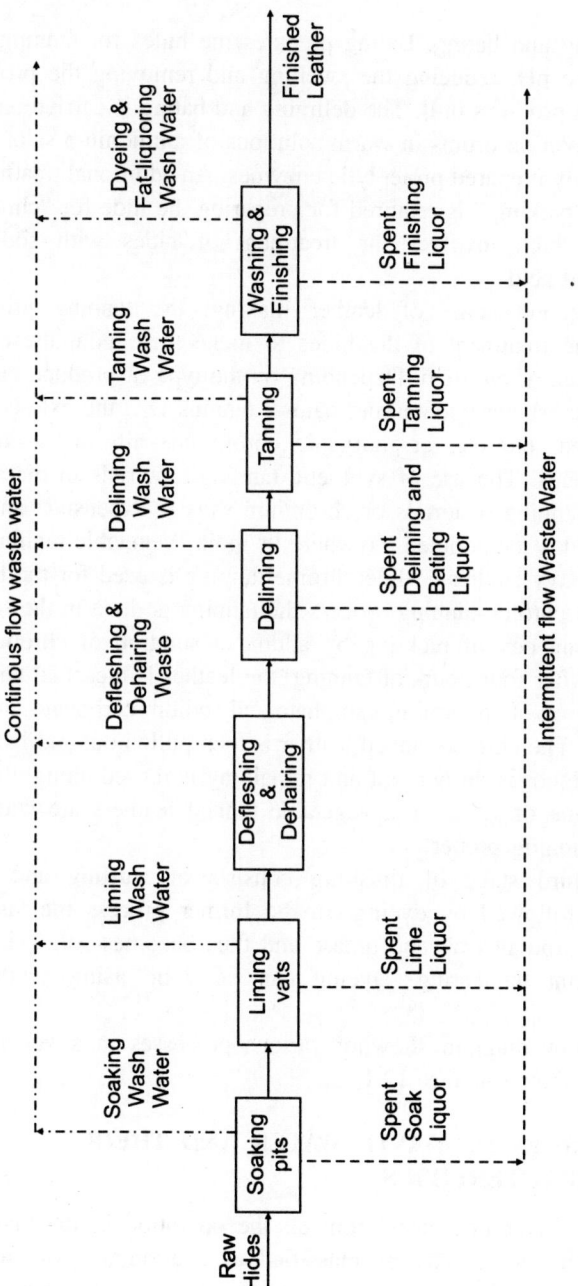

Figure 12.1: Flow Diagram of a Vegetable Tanning Process.

are deliming and bating. Bating prepares the hides for tanning by reducing the pH, reducing the swelling and removing the protein degradation products in it. The deliming and bating is carried out in vertically rotating drums in warm solutions of ammonium salts and commercially prepared proteolytic enzymes. An additional treatment known as "pickling" is required for preparing the hide for "chrome tanning", which involves the treatment of hides with sodium chloride and acid.

The second stage of leather making, the tanning proper, involves the treatment of the hides to make them nonputrescible and soft even when dried. Depending on the type of product, either vegetable substances containing natural tannins *viz.,* the extracts of barks, wood, nut etc, or inorganic chromium salts are used as tanning agents. The use of synthetic tanning materials in place of vegetable tanning materials or chromium salts is expensive, and is not reported to be adopted anywhere in India. Vegetable tanning is used for heavy leathers, while chrome tanning is used for the light leathers. In chrome-tanning process the tanning is done in the same vat after one day of pickling by adding a solution of chromium sulphate. After four hours of tanning, the leather is bleached with a dilute solution of sodium thio-sulphate and sodium carbonate, in the same bath. The chrome-tanned leather is then pulled out and half of the spent liquor is thrown out and remaining is reused along with a fresh volume of water. The vegetable tanned leathers are washed after the tanning proper.

The third stage of finishing consists of stuffing and fat-liquoring, followed by dyeing. In the former process the tanned leather incorporates oil and grease and thus becomes soft, pliable, and resistant to tearing. Dyeing can be done using synthetic dyestuffs.

The flow diagram showing various processes in a vegetable tannery is shown in Fig. 12.1.

12.2 SOURCES OF WASTE WATER AND THEIR CHARACTERISTICS

The waste water originates from all the operation in the tanning process. The waste may be classified as continuous flow waste and intermittent flow waste. Continuous flow wastes consist of wash waters after various processes and comprise of a large portion of the total waste, and are relatively less polluted than the other one. Spent liquors belonging to soaking, liming, bating, pickling, tanning and finishing operations are discharged intermittently. Although these are relatively small in volume, they are highly polluted and contain varieties of soluble organic and inorganic substances.

The spent soak liquor contains soluble proteins of the hides, dirt and a large amount of common salt when salted hides are processed. The spent liquor undergoes putrefaction very rapidly as it offers a good amount of nutrients and favourable environment for bacterial growth. The growth of pathogenic anthrax bacteria in this waste is also reported.

The spent lime liquor contains dissolved and suspended lime, colloidal proteins and their degradation products, sulphides, emulsified fatty matters, and also carry a sludge composed of unreacted lime, calcium sulphide and calcium carbonate. As such, the spent lime liquor has a high alkalinity, moderate BOD, and a high ammonia-nitrogen content.

The spent bate liquor contains high amount of organic and ammonia nitrogen due to the presence of soluble skin proteins, and ammonium salts used in bating.

The vegetable tan extract contains tannins and also non-tannins. Tannins are of high COD but relatively low BOD values, while non-tannins including inorganic salts, organic acids and salts and sugar, are high both in COD and BOD. The spent vegetable tanning liquor is the strongest individual waste in the vegetable tanneries, having the highest BOD and very strong dirty brown colour. It contains organic tannic compounds, carbohydrates, albuminous compounds, organic acids, inorganic salts etc and is slightly acidic (pH 4.5 to 5.0). When mixed with spent lime liquor, the spent vegetable tanning waste yields a bulky precipitate.

The spent pickling and chrome-tanning waste comprise a small volume, have a low BOD, and contains traces of proteinic impurities, sodium chloride, mineral acids and chromium salts. Chromium is known to be highly toxic to the living aquatic organisms in the hexavalent form and somewhat less toxic in the trivalent form. However, hexavalent chromium is no longer found in the waste effluent, as the present chrome tanning practices use only trivalent chromium salts. And when this waste is mixed with spent lime liquor most of the trivalent chromium are precipitated, and the rest get diluted to a great extent in the combined waste.

The spent dyeing and fat liquoring waste has got little significance, as major part of the dye and fat liquor are made to fix to the leather.

While segregation of spent chrome-tan liquor is advised for both chemical recovery and better pollution abatement, in most of the tanneries (both chrome-tanning and vegetable tanning) the wastes from all the units (intermittent and continuous) are combined, and disposed off the plant with or without some sort of treatment. The average characteristics of different spent liquors and the 10 hr composited combined waste from a tannery processing about 18000 kg of cow hides by chrome tanning process, and about 11000 kg of buffalo hides by vegetable tanning process, are given in Table 12.1. It may be noted that a much higher value of the concentration of trivalent chromium in spent chrome tan liquor has been reported in another case (5125 mg/l). The average concentration of chromium in combined waste in one chrome tannery is reported to be 180 mg/l.

12.3 EFFECTS OF THE WASTE ON RECEIVING WATERS AND SEWERS

As stated earlier the tannery wastes are characterised by high BOD, high suspended solids, and strong colour. These wastes when discharged as such deplete the dissolved oxygen of the stream very rapidly, due to both chemical and biological oxidation of sulphur and organic compounds. A secondary pollution of the stream may occur due to the deposition of solids near the discharge point and its subsequent putrefaction. The gas evolved during this process has got a typical foul odour. Usually chlorides are refractory to water treatment processes; as such, chloride in excess of tolerance limits (usually 500 mg/l when used as raw water for domestic purposes) render the water unsuitable for future use. The chromium is toxic to aquatic life and inhibits the growth of fish in the stream. However, most of chromium is precipitated in the combined tannery waste when mixed with spent lime liquor. The

TABLE 12.1. Average Compositions of Spent Liquors and Combined Wastes (Including Wash Waters) in a Typical Indian Tannery (Chakravarty et al).

Item	Spent Soak Liquor	Spent Lime Liquor	Spent Delime Liquor	Spent Bating Liquor	Spent Vegetable Tan Liquor	Spent Chrome Tan Liquor	Spent Dyeing Liquor	Combined Waste (10 hourly composited sample)	Hourly Maximum of Combined Waste
Average flow, m³/day	176	123	73	13.4	8.2	41	16.4	1310	2240
pH	8.4	12.8	9.3	9.9	5.4	3.2	6.2	8.9	12.3
Alkalinity as CaCO₃, mg/l	600	1600	800	600	—	—	—	260	700
Acidity as CaCO₃ as	—	—	—	—	2560	5400	1000	—	—
Chloride as Cl, mg/l	16800	8900	400	240	3000	not reported*	1000	4280	10600
Total Solids, mg/l	35800	38240	27450	5000	34800	7480	4255	10505	23410
Suspended Solids, mg/l	4500	3590	445	1060	2660	705	1255	1080	3310
COD, mg/l	3584	12000	2500	2374	30240	3584	6720	3700	6675
BOD, mg/l	708	7300	775	887	16000	—	—	1725	4000
Chromium as Cr, mg/l	—	—	—	—	—	2800	—	—	—

* The Chloride content of the Spent Chrome Tan Liquor is reported to be equal to 7200 mg/l, in a different Tannery, with a Slightly different composition of waste water.

vegetable tans are reddish tan in colour and become inky blue when they come in contact with water. Presence of tannins in the raw water renders it unsuitable for use in certain industries.

Even the lagooning of the untreated tannery waste on open land may adversely affect the ground water and nearby surface water sources due to the seepage of dissolved solids (chiefly in respect to sodium chloride); this also makes the soil unsuitable for cultivation for its high salt content. In a stream used as a drinking water source, the chloride content is found to go up to a level of 1000 mg/l, only due to seepage.

The tannery waste when discharged into a sewer not only chokes the sewer due to the deposition of solids, but also reduces the cross section of the sewer arising out of the lime encrustation. Lime encrustation occurs due to the precipitation of insoluble $CaSO_4$ and $CaCO_3$. When mixed with acidic waste in sewer, the evolution of H_2S occurs resulting in the corrosion of concrete surfaces of the sewer.

Chromium compounds in excess of 10-20 mg/l disrupt the operation of the trickling filter. Sulphides are also toxic to the microorganisms; however they get oxidised in the sewer before their arrival to the sewage treatment plant. Presence of lime also inhibits biological actions in the sewage treatment plant. However a proper dilution preceeded by sedimentation may make the waste amenable to the biological treatment.

12.4 TREATMENT OF WASTE

The method of treatment of tannery waste may be classified as physical, chemical and biological. The physical treatment includes mainly screening and primary sedimentation. The physical treatment is the only treatment which is provided in most of the tanneries in India. Screens are required to remove fleshings, hairs and other floating substances. The screenings may be used in the manufacture of glue, or may be simply recovered for hair, fleshings, fat etc. A continuous flow sedimentation tank, designed on maximum hourly flow with four hours of detention, is found to be effective in 90% removal of suspended solids. However in most of the tanneries, the fill-and-draw sedimentation tanks are in use, and found to be more efficient in the removal of suspend solids.

About 98% of the chromium is precipitated in the primary sedimentation tanks, and is removed along with the sludge. The sludge is dried over sand drying beds and can be used as a good manure. No appreciable reduction of dissolved solids, BOD, COD, colour and chloride can be achieved in the physical treatment processes, but the effluent may be discharged into the sewers with less chances of sewer ehokage. But even with a primarily treated tannery effluent, lime encrustation occurs in the sewer, probably due to the precipitation of $CaCO_3$ and $CaSO_4$.

Chemical coagulation, with or without prior neutralization, followed by biological treatment is necessary for better quality of the effluent. Several coagulants like Alum, Ferric Chloride and Ferrous sulphate—have been tried for chemical coagulations. Ferrous sulphate is reported to be the best coagulant for the removal of sulphides and may be used for the effective removal of colour, chromium, sulphides, BOD and suspended solids from chrome-tan wastes. However ferrous sulphate cannot be used in the case of a vegetable tan waste due to the intensification of colour of the waste. Instead, alum may be used with prior neutralization using acid or carbon dioxide. Chemical coagulation with ferric chloride alone is reported to be quite effective in the removal of tannin and COD.

Biological treatment of the tannery waste, in activated sludge process, after mixing with municipal waste water in a suitable proportion, and using acclimatized microorganism is capable to reduce the BOD, COD and tannin by about 90%. Prior chromium reduction is, however, required in the case of a chrome-tanning waste. Trickling filters may also be used for effective removal of BOD, COD and colour. The biological treatment of tannery wastes are not difficult, but are relatively costlier due to their high BOD content, and are not within the easy reach of a small tanner. Vegetable Tanning waste can also be effectively treated by anaerobic contact filters. A field study conducted by NEERI indicates that, with a retention time of 12 hours, 89.5 to 90.8% removal of COD, 91.9 to 97.5% removal of BOD, and 73.6 to 90.5% removal of Tannin can be accomplished.

The low cost biological methods of treatment may effectively be used for the treatment of tannery wastes. Both oxidation pond

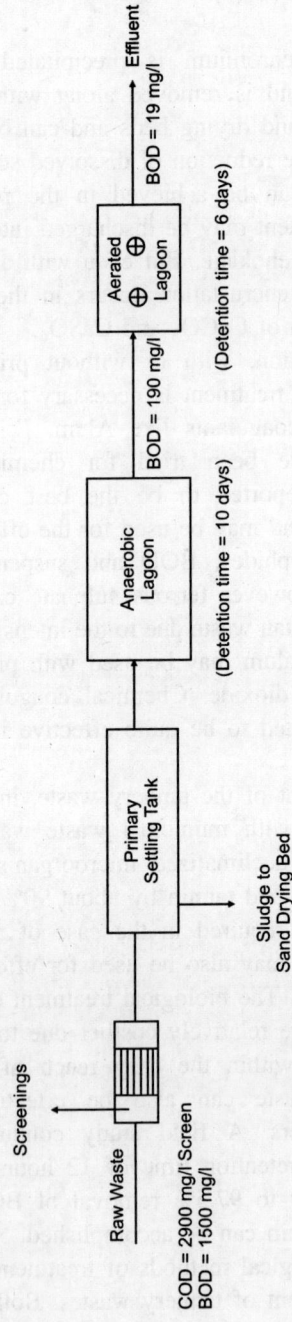

Figure 12.2: Flow Sheet for the Treatment of the Combined Waste from a Vegetable Tannery.

and anaerobic lagoons are recommended for small and isolated tanners. Anaerobic lagoons require less land area and nutrient addition compared to those in oxidation ponds.

For further improvement of the effluent quality, the anaerobic lagoon may be followed by an aerated lagoon.

A tannery waste can be successfully treated using physical and two stage biological treatment as shown in the flow sheet in Fig. 12.2. Oxidation ditches may also be used in lieu of an aerated lagoon.

As stated earlier, the major portion of the chromium in the combined chrome-tanning waste is removed in the primary settling tank. The chromium concentration in the primary effluent is further reduced by the dilution by domestic waste or river water. The residual chromium concentration usually does not impair the BOD removal in the biological processes.

Chloride, particularly sodium chloride is not removed by the conventional treatment processes. When chloride content of the receiving water courses exceeds the maximum allowable limit, several remedial measures like solar evaporation of spent soak liquor and spent pickling liquor were suggested. But ultimately, it leads to another problem of disposal of dried impure salts. Sodium chloride problem can be tackled by making reuse of spent liquors. The spent soak liquor contains more than 10% sodium chloride, but the pickling liquor requires only 8%. As such a portion of the spent soak liquor may be used as pickling liquor after settling with 0.2% alum (added to hasten settling), followed by salt percentage adjustment and acid addition. The same spent pickling liquor may be reused several times with or without make up water. Use of substances, other than salt, as preservative to the raw hides will also reduce the problem with sodium chloride. Recently it has been observed that the use of saltless short term preservatives, like Neem oil, has got several other advantages also.

The volume of the waste in any tannery can be much reduced by direct reuse of some comparatively clean wash wasters, such as bating wash water, pickling wash water, and dyeing wash water.

Segregation of spent chrometan liquor and the recovery of chromium from it is often suggested. The process involves the chemical precipitation of chromium, in the form of $Cr(OH)_3$, with lime at a pH value of 6.6, and subsequent separation or $Cr(OH)_3$ sludge either by settling or preferably by filtration, and subsequent generation of chrome sulphate solution. The filtration can be made on a sand filter bed, and the generation may be accomplished directly on the same filter bed using concentrated H_2SO_4. A laboratory study (Arumugam) indicates that 90% recovery of chromium is possible, and about 25 ml concentrated H_2SO_4 is required for each litre of spent chrome tan liquor. The recovered chrome tan liquor is very well comparable to the freshly prepared liquors, but needs only adjustment of pH by the addition of soda ash. The filtrate containing residual chromium may be sent for further treatment along with other wastes. Although the cost of recovery of the chromium is less than the cost of the fresh chrome powder, the margin of profit is not pretty attractive, and only in the order of 27% of the total cost of the chrome powder. However, the practice of recovery will considerably reduce the pollution load of the combined tannery waste, and also save some amount of foreign exchange, as the commercial chromium is usually imported from abroad.

QUESTIONS

1. What is the manufacturing process in a tannery?
2. What are the effects of tannery wastes in receiving waters and sewers?
3. How do you treat tannery wastes?

13 | Textile Mill Waste

13.0 INTRODUCTION

The fibres used in the textile industry may be broadly classified into four groups: cotton, wool, regenerated and synthetics. The characteristics of the waste from the mill depend on the type of fibre used, as different types of fibres go through different sequences of operations before the woven cloth is being sent out of the mill. The pollutants in the waste water include the natural impurities in the fibres used, and the processing chemicals.

13.1 COTTON TEXTILE MILL WASTE

An integrated cotton textile mill produces its own yarn from the raw cotton. Production of yarn from raw cotton includes steps like, opening and cleaning, picking, carding, drawing, spinning, winding and warping. All these sequences are dry operations, and as such do not contribute to the liquid waste of the mill.

The entire liquid waste from the textile mills comes from the following operation of slashing (or sizing), scouring and desizing, bleaching, mercerizing, dyeing, and finishing. A flow diagram of processing of raw cotton to finished cloth is shown in Fig. 13.1.

In slashing, the yarn is strengthened by loading it with starch or other sizing/substances. Waste originates from this section due to spills, and the floor washings at the weak end. The substitution of low-BOD sizes (such as corboxy methyl cellulose) for the high BOD starch reduces the total BOD of mill effluent by 40-90%.

After slashing the yarn goes for weaving. The prepared cloth now requires scouring and desizing to remove natural impurities and the slashing compounds. Enzymes are usually used in India to hydrolize the starch; acids may also be used for this purpose. Caustic soda, soda ash, detergents etc. are used in scouring in Kier boilers. Replacement of soap used in scouring by low-BOD detergents may reduce BOD load by 35%. About 50% of the total pollution load of the mill is contributed by this section.

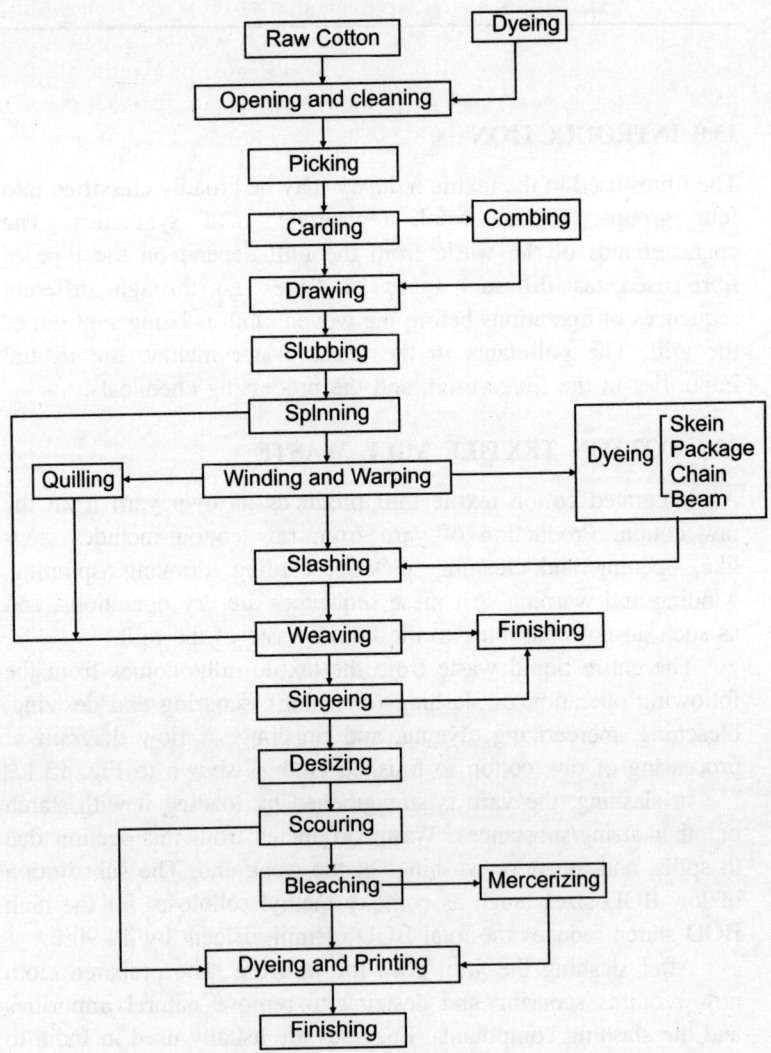

Figure 13.1: Flow Diagram of an Integrated Cotton Textile Mill.

Bleaching operations use oxidizing chemicals like peroxides and hypochlorites to remove natural colouring materials. This section contributes about 10% of the total pollution load.

Mercerising consists of passing the cloth through 20% caustic soda solution. The process improves the strength, elasticity, lustre and dye affinity. Waste from this section is recycled after sodium hydroxide recovery. Negligible waste which may come out of this section contributes little BOD but a high degree of alkalinity.

Dyeing may be done in various ways, using different types of dyes and auxiliary chemicals. Classes of dyes used include vat dyes, developing dyes, naphthol dyes, sulfur dyes, basic dyes, direct dyes etc.

Vat dyes require caustic soda and sodium hydrosulfite to reduce the dye into a soluble form. This is oxidised, after the fabric is being impregnated with it, by oxidizing agents like peroxide or chromate.

Sulfur dyes are reduced by sodium sulfide and oxidized by chromate.

Indigo dyes are also similar to vat dyes, but require only air oxidation.

When naphthol is applied first to the fabric, dried, and then passed through a developer for chemical coupling to produce dye, the process is called naphthol dyeing.

Developing dyes require acid and sodium nitrite and the colour is developed by treating with a developer.

Colour from the dyes vary widely and although those are not usually toxic, they are esthetically objectional when they impart colour in the drinking water supplies. Certain chemicals used in dyeing such as chromium are toxic and they are treated separately.

Thickened dyes, along with printing gums and necessary auxiliaries, are used for printing and subsequent fixation. After fixation of the prints, the fabric is given a thorough wash to remove unfixed dyes.

The Finishing Section of the mill imparts various finishes to the fabrics. Various types of chemicals are used for various objectives. These include starches, dextrines, natural and synthetic waxes, synthetic resins etc.

Therefore a composite waste from an integrated cotton textile mill may include the following organic and inorganic substances: Starch, carboxymethyl cellulose, sodium hydroxide, detergents, peroxides, hypochlorites, dyes and pigments, sodium gums, dextrins, waxes, sulphides, sulfates, soap etc. Depending on the process and predominant dye used, the characteristics of mill waste vary widely. The characteristics of a typical Indian mill is given in Table 13.1. (Kothandaraman, Aboo and Sastry).

TABLE 13.1. Composition of Composite Cotton Textile Mill Waste.

Characteristics	Value
pH	9.8-11.8
Total alkalinity	17.35 mg/1 as $CaCO_3$
BOD	760 mg/1
COD	1418 mg/1
Total solid	6170 mg/1
Total chromium	12.5 mg/1

13.2 WOOLEN TEXTILE MILLS WASTE

Wool wastes originate from scouring, carbonizing, bleaching, dyeing, oiling, fulling, and finishing operations.

Impurities of raw wool, consisting mainly of wool grease and other foreign matter are removed by scouring the wool in hot detergent alkali solution. Some wool are scoured by organic solvents. Various methods of grease recovery or cracking of grease containing wastes has been suggested.

Wool grease may be recovered from the scouring waste by centrifuging, coagulation or flotation, and may be processed further for the production of lanoline and potash. While many individual plants recover grease chiefly by centrifuging, coagulation either by sulphuric acid or by calcium or aluminium salts is also employed.

Carbonizing is a process in which hot concentrated acids are used to convert vegetable matter in the wool into loose charred particles, followed by mechanical "dusting" of the same.

Wool may be dyed at any stage, either as raw stock, or after spinning and weaving. Normally hot dye solutions are circulated through the wool, packed in a metal container.

In oiling, usually olive oil or a barg-oil-mineral-oil mixture is sprayed over the wool to aid in the spinning. Fulling is an operation where the loosely woven wool from the loom is shrunk into a tight closely woven cloth. To aid this process, chemicals like soda ash, soap etc., are used. Excess fulling chemicals, all of the oil etc are washed out of the fabric in a finishing process.

Waste from a dyeing and finishing process are contributed by the spent liquors and by subsequent washing of wool after bleaching, dyeing, and finishing.

Characteristics of a typical wool waste are given in the Table 13.2.

TABLE 13.2. Composition of a Typical Woolen Textile Mill Waste.

Item	Value
pH	9–10.5
BOD	900 mg/l
Total solid	3000 mg/l
Total alkalinity	600 mg/l
Total chromium	4 mg/l
Suspended Solid	100 mg/l
Colour	brown.

13.2.1 Effects of the Cotton Textile and Woolen Textile Mill Wastes on Receiving Streams/Sewers

The crude waste, if discharged into the streams, causes rapid depletion of the dissolved oxygen of the streams. The condition aggravates due to the settlement of the suspended substances and subsequent decomposition of the deposited sludges in anerobic condition. The alkalinity and the toxic substances like sulphides and chromium affect the aquatic life; and also interfere with the biological treatment processes; some of the dyes are also found toxic. The colour often renders the water unfit for use for some industrial purposes in the downstream side. The presence of sulphides makes the waste corrosive particularly to concrete structures.

13.2.2 Treatment of Cotton and Woolen Textile Mill Wastes

No treatment plant should be planned without giving serious

consideration for the reduction of the waste volume and strength, through the process chemical substitution, chemical and grease recovery and recycling of water. As stated earlier, several methods have been suggested for grease recovery; some of the chemical substitutions are also mentioned in this text. Caustic recovery, from the Kiering and mercerising wastes using dializers reduces the pollution load to a great extent.

The remaining pollution load of the waste is dealt with in the operations like seggregation, equalization, neutralization, chemical precipitation, chemical oxidation and biological oxidation. Several chemicals are used to reduce the BOD by chemical coagulation. These are alum, ferrous sulfate, ferric sulfate, ferric chloride etc; lime or sulfuric acid is used to adjust the pH in this process. Calcium chloride is found to be effective in treating wool-scouring waste. Biological treatment of Kiering or Scouring waste without any pretreatment is found difficult.

The dye wastes may be treated economically by biological methods, with prior equalization, neutralisation and chemical oxidation for certain wastes.

A composite waste, when free from toxic substances may be treated as efficiently as domestic sewage, as most of the textile mill wastes contain sufficient nutrients like nitrogen and phosphorus. For most of the wastes, pH adjustment may be required to be done. Trickling filters, activated sludge process, waste stabilisation ponds, all these types of biological treatment have been tried for the treatment of textile mill wastes and all of them are found to be very effective. Excellent results were also obtained with "Extended aeration" in treating a strong waste, even without any equalization and pre-treatment; this method eliminates the necessity of sludge digestion as well.

The Figs 13.2 and 13.3 show flow diagrams for a cotton textile mill waste treatment and a woolen mill waste treatment respectively.

13.3 SYNTHETIC TEXTILE MILL WASTE

Manufacture of synthetic fabrics involve two steps: (i) manufacture of the synthetic fibre, and (ii) preparation of the cloth. These two steps may be carried out either in one integrated plant,

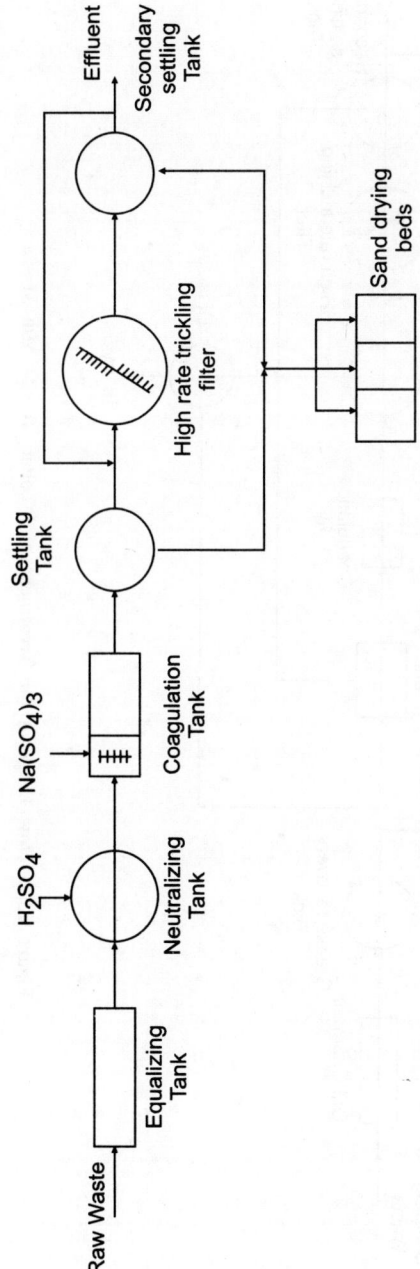

Figure 13.2: Flow Diagram for Treatment of Cotton Textile Mill Waste.

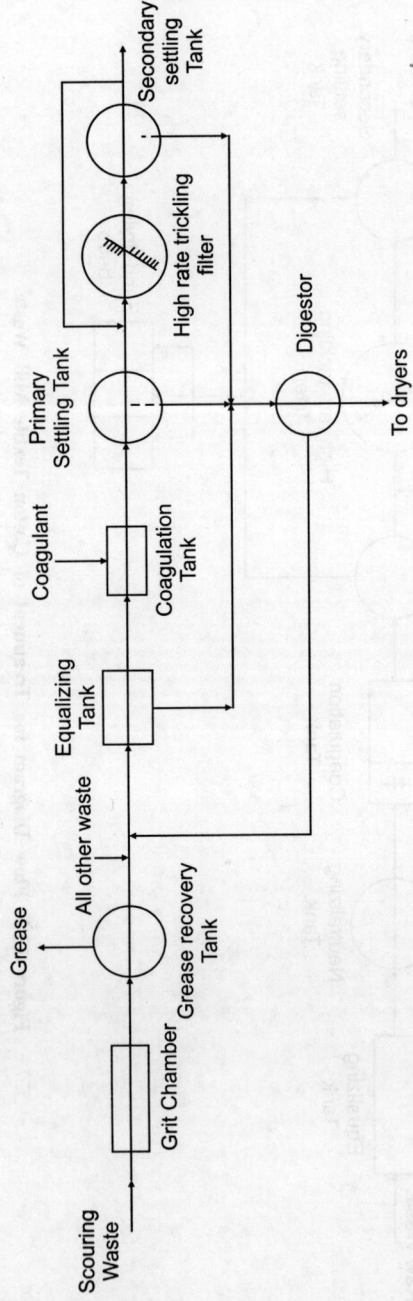

Figure 13.3: Flow Diagram for Treatment of Woollen Textile Mill Waste.

or may be separated in two different plants.

Wastes from the manufacture of the synthetic fibre resembles chemical manufacturing wastes, and depends entirely on the raw-materials used and the process adopted. A typical synthetic fibre Nylon-6 is obtained through polymerization of caprolactum, and subsequent pelletization, drying, remelting in extruders, spinning and twisting. A flow diagram of these operations is shown in Fig. 13.4.

The wastes from this manufacture are usually characterized by a colloidal type turbidity, a typical colour (a grey-yellowish colour in ε-amino caprolactum based manufacturing process), a low alkalinity (pH around 7.5), high amount of total solids (mainly organic) in the order of 2500 mg/1, and a comparatively small amount of suspended solids. The waste usually contains a large amount of nitrogen, entirely of organic origin. The waste is also characterised by a high COD value (in the order of 500 mg/1), though the BOD is found to be very low (around 50 mg/1).

The synthetic yarns are essentially composed of pure chemical compounds, so very light scouring is necessary prior to their dyeing and knitting into cloth. The processing of synthetic yarn and cloth is carried out on the conventional machinery used for cotton-textiles or woolen textiles. The major pollutants of the waste water in these mills originate from the various scouring and dyeing chemicals used to process the fibre and fabric. The characteristics of waste from processing of the synthetic fibres vary with the type of fibres being processed. Usually they are characterised by a low rate of discharge compared to the wastes from cotton or woolen textile mills.

13.3.1 Treatment of Wastes from Synthetic Textile Mills

Wastes from the manufacture of synthetic fibres are difficult to decompose biologically, as evidenced from high COD: BOD ratio. However, this can be treated biologically when diluted with municipal wastes in a ratio of at least 1 : 3. A direct aeration, preceeded by dilution (with river water say), neutralisation (if acidic), and addition of phosphoric acid, is reported to be capable of about 85% COD reduction.

The constituents of the wastes from preparation of fabric from

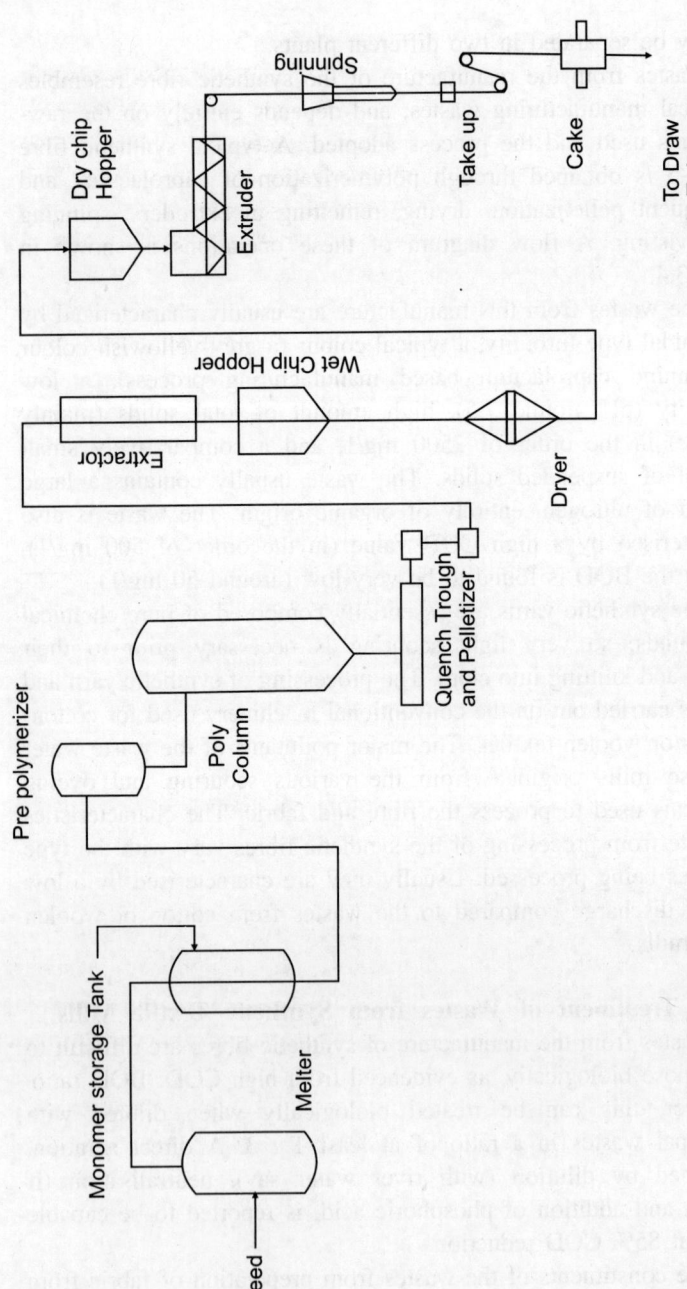

Figure 13.4: Flow Diagram of Manufacture of Nylon-6 from Caprolactum.

the drawn yarn being almost similar to those from scouring and dyeing operations in other textile mills, the treatment methods include those described earlier for cotton and woollen textile mills.

13.4 VISCOSE-RAYON PLANT WASTE

Rayon fabric is the principal regenerated or semi-synthetic fabric produced in our country. The part of the rayon yarn produced is again by what is known as viscose process. Viscose rayon is chiefly composed of regenerated cellulose. The rayons dye readily with most of the dyestuffs used for cotton textiles. However, it is not as resistant to chemicals as the cotton is. It is also highly adsorbent and takes up chemical reagents, oil etc. readily. Resistance of the rayons to acids and oxidizing bleaches decreases with higher temperature. The rayon, as such, is not given any drastic treatment in its processing. Like any other semi-synthetic and synthetic fibre, rayon has got negligible natural impurities.

The raw-material for the viscose rayon is the cellulose in the form of wood pulp. The shredded pulp is first soaked in about 18% caustic soda solution. The insoluble or undissolved cellulose, known as alkali cellulose is then separated from the spent liquor in hydraulic presses. The caustic solution is sent to dialyser for regeneration, and the alkali cellulose is then shredded and aged for certain period under controlled increased temperature.

The aged alkali cellulose is then treated with carbon disulphide under controlled temperature when an orange coloured mass of cellulose xanthate is obtained. The xanthate so formed is then dissolved in 6 to 7% caustic soda solution, yielding a viscous spinning fluid, known as "viscose". The viscose is then filtered through the filter presses, ripened and deaerated under controlled conditions.

The viscose solution is then extruded under pressure through a number of spinnerets into a continuously flowing acidic spinning bath containing 8 to 10% sulphuric acid, 13 to 20% sodium sulphate and 1 to 2% zinc sulphate. The viscose is coagulated and regenerated into filaments in the spinning bath, and are collected and wound in a bucket spinning at a high speed. A number of filaments make rayon yarn. The spinning buckets are washed either continuously or intermittently with water to remove impurities from the surface of the yarn. When intermittent washings are adopted, the bobbins of the yarns are taken out from the spin buckets after wash and are further washed with water on washing racks. The washings of the yarn, in spin buckets or on washing racks, carry most of the zinc used in the process. Intermittent wash waters of spin buckets are sent for chemical recovery along with spent spin bath.

The rayon yarn is then further treated with alkaline sodium sulphide solution to remove sulphur, deposited on the rayon fibres during the spinning operation. The yarn is then washed with water, bleached and again washed with dilute acid, detergent and plain water successively. The yarn is finally dried to proper moisture content, and sent to the weaving mill.

The spent spinning bath is sent to an evaporator for sodium sulphate recovery and then the latter is recycled to the process after being made up with requisite amount of sulphuric acid and zinc sulphate.

Further processing of the yarn for the finished cloth requires much less pretreatment particularly because it dyes easily and is not resistant to chemicals. The wastes from the rayon weaving units will not be considered in this section, as the small quantity of waste which originates there resembles that from other textile mills.

Principally two types of wastes originate in a viscose rayon plant, one being acidic and the other alkaline.

Acidic waste is made of mainly spin bucket washings and the first rayon yarn wash water. Cooling and condenser water discharged from the spent spin bath evaporators also joins the acid waste flow. Acidic waste also originates from the spin bath make up tanks.

Alkaline wastes originate from the caustic soda dializers, viscose ripening, and deaeration tanks, xanthating operations, and from the desulphurization process. Strong alkaline wastes are also obtained from the washing of filter clothes.

The sources of different types of wastes are shown in Fig. 13.5.

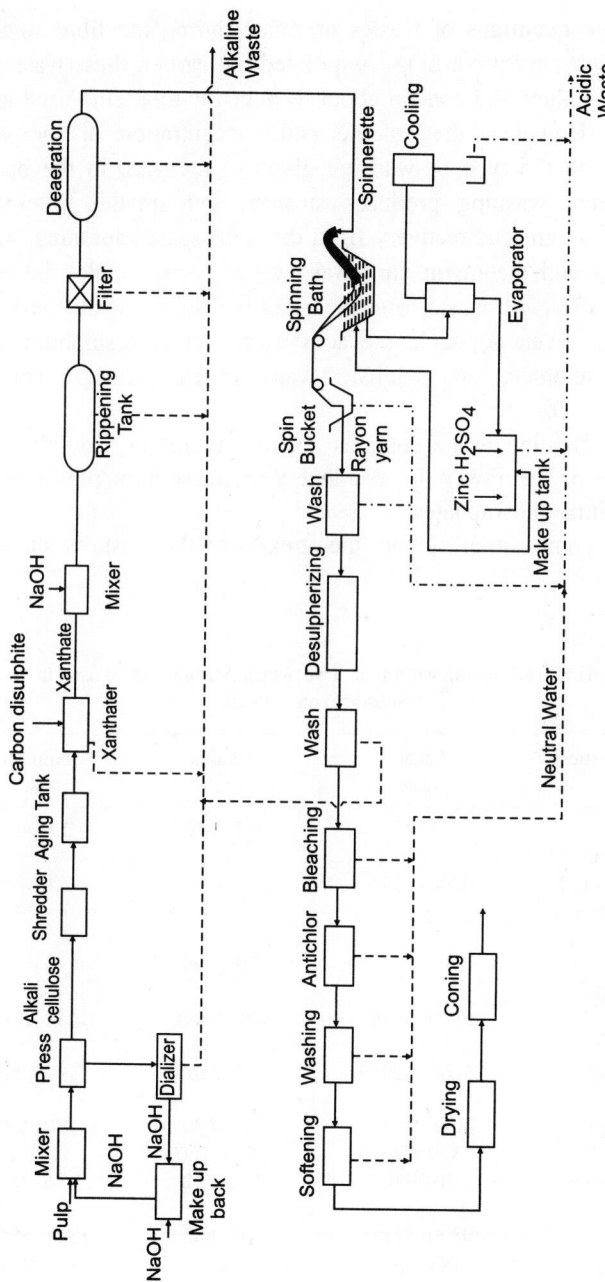

Figure 13.5: Sources of Waste Water in a Viscose Rayon Plant.

Large quantities of wastes originate during the fibre spinning and its subsequent washing. As pointed out earlier, these wastes are acidic in nature and contain about 85% of the total zinc used in the process. However, the volume and concentration of the waste depends on the type of washing given to the yarn in the bucket. Intermittent washing produces stronger but smaller volume of waste. Though zinc recovery from the seggregated spinning bucket washings and the rayon yarn wash water is preferable, the entire quantity of acidic waste from the plant is discharged out through a common sewer. As such, the acid waste contains sulphuric acid, sodium sulphate, zinc sulphate, waste fibres, hydrogen sulphide etc.

The alkaline waste contains sodium hydroxide, soluble hemicellulose of the raw pulp, residual viscose, sodium polysulphides and sodium thiosulphates.

The composition of both the streams of the waste is shown in Table 13.3.

TABLE 13.3 Composition of Two Main Streams of Waste in a Viscose-rayon Plant.

Characteristics	Acidic waste	Alkaline waste	Composite waste
pH	1–2	9.2–10.7	2.8–4.1
Total acidity as $CaCo_3$ mg/l	4200–7700	–	70–590
Total alkalinity as $CaCo_3$, mg/l	–	310–790	–
Total solids, mg/l	26800–32300	1480–2940	1630–2780
Dissolved solids, mg/l	26800–31950	960–2260	1490–2580
Suspended solids, mg/l	800–1050	520–680	190–200
COD, mg/l	390–790	154–1160	35–210
Chloride, mg/l	20–800	148–242	36.8–184
Sulphates, mg/l	20800–32188	267–683	1311–6689
Zinc, mg/l	184–203	–	6.2–18.4

13.4.1 Effect of Viscose-Rayon Wastes on the Receiving Water and Sewers

The sodium polysulphide formed during the desulphurization process is highly toxic to the aquatic life. Zinc and hydrogen sulphide are also toxic, particularly to fish. The sulphuric acid and carbon desulphide may affect the aquatic life if not adequately diluted by the receiving water. The precipitation of different compounds to the bed of the river, due to the slow chemical reactions of acidic and alkaline wastes, may continue for a long stretch of the river. The slow decomposition of the deposited flocs of hemicellulose may cause secondary pollution of the stream due to oxygen depletion. Excessive sulphate content may lead to the erosion of the river bank and corrosion of concrete hydraulic structures. The question of discharge of the untreated waste into the municipal sewers does not arise, because of its high acid content. However, this may be admitted to the sewers after proper neutralisation and hydrogen sulphide removal.

13.4.2 Treatment of the Viscose-Rayon Waste

The treatment of the waste from the viscose rayon plants should be based on segregation of different wastes and the recovering of substances used in the process as far as practicable.

As already pointed out, the recovery of caustic soda, from the spent solution separated from the alkali cellulose by presses is normally an integrated part of the plant operations. About 90%, of the caustic soda thus can be recovered by the use of dializers.

The spent spinning bath is usually treated for sodium sulphate recovery as mentioned earlier.

The recovery of zinc from the acidic waste by the commonly used method viz., chemical precipitation process, is not economically profitable, unless the initial acidity is very high (more than 3800 mg/1). However, such a recovery of zinc is justified since it leads to a saving of reasonable amount of foreign currency, zinc being an imported material. In addition to this, the acidity of the waste is completely neutralized during the process of zinc recovery. As such, the reclamation of zinc results not only in saving foreign currency, but also in a considerable reduction of pollution load.

Two methods are suggested for the recovery of zinc from such wastes. One is by chemical precipitation, as mentioned earlier and the other is by an ion exchange process.

In the chemical precipitation, a lime suspension is added to the waste in order to increase the pH of the waste to about 6.0, which is followed by settling. During settling the excess sulphuric acid is neutralized and calcium sulphate and traces of zinc are precipitated as sludge and subsequently wasted. The supernatant liquor is then transferred to another mixing tank and is treated with a requisite amount of sodium hydroxide (caustic soda), with pH adjusted to 9.2. Almost the entire quantity of zinc is precipitated as zinc hydroxide in the following settling tank. The zinc hydroxide sludge is taken out and treated with sulphuric acid to make zinc sulphate solution, which is returned to the spinning bath make-up tank. The effluent of the second settling tank is discharged as waste.

In ion exchange process, the acidic waste is directly passed through a cation exchange resin bed, when zinc is adsorbed. The spent resin bed is then eluted by a 5 to 8% solution of sulphuric acid to liberate zinc sulphate. The eluate is then returned to the process through spinning bath make up tank. Although the ion exchange method is technically feasible, chemical precipitation will still remain to be a cheaper and more efficient method of zinc recovery, unless a high capacity cation exchange resin is available. However, in the ion exchange method no strict pH control is required, and it requires less land area for its installation. The running cost can also be reduced in comparison to that in chemical precipitation, if a high capacity ion exchange resin is available.

The process of zinc recovery by chemical precipitation is shown diagramatically in Fig. 13.6.

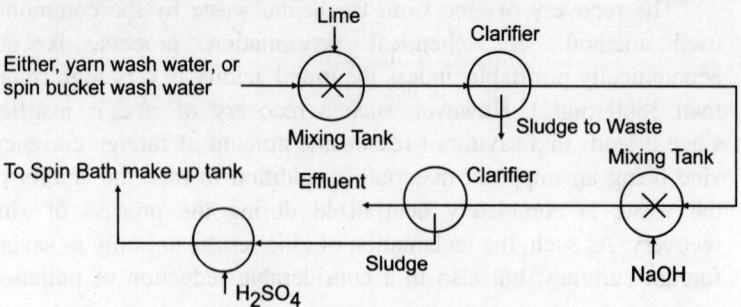

Figure 13.6: Flow Diagram for Zinc Recovery.

Further treatment of waste from viscose-rayon plant may consist of physical, chemical and biological methods. The physical methods of treatment include mixing in an equalization tank, screening, and sedimentation. The polysulphides decompose to hydrogen sulphide in the equalisation tank during mixing; the hydrogen sulphide may be removed by aeration in the said tank. The hemicellulose and sulphur settles well in the settling tank under alkaline conditions. Necessary pH adjustment using lime may be necessary for an acidic waste. Usually neutralisation, when required, is carried out after the primary settling tank, and the process is followed by another clarifier.

The residual carbon disulphides and the polysulphides which come out of desulphurization process may be removed by segregating the waste and treating them in a trickling filter. These said pollutants are oxidized to sulphates and thiosulphates during the process and are precipitated in the following clarifier. However, BOD reduction is comparatively small (about 70%) in the trickling filters.

Disposal of calcium sulphate sludge, produced either during the zinc recovery or during neutralisation, always poses a problem, as it dries only with great difficulty. Use of this sludge in any other industry is yet to be found.

QUESTIONS

1. Draw a flow diagram of an integrated cotton textile mill.
2. Explain the method of treating cotton textile mill wastes with the help of flow sheets.
3. How are the wastes from Viscose rayon plants treated?

14 | Dairy Wastes

14.0 INTRODUCTION

With increase in demand for milk and milk products, many dairies of different sizes have come up in different places. These dairies collect the milk from the producers, and then either simply bottle it for marketing, or produce different milk foods according to their capacities. Large quantity of waste water originates due to their different operations. The organic substances in the wastes comes either in the form in which they were present in milk, or in a degraded form due to their processings. As such, the dairy wastes, though biodegradable, are very strong in nature.

14.1 Sources of Waste

The liquid wastes from a large dairy originate from the following sections or plants: receiving station, bottling plant, cheese plant, butter plant, casein plant, condensed milk plant, dried milk plant, and ice cream plant. Waste also comes from water softening plant and from bottle and can washing plants.

At the receiving station the milk is received from the farms and after inspection the same is emptied into large containers for transport to bottling or other processing plants. The empty cans are rinsed, washed, sterilized and are returned to the farmers.

At the bottling plant, the raw milk delivered by the receiving station is stored. The processing includes cooling, clarification, filtration, pasteurization, and bottling.

In the above two sections, the liquid wastes originate out of rinse and washings of bottles, cans and equipments, and thus contain milk drippings and chemicals used for cleaning containers and equipments.

In a cheese plant, the milk (whole milk or skimmed milk) is pasteurized and cooled and placed in a vat, where a starter (lactic acid producing bacterial culture) and rennet are added. This separates the casein of the milk in the form of curd. The whey is then withdrawn and the curd compressed to allow excess whey to drain out. Other ingredients are now added, and the cheese blocks are cut and packaged for sale. Waste water from this plant include mainly the discarded whey and the wash water used for cleaning vats, equipments, floors etc.

In the creamery process, the whole milk is preheated to about 30°C to separate the cream from the milk. In a butter plant the cream is pasteurized and may be ripened with a selected acid and a bacterial culture. This is then churned at a temperature of about 7-10°C to produce butter granules. At a proper time the butter milk is drained out of the churn and the butter is washed and after standardisation, packaged for sale. Butter milk and wash waters used to clean the churns, and small quantity of butter comes out as the waste from the butter plants.

The skimmed milk may now be sent for bottling for human consumptions, or for further processing in the dairy for other products like non fat milk powders. Milk powders are produced by evaporation followed by drying by either roller process or spray process. The dry milk plant wastes consists chiefly of wash waters used to clean containers and equipments.

The soured or spoiled milk, and sometimes the skimmed milks are processed to produce caseins used for preparation of some plastics. The process involves the coagulation and precipitation of the casein by the addition of some mineral acids. The waste from this section includes whey, washings and the chemicals used for precipitation.

Very large dairies also produce condensed milk and Ice-creams.

In addition to the wastes from all the above milk processing units, some amount of uncontaminated cooling water comes as wastes; these are very often recirculated.

The dairy wastes are very often discharged intermittently; the nature and composition of waste also depend on the types of products produced, and the size of the plants. The Table 14.1 gives the characteristics of the waste (composite) of a typical Indian Dairy, handling about 300000 to 400000 litres of milk in a day.

TABLE 14.1. Composition of the Waste Water of a Typical Dairy.

Item	Value
PH	7.2
Alkalinity	600 mg/l as $CaCO_3$
Total dissolved solid	1060 mg/l
Suspended solid	760 mg/l
BOD	1240 mg/l
COD	84 mg/l
Total nitrogen	84 mg/l
Phosphorous	11.7 mg/l
Oil and grease	290 mg/l
Chloride	105 mg/l

14.2 EFFECTS OF THE WASTES ON THE RECEIVING STREAMS/SEWERS

As observed from the above table the waste is basically organic in nature. This is also slightly alkaline when fresh. When these wastes are allowed to go into the stream without any treatment, a rapid depletion of the dissolved oxygen content of the stream occurs, along with growth of sewage fungi covering the entire bottom of the stream and the submerged parts of the hydraulic structures within it. The waste is said to carry, occasionally, the bacteria responsible for tuberculosis. Though alkaline in fresh condition, the milk waste becomes acidic due to the decomposition of Lactose into lactic acid under anaerobic condition, particularly after complete oxygen depletion of the stream. The resulting condition precipitates casein from the waste, which decompose further into a highly odourous black sludge. At certain dilutions the dairy waste is found to be toxic to fishes also.

As the dairies are usually situated in rural areas or in small towns, the question of discharging the dairy waste into the sewers does not arise. In large cities, combined treatment of domestic sewage and dairy waste may be considered if the latter constitutes only 10% in volume of the former. In that case the dairy waste should be discharged in a fresh condition, as a putrefied waste may cause corrosion of the sewers.

14.3 TREATMENT OF THE DAIRY WASTES

As evident from the low COD: BOD ratio, the dairy wastes can be treated efficiently by biological processes. Moreover, these wastes contain sufficient nutrients for bacterial growth. But for economical reasons, attempt should be made to reduce volume and strength of the waste. This may be accomplished by (i) the prevention of spills, leakages and dropping of milks from cans, (ii) by reducing the amount of water for washes, (iii) by segregating the uncontaminated cooling water and recycling the same, and (iv) by utilising the butter milk and whey for the production of dairy by-products of good market value.

Due to the intermittent nature of the waste discharge, it is desirable to provide Equalization tank, with or without aeration, before the same is sent for biological treatment. A provision of grease trap is also necessary as a pretreatment to remove fat and other greasy substances from the waste. An aeration for a day not only prevents the formation of lactic acid, but also reduces the BOD by about 50%.

Both high rate trickling filters, and activated sludge plants can be employed very effectively for a complete treatment of the dairy waste. But these conventional methods involve much maintenance, skilled personnel, and special type of equipments. On the other hand the low cost treatment methods like oxidation ditch, Aerated Lagoon, Waste Stabilisation Pond etc. can be employed with simpler type of equipments and less maintenance.

Oxidation ditches in India may be designed with a low organic loading (about 0.2 kg/kg of MLSS), high biological mass concentration (in the order of 4000 mg/l), and extended period of aeration (in the order of 1.5 days), for BOD reduction of about 95 to 98%.

In a waste stabilisation pond, a BOD reduction of 52 to 74% could be achieved after 12 days of retention and an organic loading of 550 to 585 kg/hectare/day.

BOD reduction of about 90% may be obtained with a retention time of 7 days and a depth in the order of 3 m, in an anaerobic lagoon. An organic loading in the order of 0.48 kg/m^3/day is suggested for the above.

Use of dairy waste for Irrigation after primary treatment in an Aerated lagoon may also be good answer for the disposal of Dairy waste.

QUESTIONS

1. What are the effects of dairy wastes on receiving streams and sewers?
2. How do you treat Dairy wastes?

15 | Fertilizer Plant Waste

15.0 INTRODUCTION

The fertilizers produced in India can be classified broadly into two groups *viz.*, nitrogenous fertilizers, and phosphatic fertilizers. Plants may be producing only nitrogenous fertilizers like Urea, Ammonium Sulphate, Ammonium Nitrate, Ammonium Chloride, or only phosphatic fertilizers like super phosphates; there are plants where complex fertilizers containing both nitrogen and phosphates like Ammonium phosphate and Ammonium sulphate phosphate are produced. Another fertilizer industry, the potash industry will not be considered here; because this industry is quite simple in comparison to the others, and involves nothing but mining only. Potash ore, mainly potassium chloride, is mined, beneficiated, and sent for use without any chemical processing.

15.1 MANUFACTURING PROCESSES

Ammonia is the principal intermediate in the manufacture of all nitrogenous fertilizers. So, except when the by-product ammonia will be available from a coke-over, raw materials for nitrogenous fertilizer production is the carbonaceous materials, which are required for making ammonia. So all the nitrogenous fertilizer plants will have essentially an Ammonia production unit and a reactor where the synthetic ammonia will be reacted with other chemicals to produce the final product. The plant may have auxiliary units to produce the reacting chemicals also.

Basic process steps in the manufacture of Urea, from carbonaceous raw materials like naphtha, are as follows:

(*i*) reaction of the carbonaceous materials with steam and air to form a mixture of hydrogen and carbon monoxide, known as systhesis gas,

(ii) reaction of the carbon monoxide with steam over a catalyst to form more hydrogen, and carbon dioxide,

(iii) separation and purification of carbon dioxide,

(iv) removal of residual carbon monoxide from gas mixture,

(v) synthesis of ammonia by reacting hydrogen and nitrogen over a catalyst. (Nitrogen is supplied as air in an earlier step), and

(vi) synthesis of urea by treating ammonia with carbon dioxide in a reactor at higher temperature and pressure.

Different stages in the manufacture of urea, and unit operations involved in each stage is shown diagramatically in Fig. 15.1.

Plants using by product ammonia from other manufacturing plants (like coke oven) have to produce carbon dioxide separately for the production of urea.

Ammonium sulphate may be produced by reacting anhydrous ammonia with sulphuric acid, usually obtained as by-product sulphuric acid from other manufacturing plants. Ammonium sulphate may also be manufactured from gypsum or from calcium sulphate sludge, obtained from the phosphatic fertilizer plant, using ammonia and carbon dioxide obtained from ammonia plant. In this process calcium sulphate is reacted with ammonium carbonate solution to produce ammonium sulphate.

Ammonium Nitrate is produced when ammonia is reacted with Nitric acid. Normally the required quantity of Nitric acid is produced in the same plant, by oxidizing the ammonia.

Super phosphate is produced by merely mixing the phosphate ore (commonly known as phosphatic rock) with sulfuric acid to convert the phosphate to "monocalcium phosphate". The by-product calcium sulphate of this process may be used in the manufacture of Ammonium sulphate.

Ammonium phosphate is made by treating phosphoric acid with ammonia; the phosphoric acid production process involves, the following steps:

(i) dissolving phosphate rock in enough sulphuric acid,

(ii) holding the mixture until the calcium sulphate crystals-grow to adequate size,

(*iii*) separating the phosphoric acid and calcium sulfate by filtration, and

(*iv*) concentration of acid to the desired level.

15.2 WASTE WATER FROM FERTILIZER PLANTS

A variety of wastes are discharged from the fertilizer plants as water pollutants, in the form of

(*i*) processing chemicals like Sulphuric acid,

(*ii*) process intermediates like Ammonia, Phosphoric acid etc.,

(*iii*) final products like Urea, Ammonium sulphate, Ammonium phosphate etc.

In addition to the above, oil bearing wastes from compressor houses of ammonia and urea plants, some portion of the cooling water and the wash water from the scrubbing towers, for the purification of gases, also come as waste.

Wash water from the Scrubbing towers may contain toxic substances like Arsenic, Monoethanolamine, Pottassium carbonate etc. in a Nitrogenous fertilizer plant, while that in a phosphatic fertilizer plant may contain a mixture of carbonic acid, hydrofluoric acid and flousilicic acid. Both alkaline and acidic wastes are also expected from the boiler feed water treatment plant, the wastes being generated during the regeneration of anion and cation exchanger units.

Previously the waste cooling water was containing toxic elements like chromates, zinc, etc., which were used for corrosion control. The development of non-chromate technology using quaternary ammonium compounds has eliminated these toxic substances from the waste cooling water.

Additional pollutants like phenol and cyanide will be introduced in the list of pollutants in the fertilizer plant where ammonia is derived from the waste ammoniacal liquor of the coke ovens.

Average characteristics of the waste water from a typical Indian fertilizer plant producing both nitrogenous and phosphatic fertilizer is given in the Table 15.1.

TABLE 15.1. Composition of Wastes of a Complex Fertilizer Plant.

Item	Values
PH	7.5 to 9.5
Total solids	5400 mg/1
Ammonia Nitrogen	700 mg/1
Urea Nitrogen	600 mg/1
Phosphate	75 mg/1
Fluoride	15 mg/1
Arsenic	1.5 mg/1

15.3 EFFECTS OF WASTES ON RECEIVING STREAMS

All the components of the waste from the fertilizer plants induces adverse effects in the stream. Acids and Alkalis can destroy the normal aquatic life in the stream. Arsenic, Fluorides, and Ammonium salts are found to be toxic to the fishes. Amines are not only toxic to the fishes but also exerts a high oxygen and chlorine demand. Presence of different types of salts renders the stream unfit for use as a source of drinking water in the down stream side. Nitrogen and other nutrient content of the waste encourages growth of aquatic plants in the stream.

15.4 TREATMENT OF FERTILIZER WASTE WATER

Major pollutants in the fertilizer waste water, for which the treatment is necessary include oil, arsenic, ammonia, urea, phosphate and fluoride. Oil is removed in a gravity separator. Arsenic containing waste is segregated and after its concentration the solid waste is disposed off in a safe place. Phosphate and Fluoride bearing wastes are also segregated and chemically coagulated by lime; clarified effluent which still contains some amount of posphate and fluoride is diluted by mixing with other wastes. Several alternatives are there for the treatment of ammonia bearing wastes, *viz.,*

   (*i*) steam stripping
   (*ii*) air stripping in towers
   (*iii*) lagooning after pH adjustment, and
   (*iv*) Biological Nitrification and Denitrification.

For all practical purposes, "steam stripping" for the ammonia removal from fertilizer wastes have been found to be

263

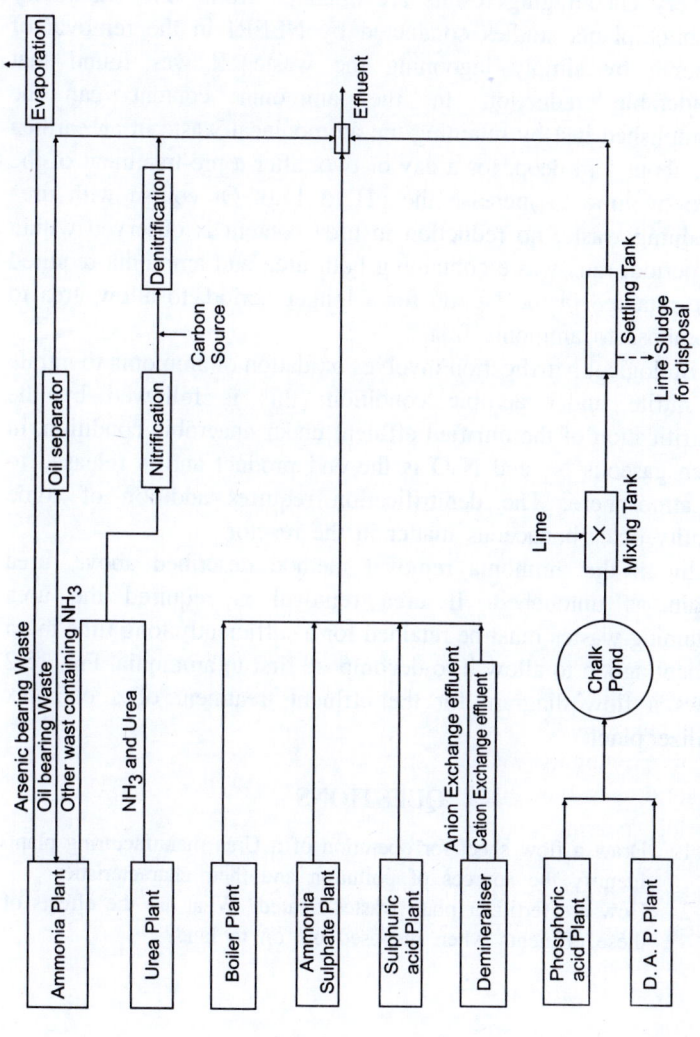

Figure 15.2: Flow Diagram for Effluent Treatment for a Complex Fertilizer Plant.

uneconomical. Removal of ammonia gas from the solution in an Air stripping tower, packed with red wood stakes, is found to be a very efficient method.

Very encouraging results are obtained from some laboratory and pilot plants studies conducted by NEERI in the removal of ammonia by simply lagooning the waste. It was found that considerable reduction in the ammonia content can be accomplished just by retaining the ammoniacal waste in an earthen tank, about 1 m deep, for a day or two, after a pre-treatment of the waste by lime to increase the pH to 11.0. Of course with urea containing waste, no reduction in urea content is observed within this period; thus waste containing both urea and ammonia required to be retained in the lagoon for a longer period, to allow urea to decompose to ammonia first.

Biological Nitrification involves oxidation of ammonia to nitrate via nitrite under aerobic condition; this is followed by the denitrification of the nitrified effluent under anaerobic condition, in which gaseous N_2 and N_2O is the end product and is released to the atmosphere. The denitrification requires addition of some quantity of carbonaceous matter in the reactor.

In all the ammonia removal method described above, urea remain as untouched. If urea removal is required the urea containing wastes must be retained for a sufficiently long time in an earthen lagoon to allow it to decompose first to ammonia. Fig. 15.2 shows a flow diagram for the effluent treatment of a complex fertilizer plant.

QUESTIONS

1. Draw a flow sheet for operation of a Urea manufacturing plant. Identify the sources of pollution and their characteristics.
2. How are fertilizer plant wastes treated? What are the effects of these effluents when disposed off on to land?

16 | Sugar Mill Wastes

16.0 INTRODUCTION

In countries like Cuba, Jamaica, and India, the sugar is produced from sugarcanes, while in many other places beetroots are used as the raw material for the sugar production. In India most of the sugar mills are situated in the countryside and operate for about 4 to 8 months just after the harvesting of the sugar canes. A large volume of waste of organic nature is produced during the period of production, and normally they are discharged onto land or into the nearby water courses, usually small streams, practically without pretreatment. Condition becomes worse as the stream flow reaches a very low level and eventually when enough dilution water is not available during the period of operation of the sugar mills (early November to end of May or June). Putrefaction of the polluted stream water caused by the heavy discharge of organic waste, resulting in the odour nuisance near the sugar mills is a very common phenomenon. In fact, all the concerned bodies, both sugar industry and pollution control agencies are aware of these problems and are trying to find an economical means to stop the nuisance created by the sugar mill effluents.

16.1 MANUFACTURING PROCESS

The sugar cane is normally harvested manually in India, which eliminates the carriage of soil and trashes to the factory along with the sugar canes. The sugar canes are cut into pieces and crushed in a series of rollers to extract the juice, in the 'mill house'. The milk of lime is then added to the juice and heated, when all the colloidal

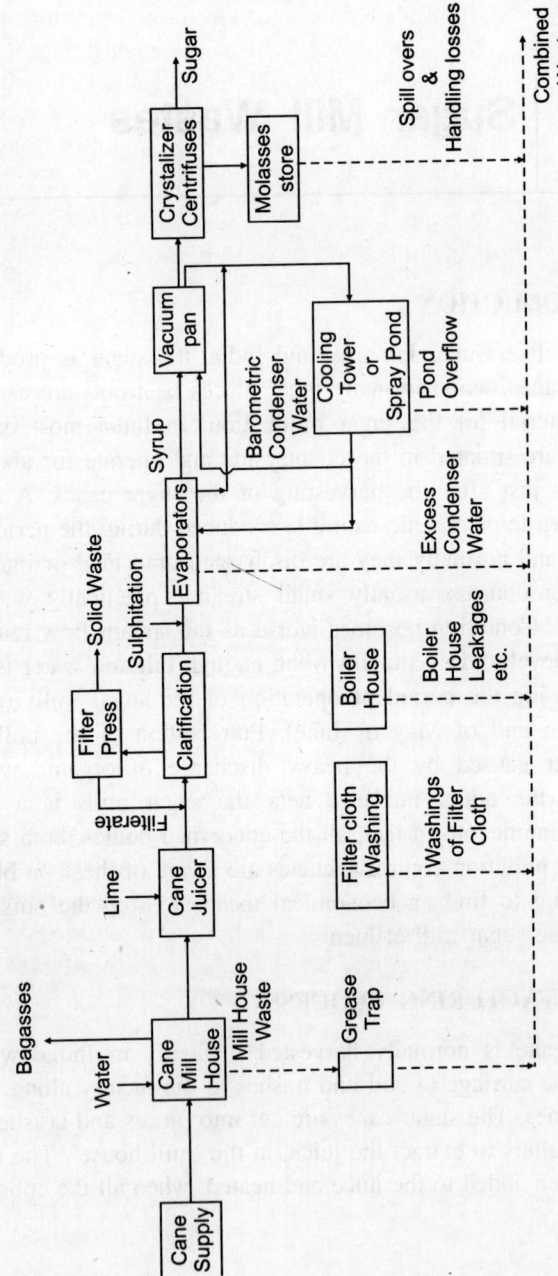

Figure 16.1: Flow Diagram for the Sugar Manufacturing Process.

and suspended impurities are coagulated; much of the colour is also removed during this lime treatment. The coagulated juice is then clarified to remove the sludge. The clarifier sludge is further filtered through filter presses, and then disposed off as solid waste. The filterate is recycled to the process, and the entire quantity of clarified juice is treated by passing sulphur dioxide gas through it. The process is known as "sulphitation process"; colour of the juice is completely bleached out due to this process.

The clarified juice is then preheated and concentrated in evaporators and vacuum pans. The partially crystalized syrup from the vacuum pan, known as 'massecuite' is then transferred to the crystallizers, where complete crystallisation of sugar occurs. The massecuite is then centrifuged, to separate the sugar crystals from the mother liquor. The spent liquor is discarded as 'black strap mollases'. The sugar is then dried and bagged for transport.

The fibrous residue of the mill house, known as 'bagasses' may be burnt in the boilers, or may be used as raw-materials for the production of paper products. The black strap mollases may be used in the distilleries.

A flow diagram of the process of manufacturing in a typical sugar mill is given in the Fig. 16.1.

16.2 SOURCES OF WASTE WATER AND THE CHARACTE-RISTICS OF THE WASTES

Wastes from the mill house include the water used as splashes to extract maximum amount of juice, and those used to cool the roller bearings. As such the mill house waste contains high BOD due to the presence of sugar, and oil from the machineries.

The filter cloths, used for filtering the juice, need occasional cleaning. The wash water thus produced though small in volume, contains high BOD and suspended solids.

A large volume of water is required in the barometric condensers of the multiple effect evaporators and vacuum pans. The water is usually partially or fully recirculated, after cooling through a spray pond. This cooling water gets polluted as it picks up some organic substances from the vapour of boiling syrup in evaporators and vacuum pan. The water from spray pond when overflows, becomes a part of the waste water, and usually of low

BOD in a properly operating sugar mill. But because of poor maintenance and bad operating conditions, a substantial amount of sugar may entrain in the condenser water; this polluted water, instead of being recirculated, is discarded as excess condenser water. These discharges contribute substantially to the waste volume and moderately to BOD in many sugar mills.

Additional waste originates due to the leakages and spillages of juice, syrup and mollases in different sections, and also due to the handling of mollases. The periodical washings of the floor also contributes a great lot to the pollution load. Though these wastes are small in volume and are discharged intermittently, they have got a very high BOD.

The periodic blow off of the boilers produce another intermittent waste discharge. This waste is high in suspended solids, low in BOD, and usually alkaline.

Characteristics of wastes from the different sections of common sugar mills, and that of the combined waste, are given in Table 16.1. Composition of the waste and its volume varies widely

TABLE 16.1. Characteristics of Sugar Mill Wastes (as given by ISI)

Characteristics	Mill-house waste	Filter cloth washings	Condenser water	Boiler house & floor washings	Combined waste (excluding condenser water)
Rate of flow, litres per tonne of cane crushed	730	360	1640	230	–
pH	6.7	9.5	–	7.2	4.6-7.1
COD, mg/l	–	–	–	–	600-4380
BOD, mg/l (5 day 20°C)	210	1765	–	5150	300-2000
Total solids, mg/l	1760	6970	–	5130	870-3500
Total volatile solids, mg/l	–	–	–	–	400-2200
Total suspended solids, mg/l	910	4000	–	120	220-800
Total nitrogen mg/l	–	–	–	–	10-40
COD/BOD ratio	–	–	–	–	1.3.2.0*

*Average value reported by Arora et al.

from mill to mill depending upon the availability of water. The figures indicated in Table 16.1 represent the average values of wide range of observations made in Uttar Pradesh and Bihar.

16.3 EFFECTS OF THE WASTE ON RECEIVING WATER

The fresh effluent from the sugar mill, decomposes rapidly after few hours of stagnation. It has been found to cause considerable difficulties when their effluent gets an access to the water courses, particularly the small and non-perennial streams in the rural areas. The rapid depletion of oxygen due to biological oxidation followed by anaerobic stabilization of the waste causes a secondary pollution of offensive odour, black colour, and fish mortality.

No question of the discharge of this waste into the seweres arises, as most of the sugar mills are situated in the unsewered rural areas.

16.4 TREATMENT OF THE WASTES

Like any other industry, the pollution load in sugar mills can also be reduced with a better water and material economy practiced in the plant. Judicious use of water in various plant practices, and its recycle, wherever practicable, will reduce the volume of waste to a great extent. Volume of mill house waste can be reduced by recycling the water used for splashing. Dry cleaning of floors or floor washings using controlled quantity of water will also reduce the volume of waste to certain extent.

The organic load of the waste can only be reduced by a proper control of the operations. Overloading of the evaporators and the vacuum pans, and the extensive boiling of the syrup lead to a loss of sugar through condenser water; this in turn increases both volume and strength of the waste effluent.

Disposal of the effluent on land as irrigation water is practiced in many sugar mills, but it is associated with odour problem.

The reasonable COD/BOD ratio of the mill effluents indicate that the waste is amenable to biological treatment. However, generally it is found that, the aerobic treatment, with conventional activated sludge process and trickling filters, is not too efficient, even at a low organic loading rate. A maximum BOD reduction of

270

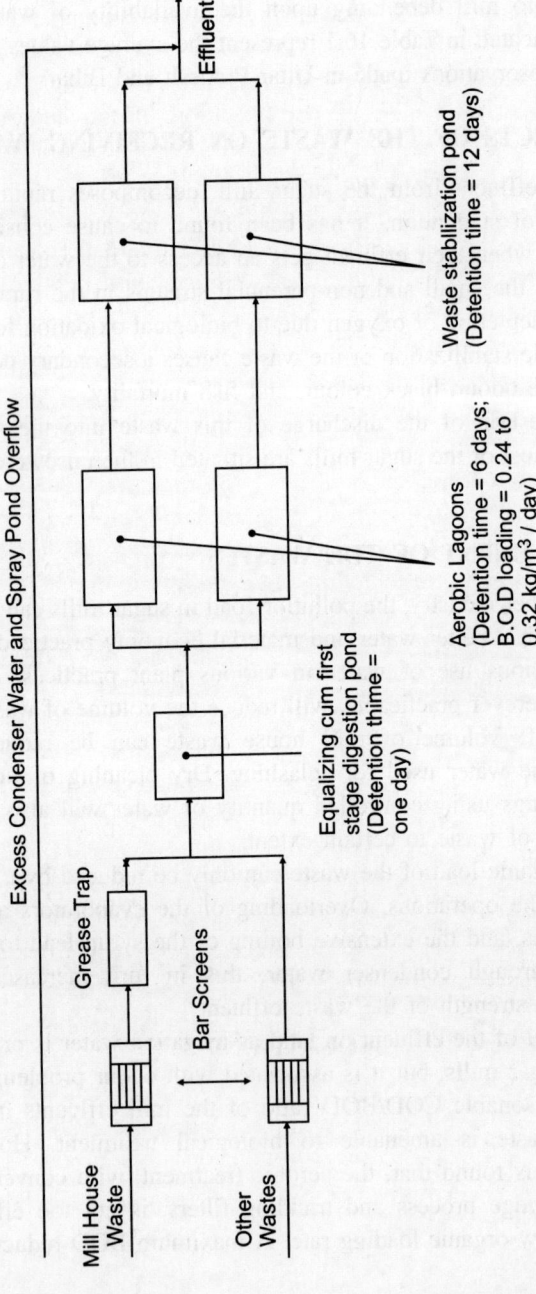

Figure 16.2: Flow Diagram for Complete Treatment of Sugar Mill Waste.

51% is observed in a pilot plant study at Kanpur, where both Trickling filter and activated sludge process were tried. In view of the high cost of installation and supervision of the treatment units, and the seasonal nature of the operation of this industry, it is generally observed that the conventional aerobic treatment will not be economical in this country.

Anaerobic treatment of the effluent, using both digesters and lagoons, have been found to be more effective and economical. A BOD reduction of about 70% was observed in a pilot plant study with an anaerobic digester, where BOD loading was 0.65 kg/m^3/ day with a detention time of 2.4 days at a controlled temperature of 37°C. In the same study, the BOD reduction was found to be 60% in an anaerobic lagoon, where a BOD loading of 0.23 kg/m^3/day and detention time of 7 days was provided. In another laboratory study, a BOD reduction of 88.9% was observed in an anaerobic digester with a detention time of 2 days and controlled temperature of 36 ± 1°C, at a higher BOD loading of 1.79 kg/m^3/day. The effluents of the anaerobic treatment units are found to contain sufficient nutrients (nitrogen and phosphorous). As such further reduction of BOD can be accomplished in aerobic waste stabilisation ponds.

Where sufficient land is available, a two stage biological treatment, with anaerobic lagoons followed by aerobic waste stabilisation ponds, is recommended for Indian conditions. The mill effluent, however, is to be pretreated primarily in bar screens and grease trap. The flow sheet for the complete treatment of the sugar mill waste is shown in Fig. 16.2. With the process parameters as indicated in Fig. 16.2, it is expected that the BOD reduction in the anaerobic process will be in the order of 60%, while overall BOD reduction may be in the order of 90%.

QUESTIONS

1. How is sugar manufactured?
2. What are the characteristics of sugar mill wastes?
3. Draw a flow diagram showing the complete treatment of sugar mill wastes.

17 | Steel Plants Wastes

17.0 INTRODUCTION

Integrated steel plants in India, as it stands now, are probably the largest industrial complexes in the country, with a large number of their subsidiaries and supporting industries. The production of steel in India is not so high, compared to that in other advanced countries. But the use of large volume of water and the consequent waste discharges, induces a tremendous effect on the water resources of the locality. A large volume of waste water, originating from different unit operations and containing mostly suspended solids, lubrication oils and several toxic substances are produced from an integrated steel plant. It is very important therefore to have the knowledge of the manufacturing processes and the type of waste originating from those processes, for establishing an effective and economical means for pollution control.

Integrated steel plants usually consist of five main units, *viz.,* coal washery, coke oven, blast furnace, steel melting shop, and rolling mills (hot and cold). In addition to the above the plants may have auxilliery units like oxygen plant and power plant for their own uses. A flow diagram of an integrated steel plant is given in Fig. 17.1.

In this chapter all the major units along with the sources of the waste water will be described briefly. It will be followed by a detailed description of the waste characteristics, and the suggested methods of effluent treatment.

17.1 COAL WASHERY AND ITS WASTE WATER

The coal needs some processing to make it suitable for use in coke

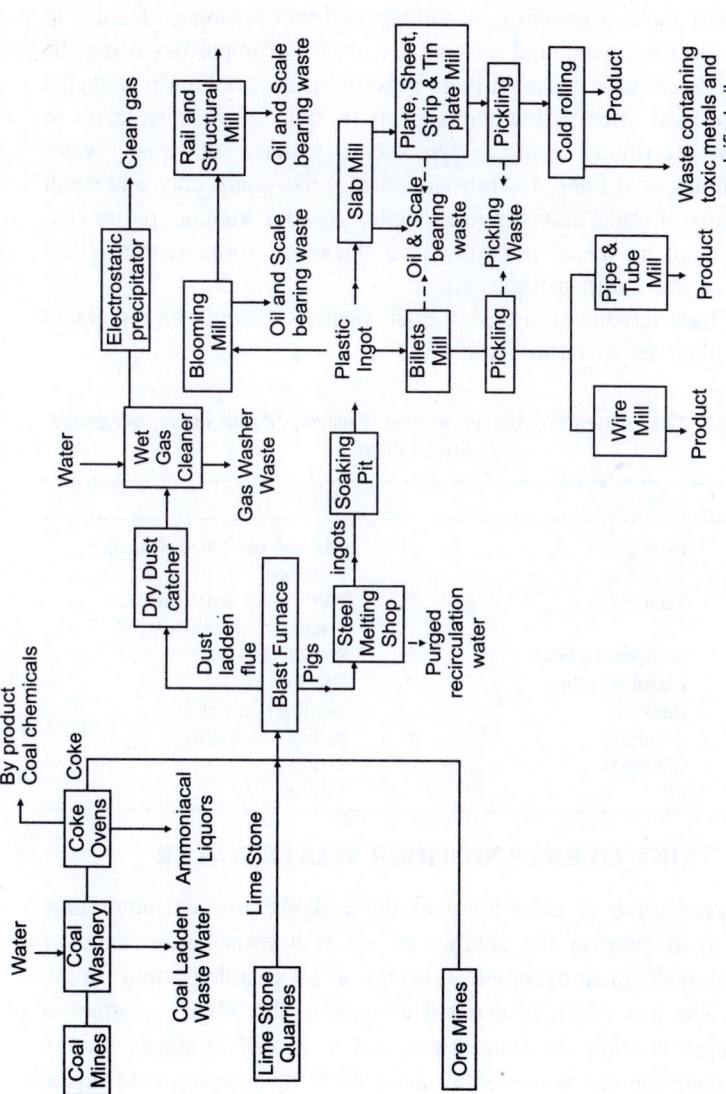

Figure 17.1: Flow Diagram for Major Operations in an Integrated Steel Plant.

274

ovens. The main objective of such treatment is the removal of solid foreign matter present in the coal. Generally the processes in a coal washery include crushing, screening and wet washing of coal. In the wet process the coal is separated from the impurities using the principle of differential settling. Water used for washing is recycled and re-used after sedimentation. But in spite of all care taken to ensure maximum reuse, appreciable quantity of wash water containing coal fines, and other impurities like shale, clay, and small amounts of other minerals like calcite, gypsum, kaoline, pyrite etc., comes out as waste, normally in a thickened form as the underflow of the sedimentation tank.

Characteristics of a typical coal washery waste in an integrated steel plant are given in Table 17.1.

TABLE 17.1. Characteristics of a Coal Washery Waste in an Integrated Steel Plant.

Items	Values
Flow	0.18 m^3 per tonne of coal washed.
Total solids	1000-25000 mg/l (normal range 5500-6000 mg/l)
Suspended solids	800-24700 mg/l
Dissolved solids	200-300 mg/l
Hardness	230 mg/l as CaCO$_3$
Alkalinity	86 mg/l as CaCo$_3$
Chlorides	13 mg/l
pH	7.4-7.8

17.2 COKE OVENS AND THEIR WASTE WATER

The production of coke involves the carbonization of bituminous coal by heating in the absence of air at a temperature range of 900°-1100°C in an oven, which drives off all volatile portions in the coal. The gas which is evolved containing the volatile matters is collected through the stand pipes and is cooled in stages. In the first stage the gas is cooled to about 80°C by spraying cold liquor over the gas, thereby producing mainly tar as the condensate. In the second stage by a further cooling to about 30°C, condensate containing additional tar and ammonia liquor is produced. These two condensate liquors after the separation of tar in a tar-decantor, are recycled as sprays in the first stage. The excess liquor known as "ammonia liquor", containing mainly ammonia and various other compounds is subjected to distillation for the recovery of ammonia; the waste is sent for further treatment or other chemical recovery. After the second stage of cooling, *i.e.*, in the third stage, the gas is compressed and cooled for further recovery of chemicals. Besides the arrangement for separation of Tar and Ammonia, this stage may include a Benzol washer for the recovery of Light Oils. The remaining gas may be used or sold as fuel. The crude light oils are either sold as such or are further refined for the recovery of Benzens, Xylene, Toluene and solvent Naptha.

The coal after being carbonized, is removed from the oven and quenched by cold water. About 30% of the quenching water is evaporated while the remaining water containing coke fines comes out as waste. This waste water is usually recirculated through breeze settling ponds and does not present any pollution problem.

The largest single source of waste water from coke oven plant, having the highest pollution potential in an integrated steel plant is the Ammonia still from where the Waste Ammoniacal Liquor comes out. The second source of waste water is the Benzol plant, the pollution potential and the volume of waste from which are much smaller compared to the first.

The characteristics of the typical spent Ammoniacal Liquor, and decanted liquor from the light oil refining process in the Benzol Plant, are given in Table 17.2 and Table 17.3 respectively.

17.3 BLAST FURNACE AND ITS WASTE WATER

Blast furnace is a basic unit in an integrated steel plant. Essentially the blast furnace process consists of charging iron ore and coke as fuel, limestone and dolomite as fluxing material into the top of the furnace and blowing heated air (blast) into the bottom. Pig iron is the metallic product of this unit.

Appreciable quantity of water is used in blast furnace for the purpose of cooling and gas cleaning operations. However, the cooling water normally remains uncontaminated and is reused after

TABLE 17.2. Characteristics of a Typical Spent Ammoniacal Liquor, (*i.e.* after Recovery of Ammonia).

Item	Values
Flow	0.14 m³/tonne of coal carbonized
pH	7.5–8.0
Total Free Ammonia	300–350 mg/1
Total phenol	900–1000 mg/1
Cyanides	10–50 mg/1
Thio-cyanates as CNS	50–100 mg/1
Thio-sulphates	110–220 mg/1
Sulphides	10–20 mg/1
Chlorides	4000–4200 mg/1

TABLE 17.3. Characteristics of a Typical Benzol Plant Waste.

Item	Value
Flow	0.02 m³/tonne of coal carbonized.
pH	6–8.5
Phenols	30–150 mg/1
Ammonia	5–30 mg/1
5 day BOD	300–800 mg/1

cooling. The entire quantum of waste water originates from the gas cleaning operations.

The blast furnace gas, which is heavily loaded with flue dust, is cleaned in a three stage process. The major portion of the flue dust which comes out along with blast furnace gas is recovered by the dry dust catchers. The remaining is removed by washing with water by "wet scrubbing". The portion which escapes wet scrubber may be removed by electrostatic precipitator. In wet scrubber, the downflow water sprays clean the dust from the upflowing gases, and the waste water contains 1000 to 10000 mg/1 of suspended solids. The wash water used in electrostatic precipitators adds very little to the waste volume, but increases the concentration of finer particles in the flow. Characteristics of a typical Blast Furnace waste water is given in Table 17.4.

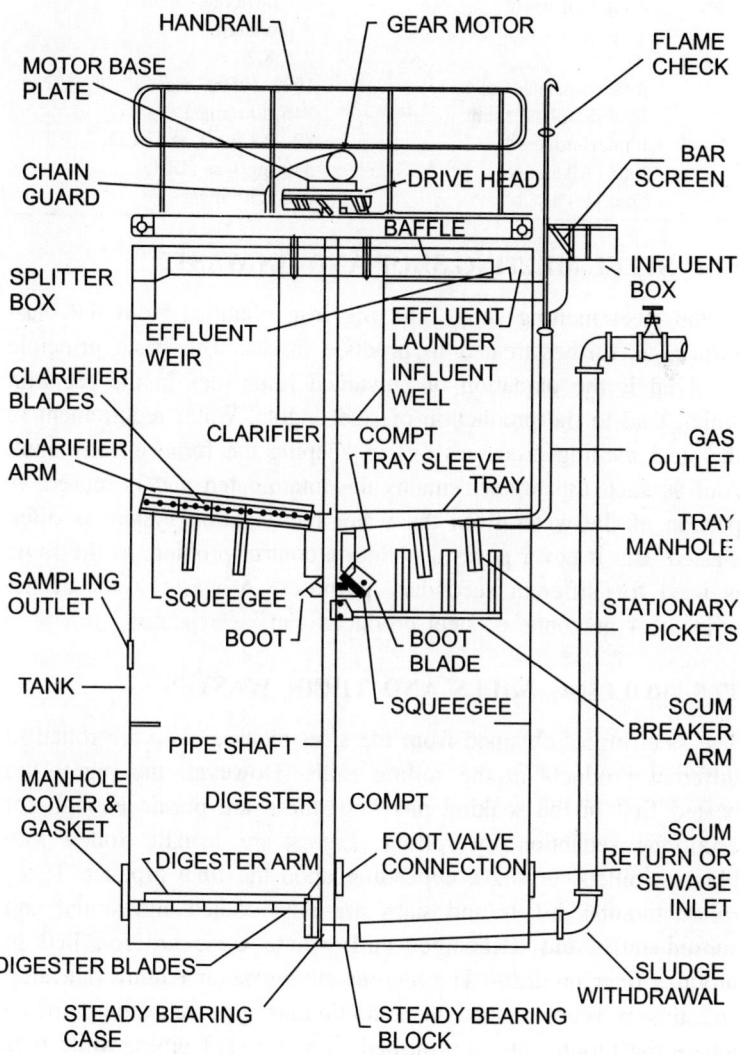

Figure 17.2: Dorr Clarigester.

Table 17.4. Characteristics of Typical Blast Furance Waste Water.

Items	Value
Volume of waste	0.5 m³/tonne of iron produced
pH	7.3–8.2
Total suspended solids	1000–10500 mg/1
Total dissolved solids	346–500 mg/1
Total Hardness	80–118 mg/1 as $CaCO_3$
Total Alkalinity	380 mg/1 as $CaCO_3$
Chlorides as Cl	210–250 mg/1

17.4 STEEL MELTING SHOP AND ITS WASTE

In the steel melting shops, the pig iron obtained from the blast furnace is further treated to produce ingots. The basic principle involved is the oxidation of unwanted impurities in the pig iron which lead to the production of steel ingots. Water requirement in the steel melting processes lies in keeping the furnace body cool. And as such this water remains uncontaminated and is reused. A portion of the water from the water recirculation system is often wasted. But it never poses a pollution control problem as the same is used for different secondary purposes. A waste water of the order of 4 m³/tonne of steel produced can be expected.

17.5 ROLLING MILLS AND THEIR WASTES

The steel ingots obtained from the steel melting shop are rolled to different products in the rolling mills. However, the ingots are heated first in the soaking pits until these are plastic enough for economic reduction by rolling. Ingots are usually rolled into blooms, billets, or slabs, depending upon the final product. These rolled blooms, billets and slabs are then cooled and stored and subsequently sent to another mill, where these are re-rolled to produce finer products. The blooms etc. however require reheating and this is accomplished in a continuous type reheating furnace where the blooms etc. are pushed over the skid pipes, while it is heated and ultimately placed over the roll table.

During the processes of rolling of ingots, blooms, billets and slabs, lots of scales are given off and are collected in the scale flume, below the roll tables. These scales are flushed down with high pressure water and are collected at the scale pit.

The rolls also get heated during the process, and are cooled with liberal supply of water. This water also joins the waste water flow through the flume. Naturally it carries a lot of oil and grease, used in different bearings.

Apart from the above, a reasonable amount of water is required to cool the skid pipes and also for furnace cooling. This cooling water is either directly sent for cooling tower, or is sent to the common scale pit, where from it is sent for cooling tower after the necessary clarification.

Water is also used, sometimes, for descaling of fine end products which eventually increases the volume of the scale bearing waste.

Whenever finer products like sheets, strips etc are required the products from hot rolling mills are normally subjected to cold rolling. For the production of cold-reduced steel sheets, it is essential however to remove the surface oxides of the steel slabs before they are subjected to rolling. The removal of oxides is essentially done by a process known as "pickling" at an earlier stage. The cold rolling however generates heat which is dissipated by the application of a rolling solution. The rolling solutions are usually oils emulsified in water, and are directed in jets, towards the rolls and the steel surfaces during rolling. These rolling solutions are recycled with occassional discharges as waste; this waste contains significant amount of oil. In addition to the above, the waste from the cold rolling mill originates from the cleaning lines, containing detergents.

The pickling, as mentioned earlier, is essentially a process in which the hard black oxides formed on the surface of the products during the hot rolling is removed by immersion in acid. Normally either sulphuric acid or hydrochloric acid is used for this purpose. The acid reacts with the iron oxide scales, and forms ferrous sulphates. Where sulphuric acid is used for pickling, the process can be continued until the build up of ferrous sulphate reaches to a strength of 25% and the free acid content reduces to less then 5%. The used pickling liquor is then discharged and a fresh batch of acid is taken up. The volume of waste depends on the method of pickling used.

After pickling and cold rolling the finer products are subjected to various finishing operations, *viz,* tinning, plating, galvanizing, wire drawing etc. The waste generated in these operations may be contaminated with chromate, cyanide, fluoride, zinc, tin, copper, acids and alkalis. The pollution potential of this waste is very high due to the presence of various toxic substances.

Characteristics of a typical hot rolling mill waste are given in Table 17.5.

TABLE 17.5. Characteristics of a typical hot rolling mill waste.

Item	Value
Volume of waste	7 m³/tonne of ingots rollled
pH	7.8–8.1
Total suspended solids	320–600 mg/l
Hardness	80–118 mg/l as $CaCO_3$
Oil & Grease	60–80 mg/l

17.6 OXYGEN PLANT AND ITS WASTE

Large quantity of oxygen is used for different operations in an integrated steel plant. It is estimated that, for the production of 1.6 million tonne of steel per year, about 204 tonne of oxygen will be required daily. As such, the integrated steel plants usually set up their own oxygen plant. Whatever method of oxygen production is adopted, the raw material, i.e., the air is first purified by removing carbon dioxide, moisture etc. This is usually accomplished by scrubbing with caustic soda solution in ring packed towers at about 10 atmospheric pressure. This results in the production of waste containing sodium hydroxide of N/2 strength, the volume of which is around 0.2 m³/tonne of steel produced.

However, this waste does not create any pollution problem, but in fact helps in reducing the pollution potential of other wastes when mixed with them judiciously.

17.6 POWER PLANT AND ITS WASTE

For all integrated steel plants, power plants will be associated to cater for emergency power supply to all vital units for protection against damages in the case of failure of bulk power supply system.

Moreover, the huge quantity of steam, required for almost all the units of an integrated steel plant, is supplied by the power plant steam generators.

The entire quantity of the waste water from the plant is constituted by the boiler blow down. The characteristics of a typical power plant waste is given in Table 17.6. Normally these wastes are not treated separately in any plant.

TABLE 17.6. Characteristics of a Typical Power Plant Waste (Blow Down from the Boilers).

Item	Value
Volume of waste	0.13 m^3/tonne of steel produced
pH	10–11
SiO$_2$	50–70 mg/l
Phosphate	40–60 mg/l
Chloride	70–140 mg/l

17.7 EFFECTS OF THE STEEL PLANTS WASTE ON RECEIVING WATER

Pollutants that are of main concern in integrated steel plant waste are suspended solids, cyanides, acids, ammonium compounds, phosphates, phenols, oils etc. All of them in one way or other adversely affect the environment. If the spent ammoniacal liquor is discharged into a stream without any treatment, the phenol alone can disturb the ecology of the receiving stream. It carries several elements which are toxic to the aquatic life; among them some are objectionable to the human consumption, when those substances find their way to the drinking water supply.

The phenol can be detected by taste in water even at a low concentration of 0.10 mg/l. When phenol-bearing water is chlorinated, chlorophenols will be formed. These chlorophenols are detected by taste in drinking water even at a concentration of 0.005

mg/1. The black suspended solids of an untreated waste discharge pollute the bed and the banks of the stream with a thick deposit. The reason for eutrophication in stream is generally attributed to the presence of excess amount of phosphates in the effluent, which act as nutrients.

17.8 TREATMENT OF COAL WASHERY WASTE

The major pollutant of the coal washery waste is the suspended solids. As such this waste is usually treated in a clarifier with or without coagulation. However the addition of coagulant reduces both the detention time and surface area of the tank. Several coagulants like starch, lime and indegenous coagulants like Nirmali seed (Strychuos potatorum Linn) extracts can be used effectively for the clarification of coal washery wastes. The clarified effluent is either recycled or discharged as waste.

The "froth floatation" is also suggested as a very efficient method of treatment for the removal of ultrafine coal particles.

Later it will be shown how these coal washery wastes can be used to reduce the pollution potential of other wastes in an integrated steel plant.

17.9 TREATMENT OF COKE OVEN WASTE

Though all the pollutants of the spent ammoniacal liquor, as listed in Table 17.2, can affect the ecology of the waste-receiving water course, the phenol is considered to be the most objectionable pollutant. The other objectionable substances include thio-cyanate, thio-sulphate, cyanide etc.

In some plants spent ammoniacal liquor is utilised for quenching of hot coke; this practice destroys the toxic matters like phenols in the liquor. But as this causes heavy corrosion in the quenching cars and in other quenching equipments, the method is not generally favoured.

Phenol being a valuable chemical by-product, may be recovered instead of destroying it. Several techniques have been developed for the recovery of phenol by liquid extraction methods. Most of these processes use Benzene as solvent, to extract phenol from the crude ammoniacal liquor, before it enters the "Ammonia

Still" for ammonia stripping. Other solvents used include light oil, petroleum oil, creosote oil, etc. The extracted phenols from all absorption processes can be recovered by washing with sodium hydroxide solution; the phenol reacts with the caustic solution to produce sodium phenolate. The crude phenol is then liberated from it using gases containing carbon dioxide. Due to the economical reasons, these methods of treatment are not found acceptable in many places.

The chemical oxidation of ammoniacal liquor with potassium permanganate, sodium dichromate, chlorine, ozone, etc. have been studied on a small scale. Out of these reagents, ozone is found to be the most attractive one because the method of manufacture of ozone is very simple. The method of treatment by ozone needs no temperature or pH adjustment, and does not produce any deleterious end product. However, the cost of chemical oxidation is very high and may be applicable for secondary treatment of waste, containing low concentration of phenol.

Certain microorganisms, both bacteria and yeast, are identified which can oxidise biologically phenols, thio-cyanates, thio-sulphates, and ammonia. When optimum pH and temperature are maintained, sufficient nutrients are added, and the reactor is suitably seeded, the proper loading of this phenolic substrate to the reactor may result in the desirable reduction of the pollution load of the waste. Phenol in concentrations as high as 800 mg/1 may be treated biologically. But it should be mentioned that high concentration of phenol prohibits the metabolism of thio-bacilli (microorganisms responsible for oxidation of thio-cyanates and thio-sulphates), and so thio-sulphates and thio-cyanates are oxidized only when phenol concentration reduces to an appreciable low value. Both activated sludge process and Trickling Filter have been tried for the biological oxidation of the phenolic wastes; waste stabilisation ponds may also be used for this purpose in places like India. But great care must be taken, in all the cases, to acclimatize the microorganisms when treating a strong waste.

In all practical cases, the phenol concentration in the waste ammoniacal liquor is too high to be treated directly by biological means. So these are either pretreated by physical or chemical methods, or are diluted using other waste to reduce the concentration of phenols. Dilution of phenolic waste by the domestic sewage is a common practice.

When coal washeries waste is available it has been shown that considerable reduction in the strength of ammoniacal liquor can be accomplished when equal volumes of coal washeries waste and ammoniacal liquor is mixed and flocculated with lime in a clariflocculator. Effluent of the clariflocculator may then be sent for activated sludge treatment.

The biological treatment of the coke oven effluents can be carried out economically in a three stage process, with isolated or cultured bacteria, appropriate in each stage. In the first stage the phenol is oxidized, in the second stage the thio-sulphate and thiocyanate are reduced, and in the third stage the ammonia is oxidized. In any case, the biological reactors should be of complete mix type for a better performance.

17.10 TREATMENT OF BLAST FURNACE WASTE

The blast furnace waste contains about 40% of the total dust coming out of the blast furnace along with the flue gas. Iron oxide and silica comprise about 70% and 12% respectively of the flue dust content. The waste can be treated in a clariflocculator even without the addition of coagulants. However the flocculation time can be reduced to a great extent when certain coagulants like alum or lime is added. The efficiency of the clariflocculator can be increased alternatively by a judicious mixing of this waste with the other wastes of the steel plant. The oxygen plant waste containing sodium hydroxide, may be used for this purpose. Ferrous sulphate, when recovered from the pickling waste, may also be used as a coagulant.

17.11 TREATMENT OF THE SCALE-PIT EFFLUENT

The primary treatment of the scale bearing waste from the rolling mills are offered by the scale pits associated with different sections of the mill. The scales produced in primary mills, like blooming mills etc. are coarse and mostly settle in the scale pit itself. The substantial quantity of finer scales produced in the process of rolling the billets, slabs etc. are also settled in the associated scale pits.

The effluent from the scale pits still contains considerable amount of fine scales, oil and grease and requires secondary treatment.

The above effluent contains about 170 to 440 mg/l of suspended solids, and can be treated well in the clariflocculators using either sodium hydroxide or the oxygen plant waste as coagulant.

The iron content of the clarifier sludge is as high as 46%. As such, the sludge is thickened, dewatered using vacuum filters and then sent to the sintering plant, so that it can be fed back to blast furnace. Thickening can accomplish almost complete removal of oil and grease from the sludge and thus prepares it for dewatering and caking in vacuum filters.

17.12 DISPOSAL AND TREATMENT OF PICKLING WASTE

The disposal of the pickling waste has always been a problem. The pollution due to this waste can be avoided if complete recovery of the waste is adopted. Alternate method of treatment of this waste include neutralization, using lime; but the cost of this treatment will be very high. As such, the ferrous sulphate recovery from the spent liquor is considered to be the most suitable means for pollution control.

The ferrous sulphate can be recovered in one easy method where sufficient amount of scrap iron and iron oxides are added to the waste pickling liquor when the free sulphuric acid is converted to ferrous sulphate, reducing the acid content to about 0.1%. The resulting solution may be settled and used as liquid coagulant after proper dilution, or the ferrous sulphate crystals may be recovered from the solution by evaporating the solution to dryness.

Alternatively both the acid and ferrous sulphate may be recovered from the pickling waste by extracting with 1 : 1 Acetone. The 85% of the ferrous sulphate precipitates after mixing and is crystalized; the acid concentration at the same time rises to about 71%. About 97% of the used acetone can be recovered for further use.

Various other sophisticated methods for the recovery of hydrochloric acid from the hydrochloric acid-pickle waste, and ferrous sulphate and sulphuric acid from the sulphuric acid-pickle waste are also available.

The process of electrolysis of spent liquor is gaining importance in the matter of regeneration of acid and recovering of iron. This iron in the form of sheet or as powder has good demand in the growing industry of power-metallurgy.

QUESTIONS

1. What are the major operations in an integrated Steel Plant?
2. How are the wastes from the following units in a steel plant treated? (a) coke ovens, (b) oxygen plant, (c) rolling mills, (d) power plant, (e) pickling bath.

18 | Oil Refineries Waste

18.0 INTRODUCTION

The number of oil refineries in India has increased substantially after the independence. But, while most of the earlier refineries are located near the sea-shore, many of the new installations are located far away from the sea, and discharge their waste to the inland water courses. As such, the pollution potential of the refinery wastes has gone up considerably and demands a careful attention. However, most of the government owned refineries have undertaken antipollution measures, and are designed to discharge a waste of acceptable quality.

The large refineries practice a number of enormously complicated basic and auxiliary operations, description of which is beyond the scope of this book. In this chapter, the basic operations and the most important auxiliary processes will be described in brief to give some idea of the refinery practices. This will be followed by a description of the sources of the waste water, composition of wastes and the available treatment methods.

Other processes in large refineries in relation to the production of different petrochemicals will be described in the next chapter.

18.1 BASIC AND AUXILIARY REFINERY OPERATIONS

Crude oils are complex mixtures of hydrocarbons of varying molecular weight and structure. These hydrocarbons range from simple highly volatile substances to complex waxes and asphaltic compounds. The final petroleum products are obtained from the crude oil through a series of operations, *viz.,* topping, thermal cracking, catalytic cracking, catalytic reforming etc. These basic

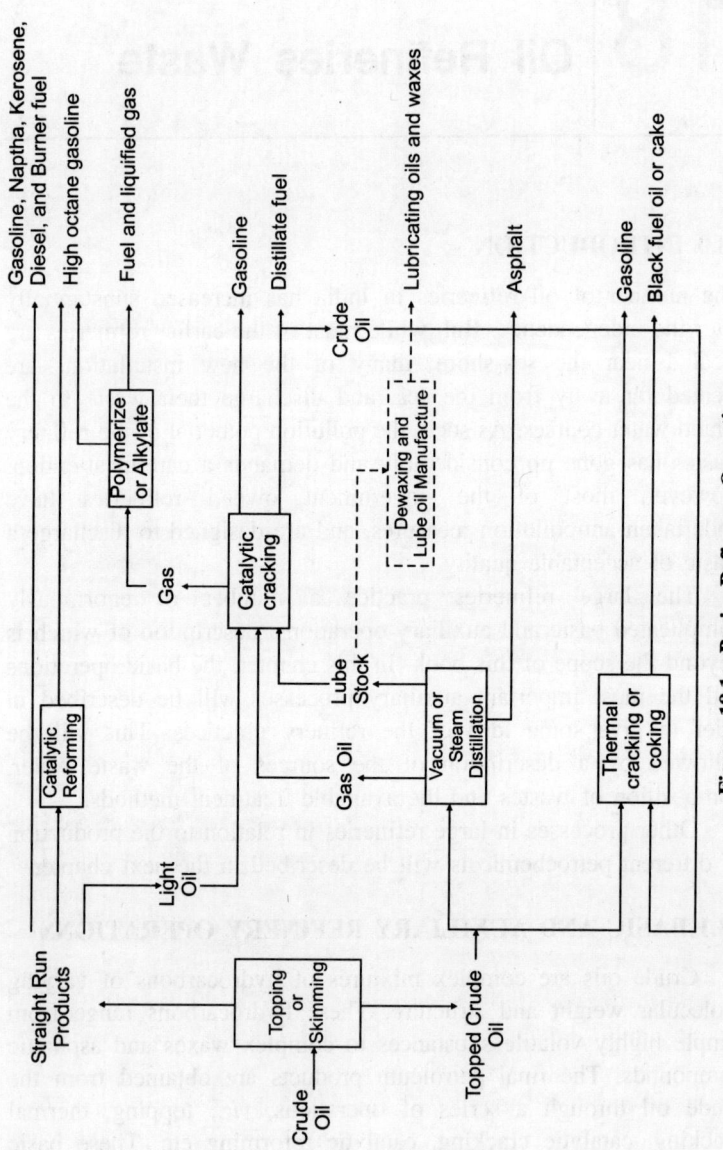

Figure 18.1: Basic Refinery Operations.

operations however vary from plant to plant depending upon the properties of the crude oil and the desired grade of the final product. The basic refinery operations as mentioned above are shown diagrammatically in Fig. 18.1.

In general, the crude oil is first subjected to fractional distillation in the process known as "topping". The products obtained are called raw products and include raw gasoline, raw naptha, raw kerosene, gas oil, fuel oil etc. A simplified flow-diagram for the process is shown in Fig. 18.2.

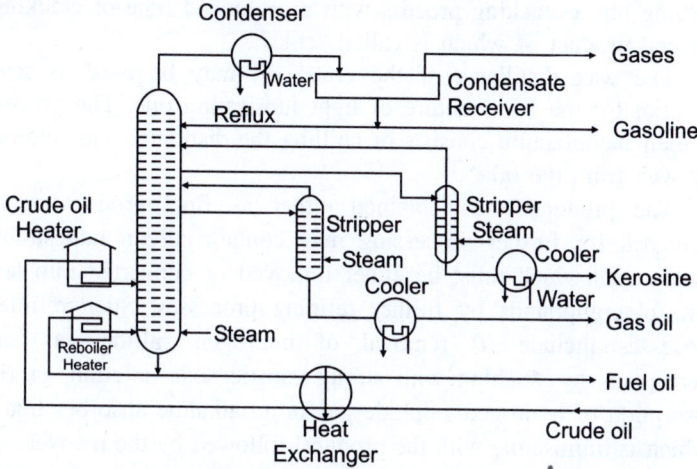

Figure 18.2: Simplified Flow Diagram for the Distillation Process.

Now these intermediate refinery products are again treated to yield various finished market products as per the requirements. The operations practiced include "catalytic cracking" or "thermal cracking", and further purification processes like "acid treatment", "sweetening treatment", "hydrodesulphurization" etc.

The decomposition of heavy or high boiling petroleum distillates like gas oil and fuel oil to lighter products like gasoline is called cracking. These reactions take place practically at the atmospheric pressure and at a high temperature. In catalytic cracking, however, an acid type solid catalyst (such as synthetic silica alumina) is introduced into the reactor.

A residue of heavy black material, known as coke, is obtained out of the process of cracking. In the catalytic cracking process, however, the asphaltic or tar-like products get adsorbed on to the surface of the catalysts in the form of coke.

When "high octane" fuels are in demand, much of the naptha is "catalytically reformed" into high octane gasoline; the process involved is another form of cracking.

Again when market is available the most degraded carbonaceous high boiling parts of the crude oil distillates are subjected to the "coking process". This "Coking process" is nothing but a cracking process with a prolonged time of cracking, the end product of which is called "coke".

The wax distillates of the crude oil may be used as raw-material for the manufacture of light lubricating oils. The process of their manufacture consists of chilling the distillates and filtering the wax from the oil.

The products thus obtained either as final product or as feedstock for further processing may contain certain undesirable constituents which must be either removed or converted into less harmful compounds by further refining processes. Such refining processes include (i) removal of hydrogen sulfide gas and mercaptans by washing with strong caustic soda solution, or (ii) absorption of hydrogen sulphide gas in an alkaline absorber liquid (which is immiscible with the product) followed by the recovery of the hydrogen sulphide gas, in the process known as regenerative process, (iii) conversion of mercaptans to less harmful desulphide, employing solutions like lead oxide in caustic soda together with sulphur, or sodium hypochlorite or copper chloride, in a process known as "sweatening", and (iv) conversion of alkylsulphides and thiophenes and practically all other sulphur compounds into hydrogen sulphide in a catalytic reaction in the process known as "hydrodesulphurization"—hydrogen sulphide thus formed is recovered by the regenerative process.

The crude oil is sometimes pretreated with caustic soda along with water to remove salts from the crude.

Water may be removed by passing the product through a bed of calcium chloride or similar drying agent.

The brine, which comes out of the desalting operation, and the spent drying agent, used for the removal of the water goes to the sewer. But wherever possible, the removed impurities are recovered and sold or are used for manufacture of other products; the process chemicals are reused after the recovery of the impurities. As such, these impurities do not join the waste water stream directly, but may be introduced to the flow through the condensate and wash waters which come into contact with products at some stage of the processing.

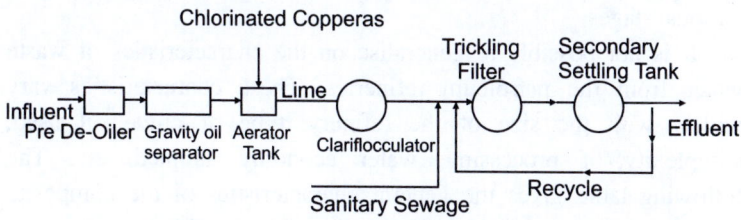

Figure 18.3. Flow Diagram for the Treatment of a Typical Petroleum Refineries Waste.

18.2 SOURCES OF WASTE WATER

A large amount of water is used in the refinery processes, and a big fraction of it comes out as waste after getting polluted by oil and other toxic substances.

Major amount of water used in a refinery, is for cooling. Like any other industry, the extent of recirculation and reuse of this water will determine the volume of the effluent discharge.

Second to cooling, the major use of water is for boiler feed. The steam obtained is utilised for different operations like desalting of crude oil, distillation and steam-stripping in topping process, and in catalytic cracking. Steam is also used for the stripping of spent catalyst before the later is sent for regeneration.

Water is also required to wash the products like gasoline to remove the traces of chemical reagents used in earlier operation. The process waste water includes the condensates and wash waters which come into contact with the petroleum products at some stage of the processing. So the waste water contains free or emulsified oil, spent caustic and acid solutions, impurities of petroleum products like hydrogen sulphides, ammonia, mercaptans, phenols, and spent catalysts.

Again where semicontinuous delayed coking is adopted and hydraulic method is used for decoking the chamber, water in the form of water jets is used for cutting and breaking down the cokes from the chamber walls. The next operation of dewatering and drying of coke produces a waste containing large amount of coke fines.

Oil enters into the waste water from leaks, spills etc., at various stages.

It is not possible to generalise on the characteristics of waste water from the petroleum refineries. These characteristics vary widely with the size of the refinery, type of crude oil used, complexity of processing, water economy adopted, etc. The following table gives the general characteristics of the composite waste from a typical Indian oil refinery (Chakravarty and Bhaskaran):

TABLE 18.1: Characteristics of Oil Refineries Wastes.

Free oil	2000–3000 mg/l
Emulsified oil	90–120 mg/l
H_2S and RSH	10–220 mg/l
Phenolic compounds	12–30 mg/l
5 day 20°C BOD	100–300 mg/l
Suspended solids	200–400 mg/l

18.3 EFFECT OF WASTES ON RECEIVING STREAMS/ SEWERS

Even a negligible amount of petroleum in the order of 0.5 mg/l can impart a bad taste to water. Napthalein also imparts bad odour and becomes toxic at a concentration of 5 mg/l. Mercaptans produces a very strong and unpleasant odour at a very low concentration. The odourous compounds of the waste impair a bad taste of the fish living in that stream. Again as the waste contains various organics, it exerts an oxygen demand and on the other hand, the oil film on the surface of the water prevents reaeration of the stream. Cattle drinking such a polluted water, frequently get attacked by some lethal diseases.

18.4 TREATMENT OF REFINERY WASTE WATER

In general, the treatment of refinery wastes is carried out in three steps:

(*i*)   Physical separation of free oil.

(*ii*)   Chemical coagulation of emulsified oil.

(*iii*) Biological treatment for the removal of BOD, phenol and other toxic materials.

The physical separation of oil is accomplished in "gravity separators". These are essentially rectangular tanks with the arrangements for oil skimming and sludge scraping. The design of gravity separator is based on fundamental principles of sedimentation, and it is assumed that all oil globules of 0.015 cm in diameter and above will be removed in the separator. A factor of safety of 1.2 is applied to the surface area to allow for short circuiting. Whatever may be the theoretical efficiency of this type of gravity separator, in practice, the efficiency reduces due to the fact that certain portion of the free oil forms agglomerates with the suspended solids present in the waste, and are not separated in the tank.

Dissolved air floatation with or without flocculation is also tried to remove the oil content of refinery wastes, and is found satisfactory in some special cases.

Effluent of the gravity separator is given a chemical treatment, consisting of coagulation, flocculation and sedimentation, to remove emulsified oil and certain other chemicals, and also to condition the waste for final treatment of the waste in biological reactors. Various coagulating agents like ferrous sulphate, ferric sulphate, chlorinated copperas, calcium chloride, calcium carbonate and hydrated lime have been tried to de-emulsify an oil-separator effluent. Most suitable type of coagulant, or the combination of coagulants, and their dosage are to be determined by laboratory tests. The flocs formed in the flocculator adsorbs the oil, and becomes considerably buoyant. Therefore the following sedimentation tank must be designed with low surface settling rate and large surface area. It has been observed that the size of the sedimentation tanks can be considerably reduced by adding coagulant aids like Fuller's earth during flocculation, which almost doubles the settling velocity of the floes. The chemical treatment not only produces effluent of low oil content, but reduces the sulphides content considerably. Chakravarty and Bhaskaran have shown that much improved quality of the effluent may be obtained if the waste is also aerated along with the chemical treatment.

The "waste decoking water" containing large amounts of coke fines is subjected to segregation and is treated separately. Coke fines are not easily wetted by water; usually the wetting is accomplished by air blowing, subsequent to which these are settled in a settling tank. As an alternative device of treatment of coke containing waste, Chakravarty and Bhaskaran reported that, treatment of the waste with a coagulant aid (2 mg/1 of coagulant aid CA-15 developed by NEERI) with flocculation (1 minute) and sedimentation (15 minutes) may produce an effluent of very high quality. By this method, not only suspended solids, but certain percentage of oil content can also be removed.

Biological treatment methods include waste stabilization pond, Aerated Lagoon, Trickling Filter, and Activated Sludge Process. Waste stabilisation ponds require large area, but are capable of reducing BOD by 43%–96% and phenol by 61–99% with a retention period of 60 days. In Aerated Lagoon, with mechanical surface aerator, a retention period of 6 days may be provided but the microbial mass produced in the lagoon settles so slowly that good effluent quality cannot be obtained unless a number of ponds are provided following the lagoon in series.

Many refinery engineers favour the trickling filter as a biological treatment method. Trickling filters can absorb shock loads of toxic matter and do well without any retardation even with an influent oil concentration of upto 100 mg/1. Deep bed high rate trickling filters have been found to be very efficient in the reduction of phenols (Rao et al). All the sulphides are removed in the Trickling Filters. A flow diagram for treatment of a typical refinery employing Trickling Filter is shown in Fig. 18.3.

Treatment of the refinery waste by Activated Sludge process is affected when the oil concentration exceeds about 30 mg/1. Chakravarty and Bhaskaran observed that activated sludge treatment of chemically treated refinery wastes results in very poor removal of BOD and phenol. To tackle this situation they developed a modified method designated as "Ferro-biological" process. In this method both microbial mass and ferric hydroxide are used simultaneously in the aeration tank to flocculate BOD, phenol and oil. In view of the toxic nature of the refinery waste, these aeration tanks are to be designed as complete mix type.

As the biological flocs formed in the above process are adsorbed with some quantity of oil, the Final Settling Tank following the Aeration Tank must be designed with a low surface settling rate, and large surface area.

The selection of the most suitable methods and the sequences of the treatment processes depends not only on the raw waste characteristics, but also on the desired effluent quality specified by the local authority. The raw waste characteristics also vary from one refinery to another. All these aspects are to be considered in the final design of the treatment units of a refinery waste.

QUESTIONS

1. What are the sources of pollution in an oil refinery?
2. How are the liquid wastes from an oil refinery treated?

19 | Petrochemical Complex Waste

19.0 INTRODUCTION

The chemicals derived from petroleum are often referred to as "petrochemicals". The range of the petrochemical products is very wide, and they include both organic and inorganic chemicals. Some of the chemicals, derived from petroleum, may also be derived from other sources, while the others owe their existence to the petrochemical industry. The petrochemicals are usually intermediates and are used further for the productions of solvents, detergents, synthetic resins, synthetic rubbers, synthetic textile fibres, pesticides, fertilizers, and a variety of industrial chemicals.

Starting materials of the petrochemicals industry are the hydrocarbons in the different available forms. These may be naturally occuring natural gas, or the products and by-products of a petroleum refinery like refinery gas, naphtha, etc.

The principal constituent of natural gas is methane. A number of products like Carbon Black, Hydrogen, Synthesis gas (a mixture of carbon monoxide and hydrogen), Hydrocyanic acid, Acetyline, Formaldehyde etc can be obtained directly from natural gas through different types of process like incomplete combustion, cracking, catalysis, pyrolysis (a severe form of thermal cracking), and oxidation.

Refinery gas contains olefins like ethylene, propylene, butylenes etc as by-products of cracking operation in the refinery. The olefins, diolefins like butadiene, and acytylene may be derived from naphtha by cracking. The naphtha may also be converted to synthesis gas. Aromatics like benzene, toluene, (methyle benzene), and xylene (dimethyl benzene) may occur naturally in certain petroleum fractions, or may be produced by catalytic operations from naphtha.

In any petrochemical plant, at first the petrochemical fraction is processed to derive the intermediates, and then the final product is synthesised by a more complex process. Some of the primary processes are mentioned earlier. Secondary processes involve stages like reaction, recovery and purification. In the reaction stage the feed stocks are mostly catalytically reacted either in a fixed bed or in a slurry. Further processing includes condensation, scrubbing, recycling, distillation etc.

The description of all the primary and complex secondary processes involved in the production of different petrochemicals is beyond the scope of this book. Only a brief description of the processes involved in the production of very common petrochemicals is given in the next section.

19.1 MANUFACTURING PROCESSES

Ethylene

The hydrocarbon feed stock (either gas oil or naphtha) is cracked in the presence of steam in tubular pyrolysis furnace. The effluent gas is rapidly quenched and then the heavy oil fraction is removed. Finally the gas is cooled by direct water quench. Effluent gas is then compressed, dried and sent for Hydrogen removal in the chilling train. The condensed hydrocarbon is then sent to the fractionation section for separation into desired products. The section consists of demethanizer, deethanizer, ethylene fractionator, depropanizer, propylene fractionator, debutanizer etc. Ethylene, Propylene, Butadiene etc. are drawn off from the top of the reactors; the by-products of reactions like Ethane, Propane etc. are drawn off from the bottom and are either recycled to the cracking furnace, or delivered as product.

Methanols

Hydrocarbons are first removed of sulphur, impurities, and then catalytically reacted with carbon dioxide and steam, to produce the synthesis gas (a mixture of Hydrogen and carbon monoxide). The synthesis gas is then compressed to high pressure and fed into a shell and tube reactor, filled with catalysts and under high temperature. The reactor effluent gas is cooled, and the methanol is condensed and then separated in a separator. Crude methanol is then further refined to meet the required standards.

Acetaldehyde

Ethylene is oxidized in a reactor in the presence of a catalyst solution (consisting of copper chloride and palladium chloride) under slight pressure and boiling temperature. The Acetaldehyde produced is condensed and scrubbed with water from unreacted gas; the gas is then recycled. Acetaldehyde is then separated from water and by products, by a two stage distillation.

Acetic Acid

Acetic acid is synthesised by reacting Methanol and carbon monoxide in a reactor at high temperature and pressure in the presence of Cobaltous Iodide catalyst. The crude acid and the unreacted gas is withdrawn, cooled and expanded. The Methanol Iodide is recovered from the unreacted gas by washing with feed Methanol. The crude acid is separated first from high boiling by-products and then from low boiling components and cobaltous chloride. The acid is then dehydrated and purified further.

Ethylene Oxide and Glycols

Ethylene is converted into Ethylene oxide in the presence of silver based catalysts, in a reactor. The Ethylene Oxide is separated from the reactor effluent gas by water in an absorber. Unabsorbed gas is recycled to the reactor. Ethylene dioxide-rich water stream is then fed into a stripper, where Ethylene oxide is recovered. The water from the stripper is cooled and recycled to the absorber. The stripper overhead effluent containing crude Ethylene Oxide is further purified in fractionator and refiner.

For the production of Glycols, the Ethylene Oxide is fed into a reactor along with water. The reactor effluent is then fed to separating and purifying systems to yield Glycols.

Polyethylene

Ethylene, along with a catalyst, a diluent (usually a Hexane), Hydrogen, and small quantities of propylene or Butylene, is

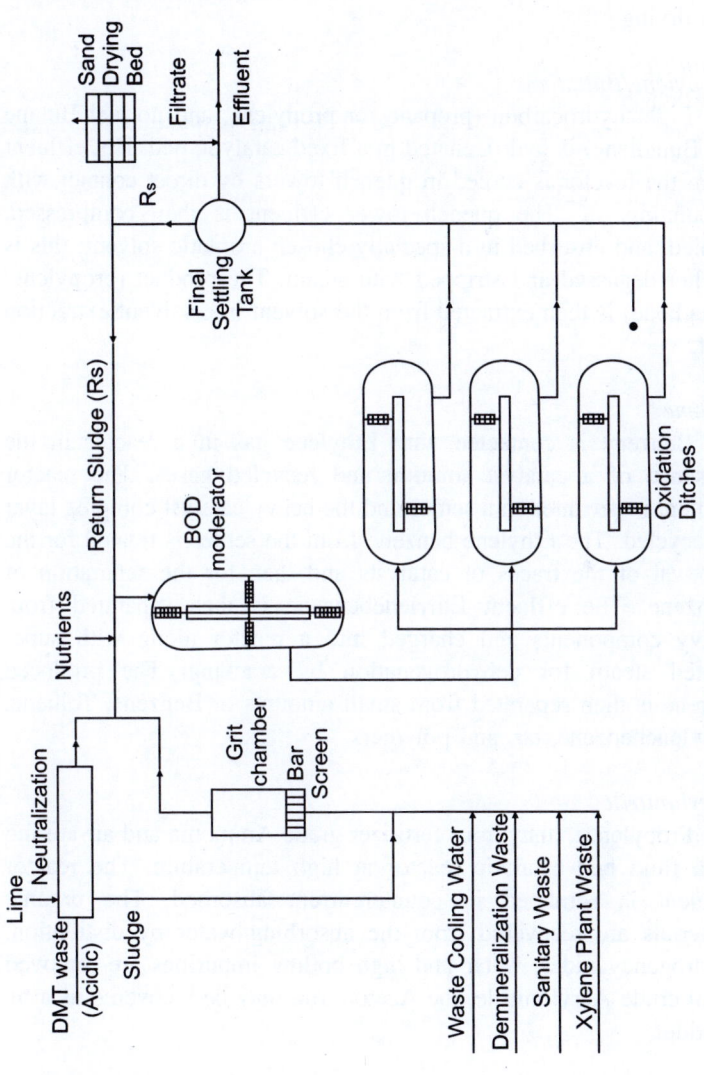

Figure 19.1: Flow Sheet for an Aromatic Waste Treatment Plant.

continuously fed in a reactor. The reaction takes place in a slurry. The polymer is obtained in the form of powder, after a series of operations like Hexane separation by steam stripping, centrifugation and drying.

Propylene/Butadiene

Light hydrocarbon (propane for propylene, and normal Butane for Butadiene) is hydrogenated in a fixed catalytic bed. Hot effluent from the reactor is cooled in quench towers by direct contact with circulating oil. The quench tower effluent is then compressed, cooled, and absorbed in a specially chosen aromatic solvent; this is further degassed and stripped with steam. The product (propylene/Butadiene) is then extracted from the solvent in a solvent extraction unit.

Styrene

Benzene is contacted with Ethylene gas in a reactor in the presence of a catalyst solution and recycled gases. The reactor effluent is decanted in a settler and the heavy catalyst complex layer is recycled. The Ethylene benzene from the settler is treated for the removal of the traces of catalysts and then for the separation of Benzene. The effluent Ethylenebenzene is then separated from heavy components and charged into a reactor along with super heated steam for dehydrogenation by cracking. The produced styrene is then separated from small amounts of Benzene, Toluene, Ethylenebenzene, tar, and polymers.

Acrylonitrile

Propylence, anhydrous fertilizer grade Ammonia and air are fed in a fluid bed catalytic reactor at high temperature. The reactor effluent is scrubbed in countercurrent absorbed. The organic materials are recovered from the absorbing water by distillation. Hydrogencyanides, water and high boiling impurities are removed from crude Acrylonitrile; the Acetonitrile may be recovered as a by product.

Butanol

Propylene and carbon monoxide are reacted in a catalytic solution containing water, Butanol and catalyst. The product

Butanol is distilled off. The overhead Butanol, water, and catalyst are separated from the gas stream by condensation and the gas is recycled. The by-product carbon dioxide, hydrogen, and propane are removed by passing a portion of recycle gas through gas scrubbers. The crude Butanol is decompressed, dehydrated and separated into normal-Butanol and iso-Butanol.

Caprolactum

Benzene and Synthesis gas are reacted in the presence of a Nickel catalyst to produce cyclohexane. This cyclohexane is oxidized in a reactor in the presence of octate as catalyst to produce cyclohexanone and cyclohexanol. Now cyclohexanol is separated from the cyclohexanone, and is further dehydrogeneted to cyclohexanone.

This cyclohexanone is reacted with Hydroxylamine sulphate to produce cyclohexanone oxime and the by-product Ammonium Sulphate. Oxime is then converted to caprolactum in another process in the presence of oleum. Caprolactum is recovered from the mixture by extraction and distillation.

19.2 WASTES FROM PETROCHEMICAL COMPLEX

Liquid waste from any plant, deriving chemicals from petroleum fractions, may include the raw material, the impurities in the raw material, chemicals used to remove the said impurities, catalysts, the products and by-products of different reactions and processes, and the chemicals used for regeneration of the catalysts. The plant for the treatment of boiler feed water also contributes towards the pollution. For economic reasons all these substances should not be allowed to go out as waste; but even with very good inplant control some loss is unavoidable, and this constitutes the liquid waste from the complex. Therefore, in the waste we may find hydrocarbons of different groups, the products of different operations with hydrocarbons, *viz.* alcohols, aldehydes, ketones, esters, acids, bases, salts, phenolic compounds, plus small quantities of emulsified oil.

In the study of an individual plant, it is necessary to identify the broad chemical classes (i.e. hydrocarbons or aldehydes) of these pollutants and their pollution potential. It must be recognised that though most of the pollutants are organic in nature, and need special chemical treatment, the waste may contain some inorganic chemicals also. Chemicals exhibiting toxic effects should get special attention in the study. Again some of the pollutants are not soluble in water, and may be removed by either setting or floation.

19.3 EFFECTS ON STREAMS/SEWERS

Crude waste water from the petrochemical complexes when discharged into streams or sewers, destroy the ecological balance of the stream and seriously interfere with the operation of a sewage treatment plant. The wastes not only contain toxic chemicals, but may contain surface active chemicals which result in the formation of unsightly foam on the surface of the stream. Even if the waste is discharged into a sewer, some of the surface active chemicals escape the biological treatment and cause the undesirable foaming on the surface of the stream. The inflamable constituents of the waste, if any, will cause hazards in the sewers. The low pH wastes damage the sewers and other structures in the treatment plant. Excessive suspended solids and overloading may even clog the sewers. Moreover the shock loads in the biological treatment units may reduce the efficiency of the units, and the toxic chemicals may disrupt the entire treatment operations. (Of course a sewage treatment plant may be designed to handle such heavy loads and some of the toxic chemicals ; but even then some of the refractory chemicals remain unaffected by the biological processes and come out with the effluent).

19.4 CHARACTERISTICS OF PETROCHEMICAL WASTES

Analysis of the compound or class of compound is necessary for the identification of the source of pollution and for inplant controls. The Union Carbide's petrochemical complex at Trombay has introduced Total Carbon Analyser and Gas Chromatographs for the continuous monitoring of the effluent. But, for an over-all pollution control, the environmental engineers rely upon the familiar parameters to express the pollution potential of the wastes. These include BOD, COD, Colour, Turbidity, Suspended Solids, pH, Total Nitrogen, etc.

Of course some additional quantitative information is required in regard to some specific toxic chemicals.

Table 19.1. Gives the characteristics of the waste from petrochemical complex of Gujarat State Fertilizer Corporation, which produces Caprolactum using Benzene, Ammonia, Hydrogen, Carbon dioxide, Oleum, and Sulphuric acid as raw materials and process chemicals.

TABLE 19.1. Composition of a typical petrochemicals waste (Patel and Patel).

Item	Value
pH	5.3 to 10.0
Total solids	1956 to 14478 mg/l
COD	947 to 9000 mg/l
BOD	650 to 5800 mg/l
COD/BOD	1.4 to 2.0
Total Ammoniacal-N	150 to 600 mg/l
PO_4	Traces to 15 mg/l
Oil & Grease	10–85 mg/l

19.5 TREATMENT OF THE PETROCHEMICAL WASTE

In fact, the pollution abatement in a petrochemical plant starts from the plant itself. As mentioned earlier, due to the economic reasons, several methods of recovery have been developed. Some of the wastes again instead of wasting, are utilised for the production of other substances inside or outside the plant. Common methods of recovery include stripping, scrubbing, adsorption, and distillation. As for example (i) Benzol may be stripped with Nitrogen or adsorbed on activated carbon, (ii) Butanol, Butyl acetate and Acetic acid may be adsorbed on activated carbon, (iii) wet scrubbing removes CO_2 from H_2 stream in synthesis gas plant, (iv) distillation allow the recovery of acetaldehyde from the production of vinyle acetate, etc. (Koziorowski & Kucharski). Oxidation, Extraction and Crystallization are also used frequently for the recovery of valuable pollutants in the petrochemical waste.

Multiple reuse of water, control of spills and leaks, and general good housekeeping also result in much reduction in the pollution load.

Data, in regard to the centralized treatment of the plant effluent waste, is very limited. Moreover, as the composition and the volume of waste differ from plant to plant, producing different chemicals using different processes, each should be considered separately. Laboratory-bench scale and pilot-plant studies are to be conducted to find out the most suitable and economical method of treatment of the wastes.

Simple Aeration reduces much of the BOD load of the waste from the petrochemical plants. Again any organic waste may be destroyed by Incineration; but for dilute waste this method is never an economical and practical method of treatment. Some toxic wastes may be concentrated, and incineration may be used as a means of disposal of that concentrated waste.

The organic pollutants in the waste are mostly not amenable to chemical treatment. However, chemical treatment may be necessary, as a pre-treatment in the removal of coloids and for the control of pH, prior to a Biological treatment.

Biological treatment processes have received much attention in the treatment of the waste from petrochemical complex due to the organic nature of the wastes.

All the biological treatment processes are preceded by primary treatment *viz.* sedimentation, screening, and some times by oil separation. Wide variations in the volume and strength of the waste requires equalization prior to any treatment. For some wastes, adjustment of pH prior to biological treatments may be necessary.

Activated sludge process, Oxidation ditch, Stabilization ponds, Extended Aeration Process, and Trickling filters are used for the biological treatment of the wastes. For some of the processes it requires post treatment in a secondary clarifier and an anaerobic sludge digestion tank. But it must be borne in mind that some of the wastes may be completely devoid of essential nutrients for the bacteria, and therefore Nitrogen and Phosphates are to be added to the system. Again, though some of the organics are readily assimilated by the microorganisms, many are not equally acceptable to them, and some are never utilized for the metabolism.

The common microorganisms, responsible for biodegradation in the municipal sewage treatment plants, may not be employed directly in the biological treatment of the petrochemical waste. This necessitates the Acclimatization of the microorganisms, and such a biological treatment process with acclimatized microorganisms may affect about 99% reduction in BOD value.

A certain portion of the waste may need "seggregation" for completely separate treatment, or for a special pre-treatment prior to biological treatment. As for example, Acrylonitriles require thermal reduction treatment, and wastes from Butadiene purification process which contain Cupric Ammonium Acetate, are to be treated with alkali lyes at higher temperature.

Very strong wastes may be just lagooned, or diluted with municipal sewage, if feasible, prior to any treatment.

The waste from the caprolactum plant (about 2500 m^3/day) of Gujarat State Fertiliser Company is treated biologically in an Oxidation Ditch which is preceeded by Nutrient addition, Aeration, Equalization and Dilution by water; this treatment is followed by a Secondary Settling Tank, Recycling of sludge and Sand Drying Bed. Acclimatized microorganisms are employed in the Ditch and it was found that, upto a loading of 0.50 kg of BOD or 0.745 kg of COD/day 1 kg of MLSS, at a MLSS concentration in between 3000 & 4000 mg/1 the COD and BOD reduction of about 74% and 99% respectively may be obtained. However, the effluent COD is found to be much higher than the "tolerance limit for industrial effluents discharged into inland surface waters" (500 to 610 mg/1 against the limit of 250 mg/1).

Similar observations are reported for the petrochemical complex of Union Carbide at Trombay, which produce Ethylene, Propylene, Butanes, Aromatics, Acetic Acid, Butanol, 2-Ethyl hexanol, Esters and plasticisers, starting from refinery Naphtha. In a pilot plant batch operation, COD reduction of 70% corresponding to a 81% reduction of BOD is observed at a loading of 0.16 kg of BOD or 0.32 kg of COD/day/kg of MLSS, in an aerated tank, with MLSS around 4000 mg/1. The effluent COD at this loading is 389 mg/1 which is also higher than the specified limit. The reason for the lower COD reduction is the non-biodegradability of certain organics.

Another petrochemical complex at Baroda, the Gujarat Aromatics Project produces Xylene and Dimethyl Terphthalate (DMT) (monomer for the Terylene type of polyester fibre) starting from special fraction Naptha obtained from Gujarat Refineries. The liquid waste consists of various organic chemicals like Acetic acid, Formic Acid, Formaldehyde, Methyl and Ethyl Benozates, DMT,

306

Dimethyle Isophthalate, Methyl formate & acetate, Methanol, o-,m, and p-xylene, and Potassium sulphate—mostly from DMT unit. DMT unit alone produces a very high BOD of about 35,000 mg/1 which is about 98% of the total BOD load, where as its volume of waste is only 5% of total waste flow. The combined waste includes that from DMT unit, waste cooling water, waste from demineralisation unit, sanitary waste, floor washes, and a negligible quantity of waste from xylene unit. The BOD of the composite waste is about 1900 mg/1, while COD : BOD ratio is about 1.55. The pH of the waste from DMT unit being very low it is neutralized to pH 5-6 using lime. The neutralised DMT waste along with other wastes is then treated in a two stage biological treatment plant.

The first stage of the above biological treatment consists of an oxidation ditch shaped aerator unit, where a MLSS concentration of 4000 mg/1 (MLVSS = 2400 mg/1) and a hydraulic retention time of 10 hrs is maintained at a process loading of 0.5 kg of BOD_5/kg of MLSS (or 0.64 kg of BOD_5/kg of MLVSS). The second stage consists of three oxidation ditches, where similar MLSS (MLVSS=60% MLSS) is maintained but with a process loading of 0.058 kg of BOD_5/kg of MLSS (or 0.097 kg of BOD_5/kg of MLVSS) and a hydraulic retention time of 26 hr. The treatment system is having a mean cell residence time of 24 days, and 70% of the flow is recirculated to the 1st stage inlet after the clarification in the settling tank. In both the units the aeration is accomplished by aerator rotors of capacity 2.8 kg of O_2/hr/meter length, and acclimatized microorganisms are employed. Necessary nutrients are added ahead of biological treatment. The wasting of the sludge is accomplished from the recycle line, and the sludge is dried over sand drying beds. The BOD effluent of this treatment plant is 20 mg/1. It may be added that upto 5 days of aeration the first order BOD removal rate constant for the waste is found to be 0.235/day for overall BOD and 0.272/day for soluble BOD; after 5 days it was reported to have fallen rapidly to 0.098/day and 0.085/day respectively for overall and so soluble BOD.

QUESTIONS

1. Draw a flow sheet for an aromatic waste treatment plant.
2. What are the characteristics of petrochemical wastes?
3. How are the petrochemical wastes treated?

20 | Pharmaceutical Plants Waste

20.0 INTRODUCTION

Like many other industries the pharmaceutical industry produces a wide variety of products. This industry uses both inorganic and organic materials as raw materials, the latter being either of synthetic or of vegetable and animal origin. Some of the pharmaceutical plants do not discharge liquid waste at all, some discharge very small but concentrated liquid waste, while some others discharge highly alkaline and toxic liquid wastes. Therefore it is very difficult to make any generalisation in regard to the characteristics of the pharmaceutical plant wastes.

20.1.1 Origin and Characteristics of the Antibiotic Wastes

Antibiotics and vitamins are produced by the fermentation of fairly complex nutrient solutions of organic matter and inorganic salts, by fungi or bacteria.

In the production of penicillin, molds of "Penicillium notatum-chrysegenum group" are cultured under submerged aerobic conditions on a medium consisting of corn steep liquor (nitrogen source), peanut meal, mineral salts, and lactose. After fermentation the mold mycelium is separated by filtration. The filtrate is then acidified to a suitable pH using phosphoric acid, and the penicillin is removed by extraction with amyl acetate. The solution of penicillin is further extracted with a buffered solution of sodium chloride. The isolated penicillin is finally purified by extraction with an organic solvent.

Streptomycein is produced in a similar way, using a "strep-

308

tomyces griseus" culture on a medium consisting of glucose, corn
steep liquor etc. The fermentation broth is filtered, and the filtrate is
adsorbed on charcoal or a resin. The streptomycein is eluted from
the charcoal or resin with dilute acids.

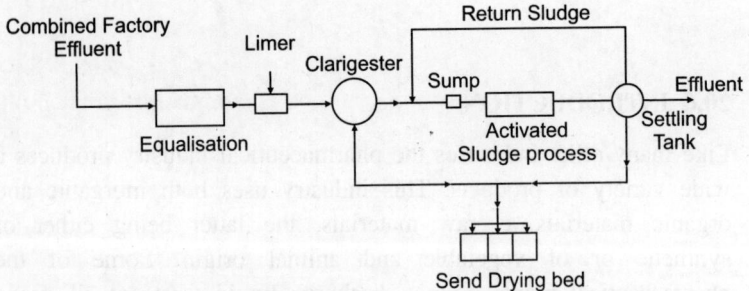

**Figure 20.1. Flow Sheet for Treatment of Combined Antibiotics and
Other Chemical Wastes.**

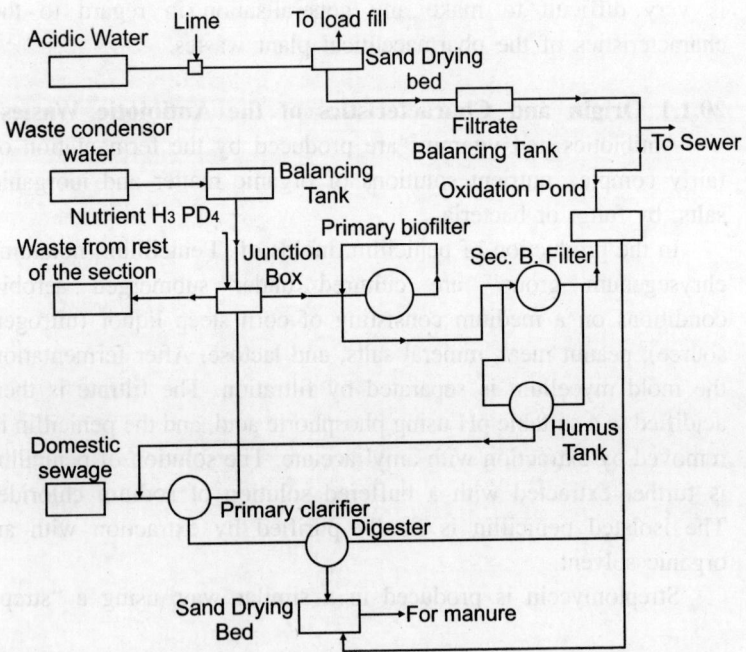

**Figure 20.2 Flow Sheet for Treatment of Waste from a Large Synthetic
Drug Plant.**

The elute is then neutralised and concentrated. The crude streptomycein is then precipitated by the addition of acetone, and further purified.

Yields from the above processes, in terms of weight, are small, and out of the raw materials used in the fermentation more than 90% appear as wastes. This waste mycelium may either be sold as manure or stock feed, or may be disposed off in any other way.

The liquid wastes from any antibiotic plant may be divided into the following groups:

    (*i*) spent liquor of the fermentation process,

    (*ii*) wash waters used for cleaning of floors and equipments,

    (*iii*) wastes containing acids, bases, and solvents used for extraction and purification of the product,

    (*iv*) "Filter Aids" used in the filtration, such as, "Diato maceous Earth",

    (*v*) Condensate from barometric condensers in evaporation and drying under reduced pressure.

The composition of the combined wastes from two typical plants producing penicillin and streptomycein respectively are given below (Mudri & Phadke):

TABLE 20.1. Characteristics of Antibiotic Wastes.

Item	Penicillin plant	Streptomycein plant
Colour	Colourless	Pale yellow
Odour	Fruity smell	Septic
BOD, mg/l (5 day, 37°C)	650–5500	500–2800
Free ammonia nitrogen, mg/l	0–5.6	0.3–18.2
Nitrate Nitrogen mg/l	0.1–0.5	0–0.8
Phosphate, mg/l	18–700	9–700
Total solid, mg/l	480–26200	960–4950
Suspended solid, mg/l	70–1080	80–1800
Total volatile solid, mg/l	200–12180	480–3070
pH	3.9–7.8	2.9–8.7

The above table is based on the analysis of the grab samples taken from the plant effluents. The wide variation is observed due to the fact that the batch processes are adopted for the production in the plant.

20.1.2 Inplant Treatment of Antibiotic Wastes

Treatment of the liquid wastes formed during the production of antibiotics, starts from the plant itself. For economic reasons, most of the organic solvents used in the process, like Amyl acetate in penicillin production and Butyl alcohol in Aureomycin production, are recovered in the plant. The recovery reduces the BOD of the wastes by a great extent.

20.1.3 Effects of The Wastes on Receiving Water Sewer

If a crude waste from an antibiotic waste is discharged into a stream, it not only imparts an objectionable odour to the stream, but also adversely affects the biological population in it. This waste should not be allowed to discharge into a municipal sewer, unless the sewage treatment plant is properly designed to handle a widely varying and concentrated waste from such a plant. Penicillin waste is found to have a disturbing effect on the process occuring within the sludge digestion tank.

20.1.4 Treatment of The Antibiotic Wastes

Antibiotic wastes can neither be clarified in settling tanks, nor can be chemically coagulated to reduce BOD (Koziorowski & Kucharski). The poor response of waste to the coagulation is due to the fact that most of the substances contributing to BOD appear in solution.

Anaerobic digestion of the waste results in a good reduction of BOD. In this process the inflow of the raw waste is controlled, and never allowed to exceed 5% of the capacity of the digester; further more the contents of the digester are mixed slowly and continuously.

Aeration of the waste, seeded by microorganisms either from domestic sewage or from garden soil, may result in some reduction of BOD. But the formation of sludge is very small due to the low concentration of colloidal and finely divided particles. Aeration, followed by the biological treatment of the diluted sewage in Trickling filter produces good BOD reduction. The coloured effluent of the trickling filter may be decolorized by chlorination.

A large pharmaceuticals complex in India producing antibiotics, vitamin C, symbiotics and fine chemicals, treats its liquid waste of almost similar composition as given in the last table and of very high acidity (about 700 mg/1) in two stages as given below (Rajagopalan et al):

(*i*) In the first stage after equalization of the waste, the high acidity is neutralized with lime and then clarified; the clarified liquor is then anaerobically digested in the digester. This results in the BOD reduction of 28-57%.

(*ii*) In the second stage the conventional biological process is employed to reduce the BOD of the waste further. In a pilot plant study, both Activated sludge process and Oxidation Ditch are found to be capable of bringing down the BOD. Same degree of BOD reduction (70-80%) is observed with the following process parameters:

		Oxidation Ditch	Activated sludge process
(a)	Organic loading kg BOD/kg MLSS/day	0.1–0.5	0.3–0.7
(b)	MLSS concentration, mg/1	3000–4000	1500–2500
(c)	Aeration time, hrs	22	6–8

20.2.1. Characteristics of Synthetic Drugs Waste

The synthetic drugs plants utilize large number of both organic and inorganic chemicals, and usually produce a variety of drugs in different sections of the plant. The volume and composition of the liquid wastes not only vary from plant to plant, but also from section to section in a plant, producing different types of drugs from different raw materials and using varieties of processes. Therefore a "typical" plant cannot be considered in the synthetic drug industry.

In general, most of the wastes are toxic to biological life, and are usually characterised by high BOD, COD, and a high COD: BOD ratio. Wastes from these plants are either highly alkaline or highly acidic. These wastes have a negligible fraction of settleable solids in their total solids content. Wastes from the manufacture of sedatives contain toxic elements like potassium cyanide. Highly alkaline wastes originate from the manufacture of sulfa-drugs and vitamin B_1. Manufacture of certain organic intermediates gives rise to a highly acidic waste consisting of both organic and inorganic acids.

20.2.2. Effects of The Synthetic Drug Wastes on Receiving Streams/Sewers

Wastes containing toxic elements like cyanides and heavy metals, if discharged without any treatment are harmful to the aquatic life in the streams. These toxic elements interfere with the biological sewage treatment units very badly. Similar effects are observed with raw acidic wastes; these wastes corrode structures in the sewerage system. Due to their high BOD content, a raw waste when discharged into stream rapidly depletes the dissolved oxygen of the stream and renders the water unsuitable for further use.

20.2.3. Treatment of Synthetic Drug Waste

Due to the great pollution potential and the diversified characteristics of the wastes from different sections of a plant, the planning for the treatment of synthetic drug wastes should be preceeded by a careful study of each waste. Segregation and equalization very often improve the overall treatment efficiency and reduce the cost of treatment. If the wastes have high COD: BOD ratios, bench scale laboratory biological treatment studies with acclimatized seed sludge are necessary for proper planning.

The acidic wastes, wastes containing toxic elements like cyanide, and those containing offensive-odour-producing compounds are usually seggregated and are treated separately. The acidic wastes may be neutralized with lime. The odour-producing compounds are usually destroyed by chlorination; compounds which are resistant to chlorine may be destroyed by heating in a furnace.

The segregated cyanide wastes may be treated with ferrous salts, where the cyanides are converted into non-toxic complex compounds. In this particular process, lime is also added along with ferrous salts to adjust the pH to an effective range. The cyanide wastes are sometimes oxidized by a strong oxidizing agent like chlorine, in an alkaline condition. This process is commonly known as "alkaline chlorination". The cyanide-containing wastes can also be treated biologically in a two stage process—the anaerobic stage followed by an aerobic stage.

Chemical treatment for other wastes is usually found to be

ineffective in BOD and COD reduction. Biological treatment employing acclimatized microorganisms, preceeded by dilution is often found to be very effective for the treatment of composite wastes, even when it contains large amounts of toxic elements.

Waste from a pharmaceutical complex, at Bangalore city, producing large number of synthetic drugs, like Dexedrine, Eskazine, Neuraphosphates, Iodex, Furasin, etc, which include many toxic compounds is found to be biodegradable when diluted with domestic sewage, and acclimatized microorganisms are used in the treatment of these wastes. The average composition of the waste after mixing with the municipal waste is tabulated in Table 20.2. The combined waste is characterised by a very high COD: BOD ratio, and with a pH in the acidic range.

TABLE 20.2. Average Composition of Combined Composite Samples of Waste from the Synthetic Drug Plant at Bangalore City and a Nearby Residential Campus.

Total solids, mg/1	1180–1242
Volatile solids, mg/1	798–890
Fixed solids, mg/1	382–406
Settleable solids, mg/1	1.25–1.5
pH	6.5–7.0
Acidity (Phenolphthalein), mg/1	25–55
5 day 20°C BOD, mg/1	475–567
COD, mg/1	4680–6800
COD: BOD	8.3–13.9
Chloride, mg/1	54–102
Nitrate Nitrogen, mg/1	28–32
Phosphate, mg/1	2–5.5
BOD : N : P	100 : (5.6-6.5) : (0.42–1.04)

A pilot plant study with the above pharmaceutical plant waste (Shivalingaiah *et al*) indicates that about 95% reduction in COD, and 88% reduction in BOD can be achieved in an extended aeration process. A process loading of 0.075 kg of BOD/kg of MLSS/day, MLSS concentration of 5000 mg/1, and a mean cell residence time of about 50 days is found to be optimum.

In another large synthetic drug plant at Hyderabad, producing a large number of synthetic drugs including Analgesics and Antipyretics, Anthelmintics, Anti-Filarials, Anti-Tuberculosis,

Diuretics, Hypnotics, Sulphas, Vitamins, and a large number of organic intermediates like Analgin, Apidin, Cemizol, Hexavit, Cebexine, Sukcee, Calmod, Compeba, Salinex, etc., different sections of the plant are found to discharge different types of waste containing about seventy chemicals including organic acids and large number of organic and inorganic toxic materials. After a careful study of the wastes from all the sections, the acidic waste is seggregated and treated separately; the wastes from the remaining sections are treated biologically. The average composition of the acidic waste, and the waste from other sections are given in the Table 20.3.

TABLE 20.3. Characteristics of Wastes from a Synthetic Drug Plant at Hyderabad.

Parameters	Acidic waste from one section	Composite waste from rest of the sections
Rate of flow, m^3/day	307	202
pH	0.60	9.30
Total solids (%)	8.27	7.73
Total volatile solids (%)	6.66	1.70
Acidity (as $CaCO_3$), mg/l	57,564	NIL
Alkalinity (as $CaCO_3$), mg/l	NIL	10,754
Chlorides (Cl), mg/l	20,500	17,000
Sulphates (SO_4), mg/l	37,000	14,800
Total Nitrogen (N), mg/l	6,202	5,166
Phosphorus (P), mg/l	NIL	NIL
COD, mg/l	13,745	28,540
BOD, mg/l	9,400	15,250

The treatment of acidic wastes consists of neutralization by lime to pH 7.0 and drying of neutralized waste over sand drying beds for 5 to 6 days, for the separation of sludge. The neutralized filtrate still contains large amount of sulfanilic acid (about 75% of that in the raw waste), but as it is found to be biodegradable it is discharged into the sewer.

The waste from other sections are diluted with domestic waste and treated in two stage biofilters with pretreatment by bar screens, grit chamber and primary clarifier. The effluent of the secondary clarifier is further treated by a number of oxidation ponds which is then discharged into the sewers.

In a pilot plant study (Mohanrao et al) with the above waste, it was found that the waste is amenable to biological treatment only when diluted to about fourteen times (about 7%) and when acclimatized microorganisms are employed. In a laboratory biological aerobic reactor, a process loading of 0.29 kg of BOD_5/kg of MLVSS, MLVSS concentration around 3500 mg/1 (MLVSS is 90% of MLSS), and a hydraulic retention time of 8 hrs. are found to be optimum for about 90% reduction in BOD. The same study indicates that the kinetic growth yield coefficient, Y, and microorganism decay coefficient, k_d, of this waste is 0.56 and 0.12/day respectively. With these values of kinetic growth coefficients and the given value of optimum process loading, U, the mean cell residence time works out to be 25 days.

As evident from Table 20.3, the waste under discussion does not need any nitrogen supplementation. But addition of phosphorus @ 0.84 kg/100 kg of BOD_5 removed was found to be necessary.

QUESTIONS

1. Trace the origin and characteristics of antibiotic wastes.
2. Draw a flow sheet for treatment of wastes from a large synthetic drug manufacturing plant.

21 Corn Starch Industry Waste

21.0. INTRODUCTION

Corn starch products are widely used in the chemical and materials field. This industry processes corn to produce starch, oil and feed. The composition of the corn kernel is approximately as follows.

Carbohydrates	—	80%
Protein	—	10%
Oil (fat)	—	4.5%
Fibre	—	3.5%
Minerals	—	2%

Clean corn is steeped in a dilute solution of Sulphuric acid in order to loosen the hull, soften the gluten, and dissolve minerals or organic matter in the Kernel. Next, the corn is ground to free, but not to crush the germ; the ground corn is mixed with water and placed in settling tanks. When the germs float to the top, they are skimmed off, and oil is pressed or extracted from them. The kernel residues are ground finely, to separate the soluble starch and gluten from the fibre and hull, which are known as "grits and bran" and are used as feed additives. The starch is separated from the gluten by settling, centrifuging, and counter-current washing. The gluten is added to feed, and the starch is filtered and washed on vacuum filters and dried. Then it is either marketed as starch, or modified starch, or hydrolyzed into corn syrup or corn sugar. Both the feed-producing and starch-manufacturing processes evolve a process-water waste containing about 3 percent of the corn in soluble form. Van Patten and McIntosh (2) present a flow sheet for wet miling of corn (Fig. 21.1).

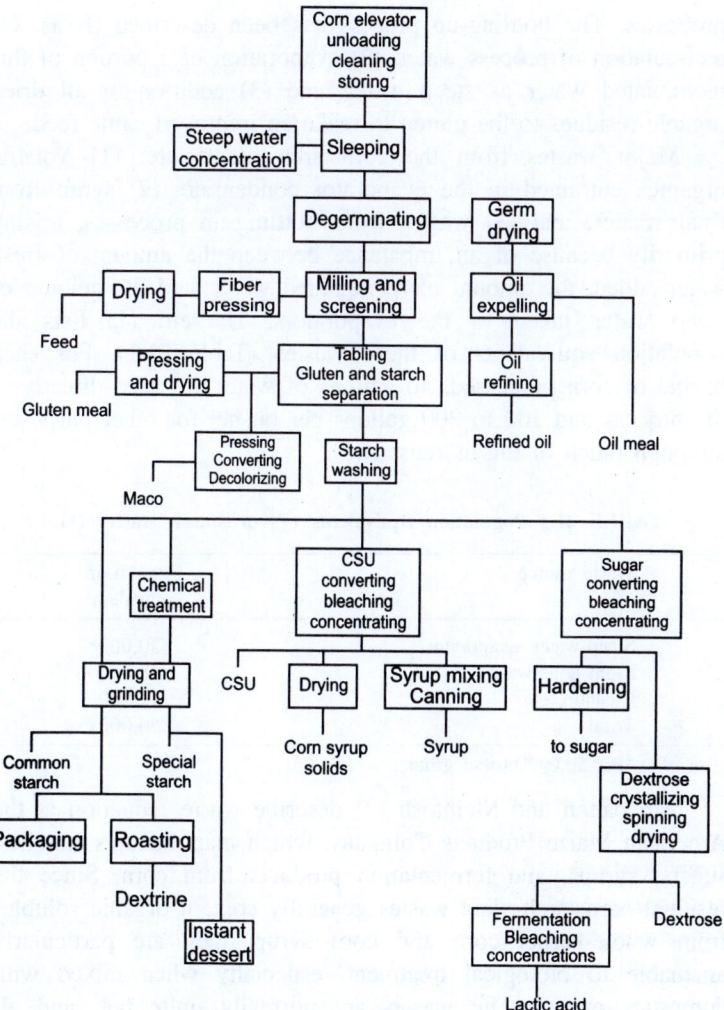

Figure 21.1: Flowsheet of Wet-Milling Process for corn. (After Van Patten and McIntosh (2).

Even before waste-treatment practices were utilized in the industry, the "bottling-up" process was common, since it was introduced to abate stream pollution, and it can now be considered as an actual part of the cornstarch plant process. The wastes from this industry consist of residues and leaks from the reuse

318

processes. The bottling-up process has been described (I) as: (1) recirculation of process water, (2) evaporation of a portion of this recirculated water as steep water, and (3) addition of all dried organic residues to the gluten to make an improved cattle feed.

Major wastes from the cornstarch plants are: (1) Volatile organics entrained in the evaporator condensate, (2) syrup from final wastes, and (3) wastes from bottling-up processes, arising primarily because of an imbalance between the amount of fresh water added, the amount of recirculated water, and the amount of steep water taken to the evaporators. Hatfield (1) lists the population equivalents of these wastes (Table 21.1). For each bushel of corn processed, 40 gallons of water are used directly in the process and 100 to 200 gallons per bushel for other purposes, although much of this is reused (3).

TABLE 21.1 Population equivalents of cornstarch wastes (1).

Waste source	Population equivalent*
Steep-water evaporators	30,000
Light bone wash	8,000
Cleanup, etc.	12,000
Total	50,000

*Per 50,000 bushel grind.

Van patten and McIntosh (2) describe waste reduction at the American Maize Products Company, which manufactures starches, sugars, syrups, and fermentation products from corn. Since the residual cornstarch-plant wastes generally contain organic solubles from whole-kernel corn and corn syrup, they are particularly amenable to biological treatment, especially when mixed with domestic sewage. The wastes are normally quite hot, and all factors directly attributable to this characteristic must be considered; for example, oxygen solubility is lowered, but at the same time digestion is enhanced. Settling is hindered by the heat of the wastes, and sewer lines tend to clog as the starch wastes cool. Equalization, to effect both cooling and homogeneity of the wastes, should thus be practised prior to biological treatment, either at the industrial-plant site or in the municipal treatment plant.

Hatfield et al. (4) used trickling filters with a 5: 1 recirculation ratio to effect a 90 per cent BOD reduction, with an influent BOD concentration of 1400 ppm or less and a loading of about 4 pounds of BOD per cubic yard per day. Control of pH and addition of nitrogen and phosphorus were necessary.

The senior author designed an anaerobic—aerobic system for treatment of corn-starch wastes, the details of which are furnished below.

21.2 TYPICAL CHARACTERISTICS OF INDUSTRIAL EFFLUENT

The waste water has the following characteristics:—

Flow	: 80 cubic metres per day in Phase I
	: 150 cubic metres per day in Phase II
Appearance	: Dirty white
Odour	: Alcoholic
pH	: 5.2
Total solids	: 7764 mg/l
Dissolved solids	: 1440 mg/l
Suspended solids	: 6324 mg/l
COD	: 3677 mg/l
BOD 5 days 20°C	: 2570 mg/l

21.3. METHOD OF TREATMENT

It is proposed to treat the effluent in anaerobic lagoons followed by aerated lagoons and a settling tank. The overflow from the settling tank will be used for irrigating 5 acres of land attached to the factory.

Design of anaerobic lagoon
Total quantity of waste water: 150 cu. metres per day
B.O.D.=2570 mg/l.

Assuming the detention period of 30 days, the capacity of lagoon would be 150 × 30 = 4500 cu. metres.

Providing a depth of 3.75 m, the surface area of pool $= \dfrac{4500}{3.75} =$ 1200 sq. metres.

2 numbers of lagoons will be provided in parallel. Size of each

lagoon = 600 sq. metres = 30 m × 20 m each.

A free board of 0.75 m above the surface will be provided with a side slope of 2 : 1. This will reduce the BOD to about 65%. Therefore, the effluent BOD after anaerobic treatment = 900 mg/1. This will be treated in aerated lagoons.

Design of areated lagoons

Flow	:	150 cu. metre per day
Influent BOD	:	900 mg/1.
Detention time	:	1 day.
∴ Volume	:	150 cu. metres.

Providing a depth of 3 m. surface area = 150 sq. mm., 2 numbers of aerated lagoons will be provided each 6 m long and 5 m wide.

Oxygen requirement

BOD load per day: 150 × 900 = 135 Kg.

Oxygen required at 1.25 Kg/Kg of BOD applied
$$= 1.25 \times 135 = 168.75 \text{ Kg./day}$$

OR $\dfrac{168.75}{24}$ = 7.03 Kg./hr.

Horse-power requirements

The surface aerators will be capable of transferring 1.35 Kg of oxygen per horse power per hour.

∴ Total HP required $= \dfrac{7.03}{1.35} = 5.20$ H.P.

2 aerators of 5 HP each will be provided – one in each tank.

Power Requirement

Volume of the lagoon = 600 cu. metres.

Power required at 1.3 K. watt/10 cu. metre of lagoon

$$= \dfrac{1.3 \times 600}{10} = 78 \text{ K.W.}$$

Efficiency of treatment	= 90%
Effluent BOD	= 90 mg/1

This effluent will go to a secondary settling tank.

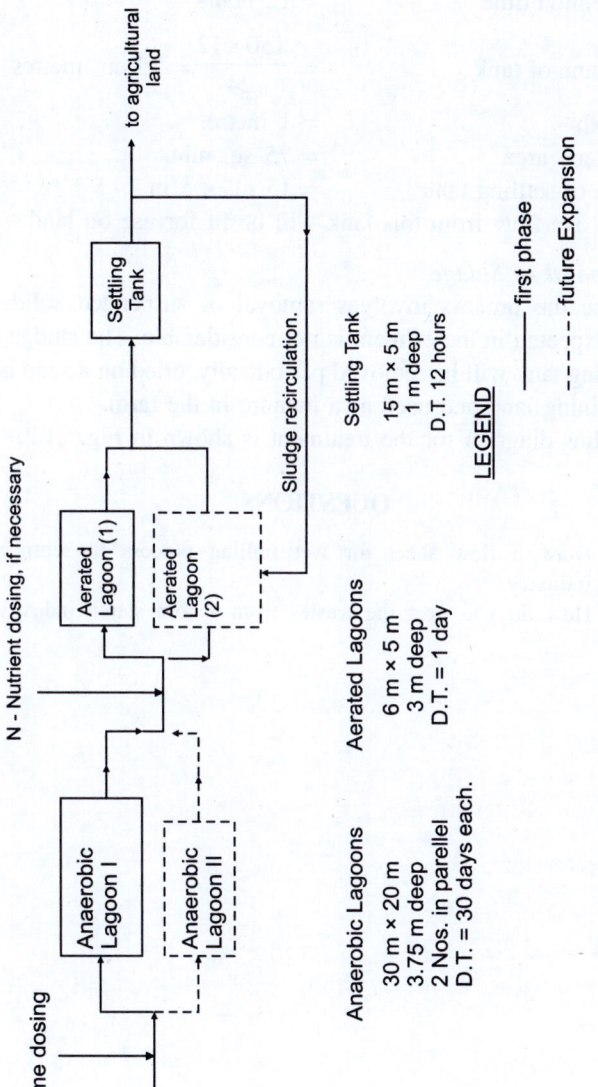

Lime dosing

N - Nutrient dosing, if necessary

to agricultural land

Anaerobic Lagoon I

Anaerobic Lagoon II

Aerated Lagoon (1)

Aerated Lagoon (2)

Settling Tank

Sludge recirculation.

Anaerobic Lagoons

30 m × 20 m
3.75 m deep
2 Nos. in parellel
D.T. = 30 days each.

Aerated Lagoons

6 m × 5 m
3 m deep
D.T. = 1 day

Settling Tank

15 m × 5 m
1 m deep
D.T. 12 hours

LEGEND

——— first phase
- - - - - future Expansion

Figure 21.2: Flow Sheet for Treatment of Corn Starch Waste.

Design of secondary settling tank

Flow = 150 cu. metres per day

Detention time = 12 hours

Volume of tank $= \dfrac{150 \times 12}{24} = 75$ cu. metres.

Depth = 1 metre.

Surface area = 75 sq. mm.

Size of settling tank = 15 m. × 5 m

The overflow from this tank will be fit for use on land.

Disposal of Sludge

Since the process involves removal of suspended solids, the sludge expected in the effluent is not considerable. The sludge from the settling tank will be removed periodically, dried on a sand bed in the adjoining land and used as a manure in the farm.

A flow-diagram for the treatment is shown in Fig. 21.2.

QUESTIONS

1. Draw a flow sheet for wet milling process for corn starch industry.
2. How do you treat the wastes from a corn starch industry?

22 | Odour and Its Removal

22.0 COMPOUNDS RESPONSIBLE FOR ODOUR

Volatile organic compounds (VOC) and inorganic compounds, that evaporate at ambient temeprature and exist in atmosphere in gaseous form, are present in waste waters and are released during their treatment. Volatile odour-producing compounds are also generated in chemical manufacture and their release through vents and stacks also contribute to odour problems.

Most of the volatile compounds are toxic. They cause repulsive and irritating feeling when inhaled and their presence in ambient air is a matter of concern. Offensive odours are complained by people before they receive any damage there-from and are often linked to air pollution.

Therefore, the presence of odour vis-vis odour-causing organic and inorganic substances in ambient air is a matter of concern and efforts are to be made to remove them to levels that are not injurious/toxic to life. Thus, odour removal or deodorisation technologies have assumed greater importance.

Examples of odour-causing inorganic and organic compounds are:

Hydrogen sulphide	Hydrocarbon solvents	Lower fatty acids
Sulphur dioxide	Benzene	Indole
Ammonia	Toluene	Methyl ketones
Mercaptans	Xylene	Aldehydes
Amines	Pyridine	Esters

22.1 REMOVAL METHODS

Deodorisation processes are broadly classified into physical, chemical and biological methods. These methods are applicable to control odours in waste gases originating from point sources. Collection of air contaminated by fugitive sources is almost impossible and hence treatment is not possible.

22.2 THE PHYSICAL AND CHEMICAL METHODS

- Wet scrubbing with water or water having chemicals to oxidise the odour-causing substances.
- Adsorption on solid material such as activated carbon
- Ozonation
- Catalytic and thermal oxidation

Water scrubbing of waste air/gases transfer the pollutants from one media into other and scrubber water requires treatment for oxidising the odour-causing compounds. Chemical scrubbers using hypochlorite, chlorine or pemanganate will leave high residues in scrubber liquor which require further treatment before disposal. These absorption systems require less space and can be effective for treating concentrated or poorly biodegradable pollutant gases.

Carbon adsorption requires regeneration to maintain adsorption capacity. This process is best suited to remove toxic pollutants from waste gas streams.

Thermal and catalytic incinerators are suitable for off-gases from industrial sources, with high concentration of flammable pollutants. Fuel cost will be high for dilute emission. Incineration is preferable where the pollutants are poorly biodegradable and are highly odorous even when they are present at ppb levels in ambient air.

Application of physico-chemical technologies for deodorization is useful for specific situations but on the whole they are expensive.

22.3 BIOLOGICAL SYSTEMS FOR DEODORIZATION

Bio-treatment of odour-causing contaminants in air streams offer an inexpensive alternative to physical and chemical methods which has recently been adopted in a number of cases.

Biofiltration is the most attractive among the available bio-processes for removal of volatile organic pollutants. It is a process in which waste gases are purified by passing through a biologically active porous medium and it offers a number of advantages in treating waste gases with concentrated VOCs.

The processes used include:
- Soil process using microorganism in soil
- Packed column process based on porous carriers allowing microorganisms to propagate
- Activated sludge/suspended growth processes in which the waste gas comes in contact with the micro-organisms in the suspended mass in water and odour ingredients are decomposed by microbes.

Packed column process has many advantages over the other two. They are
- high deodorisation capacity
- simple and compact design
- easy operation
- low running costs

The solid materials packed in the column, to propagate microorganisms, are mainly derived from peat, compost or wood chips. The materials to be used in the column should have high surface area for development of active bio-layer and for adsorption of the pollutants to be removed. The packed solid media provide the nutrient source, pH buffering and matrix for the attachment of microorganisms. The packed solids are required to be replaced once in 2 to 3 years.

Biofilters have consistently demonstrated >90% destruction efficiency for many common air pollutants including organic compounds such as alcohols, aldehydes, amines and inorganics like hydrogen sulphide and ammonia. The relative biodegradability of individual components that are present in different waste gas streams is given in the following table.

TABLE 22.1. Biodegradability of Individual and Class of Volatile Compounds

Inorganic	Organic			
Rapid	Rapid	Good	Slow	Very Slow
Hydrogen sulphide	Alcohols Methanol Butanol	Esters Ethyl acetate	Aliphatic Hydro Carbons Methane Pentane Cyclohexane	Many Halogenated Compounds 1,1,1 trichloro ethane
Ammonia	Aldehydes Formaldehyde Acetaldehyde	Ketones	–	Poly aromatic Hydrocarbons
Sulphur dioxide	Amines	Phenols Benzene Styrene	–	–
	Organic acids Butyric acid	Mercal2tans Methyl mercaptan	–	–

In the treatment of low-concentration off-gases, biofiltration tends to compete with scrubber, activated carbon adsorption and thermal/catalytic oxidation (with heat recovery). The conventional air pollution control methods can often achieve the required efficiency but can also have cost and operational drawbacks in comparison to biofiltration. Performance data on biofiltration of waste gases is presented below:

TABLE 22.2. Biofilter Performance Data

Application (Reference)	Contaminant(s)	Loading	Removal Type	Biofilter
Yeast Production Facility	Ethanol Aldehyde	35,000 cfm/500 yd^3 media, 1 g/m^3	Overall VOC Reduction of 85%	Media Filter
Plastics Plant VOC Emissions Control	Toluene, Phenol, Acetone	1,000 m^3/h	80% -95%	Media Filter
Pharmaceutical Production	Organic Carbon	1,000 m^3/h, 2,050 mg/m^3 (5,800 fig/m^3 peak)	> 98% first Stage > 99.9% overall	Media Filter (two stage)
Artificial glass Production	Monomer methyl methacrylate (MMA)	125-150 m^3/h, 50-250 mg/m^3	Biofilter: 100% MMA, 20% DCM, BTF: 95%	Media Filter Plus Biotrickling

	Dichloromethane (DCM)			Filter (BTF) in series
Hydrocarbon Emission Control Compost Plant for Garbage	Hydrocarbon solvents Odor	140,000 m^3/h, 500 mg/m^3 16,000 m^3/h, 264 m^2 (1 m deep) 60 m^3/m^{2}h, 230 mgC/m^3	95% >95%	Media Filter Media Filter
Gasoline VOCs Emission Control (Pilot scale)	Total VOCs	16 g/ft^{3}h,	90%	Media Filter
Hydrogen Sulfide Emission Control (Laboratory Scale)	H2S	1.9-8.6 mg/kg.min (25-2,651 ppmv)	93-100%	Media Filter
Styrene Removal (Bench Scale)	Styrene	Upto 22 g/m^{3}h, 0.5 min retention time	>99%	Biotrickling Filter
Styrene Removal (Bench Scale)	Styrene	Upto 100 g/m^{3}h	>95%	Media Filter (peat)
Rendering Plant	Odor	1,100 m^{3}h (650 cfm), 420 m^2	99.9%	Media Filter
Fuel-Derived VOC Emission Control	Non-methane Organic Carbon (simulated jet fuel)	500 ppm.cfm/ft^2 500-1,500 ppm. cfm/ft^2	>95%, 30%-70%	Media Filter

Thus for appropriate situations, biofiltration for treating waste gases for odour removal is typically lower in capital and operating costs, as such, it is environmentally acceptable.

QUESTIONS

1. What are the physical and chemical methods available for odour removal from effluents?
2. Explain the biofiltration technique and list its applications in odour removal

23 | Removal of Colour from Waste Waters

23.0 INTRODUCTION

Fundamentally, colour results from the absorption of visible light energy by chemical bonds in substances dissolved in waste waters. Large number of industries generate waste waters containing colour-causing substances, e.g. dye and dye intermediates manufacture, textile dyeing, pulp and paper mills, organic chemical manufacture.

Majority of compounds responsible for colour are not readily biodegradable. Secondly, biological treatment, that is typically much less costly than physico-chemical treatment, readily removes the conventional pollutants-BOD and TSS as well as some toxic compounds from waste water. However, secondary biological treatment provides very little reduction in colour.

On the other hand, colour in the waste water gets intensified after biological treatment in certain cases. The highly polymerised nature of chromophores probably accounts for their bio-refractory nature.

23.1 TREATMENT TECHNOLOGIES

Many approaches for colour removal have been attempted and only a few of them are available commercially. The available technologies are as follows.

Sorption Systems: Resin separation, Ion exchange, Activated alumina, Activated carbon.

Precipitation Systems: Coagulation with Trivalent metal salts, Lime, Poly electrolytes.

Membrane Systems: Reverse Osmosis, Ultrafiltration.

Oxidation Systems: Ozone, chlorine, Hydrogen Peroxide, UV irradiation.

Innovation Systems: Land application for crop irrigation.

Application of the above treatment systems for removal of colour from a few industrial waste waters is discussed.

23.2 ACTIVATED CARBON TREATMENT

Many industrial wastes contain refractory organics which are difficult, or impossible to remove by conventional biological treatment processes. These materials and the colour causing organics in pulp and paper, dye and dye intermediate manufacture and textile dyeing waste waters are frequently removed by adsorption on an active-solid surface. The most commonly employed adsorbent is activated carbon in granular or powder form. In general, substances of highest molecular weight are most easily adsorbed.

23.3 INFLUENCE OF MOLECULAR STRUCTURE AND OTHER FACTORS ON ADSORBABILITY

(*i*) An increasing solubility of the solute (organic molecule) in the liquid decreases its adsorbability.

(*ii*) Branched chains are usually more adsorbable than straight chains. An increasing length of chain decreases solubility (hence more adsorbable).

(*iii*) Substituent groups affect adsorbability.

Substituent Group	Nature of influence
Hydroxyl	Generally reduces adsorbability
Amino	Effect similar to hydroxyl but somewhat greater
Carbonyl	Effect varies according to host molecule
Double bonds	Variable effects as with carbonyl
Halogens	Variable effects
Sulfonic	Usually decreases adsorbability
Nitro	Often increases adsorbability

(iv) Generally, strongly ionised solutions are not as adsorbable as weakly ionised ones, i.e., undissociated molecules are in general preferentially adsorbed.

(v) The amount of hydrolytic adsorption depends on the ability of the hydrolysis to form an adsorbable acid or base.

(vi) Unless the screening action of the carbon pores intervene, large molecules are more sorbable than small molecules of similar chemical nature.

(vii) Molecules with low polarity are more sorbed than highly polar ones.

In textile industry dyeing and printing sections discharge waste waters with varying degrees of colour and the colour depends on type of dyes used in the process. The wastes are usually alkaline with high total dissolved solids and colour.

Carbon treatment for removal of colour from textile dyeing units is adopted only at the polishing stage of the effluent obtained from chemical coagulation. Efficiency of carbon treatment on textile dye house waste water has been reported to be as high as 99.4%. However carbon dose requirements will be high and adsorption using activated carbon will prove uneconomical when high degree of coloured waste waters are to be treated.

The same is true while treating waste waters from dye and dye intermediate manufacture. Carbon adsorption for treating of pulp mill waste water is tested on pilot scale after chemical coagulation and biological treatment. It is reported that carbon requirement is high and carbon regeneration is essential for economic reasons.

Resin adsorption studies with pulp mill waste water showed fouling of resin with uncertain performance.

23.4 PRECIPITATION / COAGULATION TREATMENT

High valence cations (≥ 2) are used to coagulate the colloidal substances in waste water which help in the removal of a variety of colloids and colour imparting substances such as dyes, lignin etc.

The commonly used coagulants are $Al_2 (SO_4)_3$, $Fe_2 (SO_4)_3$, $FeCl_3$ and $FeSO_4$ along with lime, Polyelectrolytes are used to improve settling and separation of the coagulated substances, for removal of colour from textile dyeing units and dye manufacturing industry. Data on extent of removal of colour and COD with different coagulants are given below.

TABLE 23.1: Coagulation of Textile Waste Water.

Plant	Coagulant	Dosage mg/l.	pH	Colour Influent mg/l.	Colour Removal %	C.O.D. Influent mg/l.	C.O.D. Removal %
1	$Fe_2(SO_4)_3$	250	7.5-11.0	0.25	90	584	33
	Alum	300	5-9	–	86	–	39
	Lime	1200	–	–	68	–	30
2	$Fe_2(SO_4)_3$	500	3-4, 9-11	0.74	89	840	49
	Alum	500	8.5-10	–	89	–	40
	Lime	2000	–	–	65	–	40
3	$Fe_2(SO_4)_3$	250	9.5-11	1.84	95	825	38
	Alum	250	6-9	–	95	–	31
	Lime	600	–	–	78	–	50
4	$Fe_2(SO_4)_3$	1000	9-11	4.6	87	1570	31
	Alum	750	5-6	–	89	–	44
	Lime	2500	–	–	87	–	44

– Colour is expressed as sum of absorbance at wave lengths of 450, 530 and 650 nm. (Ref. Olthof, M.G. and Ecken Felder, W.W., Textile Chemists and Colorist, 8, pt 7,18,1976) From the above data, colour can be removed by ~ 90% with ferric sulphate and alum. "Dose of the coagulant required varied with the colour intensity of the waste water.

Similarly data on colour removal from pulp and paper mill waste water with chemical coagulation is given below.

TABLE 23.2: Colour Removal from Pulp and Paper Mill Waste Water.

Plant	Coagulant	Dosage mg/l.	pH	Colour Units Influent mg/l.	Colour Units Removal %	C.O.D. mg/l. Influent mg/l.	C.O.D. mg/l. Removal %
1	$Fe_2(SO_4)_3$	500	3.5-4.5	2250	92	776	60
	Alum	400	4.0-5.0	–	92	–	53
	Lime	1500	–	–	92	–	38
2	$Fe_2(SO_4)_3$	275	3.5-4.5	1470	91	480	53
	Alum	250	4.0-5.5	–	93	–	48
	Lime	1000	–	–	85	–	45
3	$Fe_2(SO_4)_3$	250	4.5-5.5	940	85	468	53
	Alum	250	5.0-6.5	–	91	–	44
	Lime	1000	–	–	85	–	40

Ref. Olthof, M.G. and Eckenfelder, Water Res. 9, 853, 1975

Alum for colour removal is adopted on commercial scale. Limitations are solid/ liquid separation, high chemical requirement and sludge separation, dewatering and disposal.

Lime is also used on commercial scale. Solid/liquid separation poses some difficulty. Lime recovery is essential for economy. Since large quantity of lime is to be used, lime handling is a problem besides pH correction for final effluent before discharge.

23.5 MEMBRANE SYSTEMS

Ultrafiltration, Reverse Osmosis (RO) are used for treatment of a number of waste waters. R.O. systems are used to recover water for reuse and recycle. When used for colour removal from a waste water, the recovered water is recycled but the reject containing all the organic and inorganic pollutants requires to to treated for safe disposal. R.O. has been used economically to recover dyes and chemicals from spent baths.

Membrane separation is adopted for treating pulp mill waste water for colour removal but membrane fouling is reported. Additional provision for concentrated reject treatment is to be made.

23.6 OXIDATION SYSTEMS

Many dye containing waste waters are effectively decolorized using chemical oxidising agents like, chlorine (Cl_2, HOCl). Chlorine level of 150 mg/l in treatment of textile dyeing waste water resulted in 77% colour removal but chlorine residual of 110 mg/l of chlorine applied remained in treated waste water. When dose was below 100 mg/l no residual was detected but only 57% of colour was removed.

Use of chlorine is not desirable since chlorinated organic compounds that are generated in the process are highly objectionable as they are potential carcinogens.

Ozone is a powerful oxidant and is reported to be effective in decolorising waste waters containing reactive and basic dyes but ineffective on disperse dyes, using about 1 g/l ozone. However ozone treatment produces little reduction in TOC or DOC. Cost and efficiency are often barriers associated with ozone.

Fenton reagent (H_2O_2 and iron salt) was found to be effective in colour reduction and removal efficiency was dependent on the initial DOC in the waste water, with higher DOC resulting in less colour removal. A ratio of Peroxide to iron of 10-20 to 1 was suggested. Unlike ozone, Fenton's reagent was effective in reducing DOC from the waste water upto 60 percent. Biological treatment of waste waters was neither enhanced nor hampered by oxidative pretreatment processes, compared to control experiments.

23.7 REDUCTION SYSTEMS

Chemical reduction of azo dyes convert them to aromatic amines via reductive cleavage of arobong. In theory the aromatic amines are more amenable to aerobic biological treatment than the parent dye structure.

The most commonly used chemical reducing agent is sodium hydrosulphite (sodium dithionate). Thiourea dioxide (formamidine sulfonic acid), Sodium borohydride, Sodium formaldehyde sulfoxylate and stannous chloride are additional reducing agents for treating dye waste waters. The chemical treatment with reducing agents are effective in decolorising but the resulting effluent is reported to be inhibitory or toxic to subsequent aerobic biological treatment.

23.8 INNOVATIVE SYSTEMS

Land application of coloured waste waters particularly from pulp mills removes the colour due to lignin. The removal is due to conversion of lignin into insoluble calcium lignate which is retained by soil. In due course the lignin is slowly metabolised and is converted to humus-like matter enriching the organic matter in the soil. Field scale use of pulp mill coloured waste waters for crop irrigation is in practice in a number of mills in India as well as abroad. The percolate from the soil free from colour, recharges/increases the ground water reserve.

Soils with high cation exchange capacity are better suited for removal of colour from pulp mill waste water.

334

23.9 CONCLUSIONS

Colour in waste water is an aesthetic pollutant and is not removed in the secondary biological treatment plants. The treated effluent meeting the discharge standards with respect to all other pollutants will still catch the eyes of both regulatory authority and public.

Its removal is possible but poses an economic problem. In view of this the regulatory standards may prescribe a limit for colour and odour as "Not Objectionable".

QUESTIONS

1. What are the treatment technologies available for colour removal?
2. Explain the oxidation and membrane systems in colour removal.

24 | Industrial Waste Audit

24.0 TYPES OF ENVIRONMENTAL AUDIT

The following are the various types of environmental audits practiced

  (a) Waste audit
  (b) Energy Audit
  (c) Health and safety audit
  (d) Compliance audit
  (e) Management audit
  (f) Waste minimization audit
  (g) Liabilities definition audit
  (h) Property transfer audit

24.1 WASTE AUDIT

In a waste audit, quantities and types of wastes generated from different sources are identified and the optimum methods to minimize the quantities of wastes are evolved.

24.2 A MANAGEMENT AUDIT

A management audit, determines whether an adequate compliance management system is established, implemented and used correctly to integrate environmental compliance into everyday operating procedures.

Such an audit examines cultural, management and operational elements to include internal policies, human resources training programs, budgeting and planning systems, monitoring and reporting systems and information management systems.

A management audit detects potential systematic breakdowns that could manifest themselves as an environmental problem.

Management audit is a traditional audit evolved as a method to verify that a facility has instituted adequate procedures, dedicated sufficient resources, and installed the necessary systems to comply with environmental regulations or not.

It is a mere "snapshot" of compliance. Yet, the overall environmental objective is compliance on a continuing, long-term basis. This is a function of organization, guidance, controls, and communications. Each of these functions is the subject of a management audit whose purpose is not the identification of specific violations but the detection of activities likely to experience problems and their root causes.

Specifically, a management audit addresses a number of interpretive questions under each of several broad issues as follows:

1. Clearly defined responsibilities-How are they established and communicated? How are assignments reinforced? Are any assignments overlapping, shared or conflicting?

2. Adequate system of authority-How is it granted? Who grants exceptions? How are authorizations and exceptions recorded? Is there a real understanding of issues.

3. Division of duties-what potential exists for conflict of interest?

4. Trained and experienced personnel-What methods are used to determine who needs training? What training have personnel had to help them understand and complete their duties?

5. Documentation-What records are developed? How are records maintained? Do type and amount of documentation correspond to the importance of the activity? How are exception reports developed? Are the records reliable? Accurate?

6. Internal verification-How does the management review and evaluate environmental success? What is the accountability of managers for adherence to policies and procedures?

7. Protective measures-What engineered controls are in place to prevent non-compliance? What procedures?

24.3 WASTE MANAGEMENT CONTRACTOR AUDITS

Some generators ship significant amount of hazardous waste off site for treatment and disposal by private contractors. A fundamental principle of hazardous waste law in the United States is that of joint and several liability for management of hazardous waste the liability for remediation of problems created by a waste treatment and disposal contractor can rest with the generator.

When remediation is required, it can cost several orders of magnitude greater than the originally contracted cost for the treatment and disposal of the waste.

In addition, generators may also be financially responsible for damages to third parties derived from waste transport, treatment, and disposal operations. As a result of these potential liabilities, hazardous waste generators exercise great care in the selection of waste contractors. One tool to use in the selection of contractors, and in the verification that the selected contractor continues to operate in a prudent manner, is to audit the contractor's facilities.

A waste contractor audit follows closely a facility compliance audit. It used the same three phases (e.g., pre-audit planning). The compliance issues are the same and it places the same emphasis on document review, interviews, and inspections. However there are some key differences.

A waste contractor audit extends beyond compliance to liabilities (e.g., potential for damages to third parties and potential for remediation). A waste contractor audit also analyzes the financial strength of the company, which owns the facility; a company possessing sufficient assets and insurance probably can absorb the costs of remediation without having to pass them to generators.

Another difference between compliance and waste contractor audit is the manner in which the on-site interviews and inspections are conducted-a waste contractor audit typically provides little on going feedback to facility personnel. Unlike compliance audits, improvement is not the primary objective.

The objective is to identify risks, typically on a basis comparable with other facilities. Hence, a final difference between compliance and waste contractor audits is that the latter may tend to apply uniformly a detailed worksheet type of protocol in which the answers to a long checklist may possibly be scored.

The liability for the costs of remediating a contaminated property has surprised many landowners, tenants, and leading institutions as illustrated in the following real examples.

1. A purchaser of land finds that the property was once used for a landfill, the purchaser must pay huge amount to clean up ground water in the area.

2. A tenant moves from a site, which they had used for 6 years for an electronics manufacturing operation. Eight years later, ground water contamination is discovered. The owner successfully sues the tenant likely to have used a sufficient quality of the chemical known to have caused the contamination.

3. A bank lends money to a company. When the business started to have financial difficulties, the bank stepped in and ran part of it. The bank ultimately foreclosed. Subsequently, the property was found to be contaminated. The bank was found to be liable for a major part of the remediation. The cost of remediation was twenty-five times the amount of the original loan.

Because the value of property can be greatly diminished by the potential for having to remediate the site, the need to know the potential liabilities is crucial to both the buyer and seller of a property.

Waste minimization audits examine waste generated by a facility with the objective of identifying viable actions to reuse, recycle, or otherwise reduce the quantity and toxicity of each waste stream.

Waste minimization, is the reduction of any solid, liquid, or gaseous waste at its source of generation. It is a noble goal and certainly the preferred method for managing all waste. The first steps to successfully achieving the goal of waste minimization is an audit, consisting of planning the entire program and fully characterizing all wastes. From there, waste minimization options are identified evaluated, selected, implemented, monitored, and communicated.

24.4 LIABILITIES DEFINITION AUDIT

Another type of environmental audit is a liability definition audit. These are typically done for prospective buyers of real estate and for proposed mergers and acquisitions.

Such audits identify environmental problems that could reduce the value of a property or expose the buyer to liability.

24.5 IMPORTANT ASPECTS OF EFFECTIVE ENVIRONMENTAL AUDITING

A review of many auditing programs being conducted by both the public and the private sector has identified many common features deemed to be an important part on an effective auditing program.

- Explicit top management support for environmental auditing and commitment to follow-up audit findings;
- An environmental audit team separate from and independent of the persons and activities to be audited.
- Adequte team staffing and auditor training;
- Explicit audit program objectives, scope, resources and frequency
- A process which collects, analyzes, interprets and documents information sufficient to achieve audit objectives;
- A process which includes specific procedures to promptly prepare candid, clear and appropriate written reports on audit findings, corrective actions, and appropriate written reports
- On audit findings, corrective actions and schedules for implementation and
- A process which includes quality assurance procedures to ensure the accuracy and thoroughness of environmental audits
- These features show that conducting an audit is a process rather than a mechanical application of a rigid absolute methodology.

24.6 AUDIT PROTOCOL

The use of an audit protocol ensures consistency and

comprehensiveness in implementing the scope of the audit. It forms a record of the audit procedures, becomes a basis for development of the audit findings and also serves as a basis of any future critique of the audit. It is much more a data inventory form.

24.7 AUDIT PROCEDURE

The audit procedure includes broadly the following:
(i) Pre-audit activities
(ii) Activities at site and
(iii) Post-audit activities

24.8 PRE-AUDIT ACTIVITIES

(a) Preliminary Information

Pre-audit activities include various preparatory works. Having chosen the industry to be audited, preliminary information on the industry is to be obtained through a questionnaire.

The information includes
1. Location of the industry with surrounding land use
2. Climatic conditions
3. Products manufactured
4. Raw materials used
5. Details on water utilization, waste water generation and disposal, gaseous emissions, solid waste/hazardous waste,
6. Organizational set-up and policies of the company for Environmental Management.

(b) Audit Team

Audit team should be carefully selected to cover various aspects of the audit.

The team should include employees from production, quality control/laboratory, R & D, Pollution Control Operators, technical staff for monitoring and analysis of waste samples and an environment specialist.

The number of people may vary from 4 to 8 depending on the size and complexity of the facility being audited.

The team should be sufficiently detached to provide an independent view. The members should be so chosen that they would not hesitate to bring out even criticism, owing to obligations with supervisor.

Some times, it is advantageous to include members from the headquarters of the industry.

It is important to have well defined and systematic procedures, which are known and understood by all concerned.

24.9 STAGES OF ENVIRONMENTAL AUDIT

Regardless of the types of audit being performed, the whole programme consists of three stages:
1. Audit program planning (including pre-visit data collection)
2. On-site activities
3. Evaluation of audit data and reporting of findings.

24.10 PROGRAM PLANNING

A compliance audit program consists of many elements, including a visit to the plant. Prior to the actual visit, it is essential to plan the overall program.

Planning would include obtaining commitment by management, defining the objectives and requirements and developing an information management system.

It is also important to organize the audit team and make sure that the program has the resources and tools necessary to address the issues that need to be investigated.

In addition, one must consider the legal measures which may be appropriate to protect the results of the program.

In summary, the auditing program must be tailored to meet the needs of the facility being audited.

24.11 COMMITMENT BY MANAGEMENT

No complete audit program can be successful without the clear commitment of senior management. This commitment must come before the start of the process in two ways.

First, management must provide the program with the necessary resources and if necessary, direct plant personnel to provide access and to cooperate with the auditors during the process.

Second, it is critical that management articulates an explicit written commitment to follow-up an audit finding and correct the problems that the audit will uncover.

Undertaking an audit program without a commitment to cure the ills that it may uncover is probably worse than not conducting an audit at all. Not correcting such faults suggests to employees that compliance is not a high priority item.

Further, leaving them uncorrected, perhaps with statements in a file that they were known, is viewed by a regulatory agency as flagrant disregard for the environment and will likely result in severe penalties.

Finding a violation, correcting it, and reporting it to the appropriate agencies (i.e., self-correction) is viewed by agencies as one of the strongest benefits of a compliance audit program.

24.12 DEFINITION OF REQUIREMENTS

The process of performing an audit begins by deciding what type of audit is needed (i.e. is the goal to ensure compliance to minimize waste generation or to look for liabilities that might be there from past practices?)

After selecting the needed type of audit (in this case, a compliance audit), the auditor must define the scope of the audit, whether, are all areas of environmental management to be covered or only one specific part (e.g. air quality or hazardous waste)?

A compliance audit could cover all or just part of the various environmental management topics identified in Table 8.1. A comprehensive compliance audit would address all of these topics and perhaps related health, safety and transportation requirements.

After selecting the type of audit and its scope, other decisions need to be made. How will successful performance be measured over time? How frequently will the audits be done? How will the results be reported?

Each of these is important to evaluate prior to designing the actual audit process.

Defining these objectives in the beginning makes a major difference regarding the types of resources and techniques needed to conduct the actual audit.

The combination of tight schedules and difficult-to-define liabilities clearly reduces the certainty of the findings of a property transfer audit.

The decided advantage of having full knowledge of such liabilities during negotiations over the selling price will compel some parties to seek a thorough audit. The level of detail to which such an audit is conducted depends upon the characteristics and history of the site and time constraints set by the transaction schedule.

Underdeveloped or residential property obviously requires a lower level of scrutiny than industrial facilities. For the latter, a thorough audit will include the following issues:

1. Soil ground water contamination
2. Presence of hazardous waste and hazardous substances
3. Waste discharges to receiving waters
4. Proper permits and compliance
5. Presence of underground tanks
6. Special issues (e.g. asbestos, radon)

24.13 CONFIDENTIALITY

It may be appropriate to conduct an audit under legal arrangements to protect confidentiality. It is important to plan this carefully because such legal protections may withstand challenge only if they are built into the auditing process.

Why should one consider holding an audit confidential? The reasons go beyond the direct connection of finding violations.

For example, the audit report may describe proprietary business information. Also there may be potential for an adverse community reaction to sensitive information if not communicated clearly in an appropriate forum.

There is even the potential for self-incrimination if the disclosed violations carry criminal penalties. Many of these concerns will disappear by developing an effective strategy for correcting problems noted by the audit, Oral reporting will eliminate concern with disclosing highly sensitive issues.

24.14 ORGANIZATION OF AUDITING PROGRAM

After the objectives are defined for the audit program, the next step is to organize the program.

There are many questions to be answered: What is the frequency of reaudits, if any? How to schedule an audit of multiple sites? Will the audit program be conducted by environmental consultants? Or will it be done by the corporate staff as part of their normal jobs?

A third alternative is to have the audit managed by a small corporate oversight group with the audit function being conducted by staff from the operating divisions.

Each of these three groups (consultants, corporate staff, and operating staff) has advantages and disadvantages in its approach to various issues. Mature, successful auditing programs tend to integrate participation by all three groups.

24.15 TEAM OF AUDITING SPECIALISTS

Conducting a successful audit requires a team effort, directed by a team leader and supported by auditor specialists. This selection of the team has a major influence on the success of the audit.

The size and makeup of the team depend on the size of the facility and the complexity of the issues that must be addressed.

Auditors need to be "expert" in federal and state environmental regulations and sometimes local of country regulations.

They have to know the specific regulations that apply to the facility's particular operations and also possess knowledge of peer facilities.

For a manufacturing facility, the team needs to know production (i.e. how a facility operates, how its manufacturing processes function, where "things" could be hidden and how "things" are most likely to fail).

Auditors need to have knowledge of waste treatment and pollution control technology. The team needs to have a variety of scientific disciplines.

They need knowledge of information management systems and general management techniques for organizing staff and programs.

In summary, comprehensive audit team would offer specific experience in engineering regulatory analysis, regulatory enforcement chemistry, production and perhaps geology and community relations.

An audit could function with a team of just two experienced individuals, a regulatory specialist and an engineer. More complex facilities require additional team members.

For example, if specialized issues such as contamination of soil and ground water are to be examined, additional auditors may be required.

In addition to knowledge in specific technical areas, the auditing team should have both training and experience in conducting audits.

The auditors should be independent (i.e. free and capable of investigating the topic thoroughly and completely, probing appropriately through document reviews and interviews.)

Finally, the team members should have sensitivity, with the ability to communicate accurately and appropriately with different levels of personnel.

24.16 PRE-VISIT DATA COLLECTION

An audit usually requires facility staff to complete and return a pre-visit questionnaire that familiarizes the audit team with operations of the facility, its regulatory standing, and the environmental setting. Entering the facility without this information will result in wasting valuable on-site time instead of evaluating specific issues.

It is essential to have received a completed pre-visit questionnaire before finalizing the audit team. The responses will identify whether special expertise may be needed on the audit team (e.g. air specialists, hydrogeologists) as well as special equipment and materials.

What type of information is typically solicited by the pre-visit questionnaire? It is the critical information necessary to plan the audit including the following:

1. A management chart showing key personnel and their line and staff responsibilities.
2. A site layout showing the location of various operations, particularly the key items, which should be examined (e.g. waste storage areas, sewer locations and drainage systems, underground storage tanks and ground water monitoring wells).

3. A brief description and block diagram of operating processes and any pollution control units, showing recent or planned modifications. Block diagrams are very useful because they allow the audit team to anticipate the types of chemicals handled by the facility and the environmental emissions and wastes generated by the process. The team can preview pertinent environmental regulations and compile data on specific chemicals, appropriate maintenance programs and other information which will help appropriate maintenance programmes and other information which will help them to be more fully prepared for the on-site audit.

4. A list of all major types of waste generated and how each is managed.

5. A list of all environmental regulatory permits of the site, with copies. Permit information tells what type of reporting is required and what limits are in place. The lack of permits for certain activities would alert the auditor to potential omissions based on knowledge of similar operations.

6. All major environmental plans and policy manuals. For example, spill prevention and control plans provide relevant information about facility and the issues related to storage of waste and hazardous materials (e.g. inspection programs and structural and non-structural controls that should be in place to prevent spills).

7. Information about non-compliance history of any sort. This is particularly useful because it allows the auditor to concentrate on areas that have attracted regulatory attention and would likely receive close regulatory scrutiny in the future.

The comprehensiveness of the questionnaire should conform to the size of the facility, the sophistication of facility management, and the type of support provided by the management for the auditing efforts. In any event, the questionnaire should be of moderate length and directed only key factors.

Examples of key questions under selected headings other than hazardous waste are as follows:

24.17 GENERAL ISSUES

- What pollution control equipment is used at the plant?
- Have you been inspected by any regulatory agency in the last year?
- Have you been cited for non-compliance in the last year?
- What is the acreage of the facility?
- How long has the facility been in operation?
- What is the past use of the site?

24.18 HAZARDOUS MATERIALS

- What is the latest hazardous materials inventory?
- What quantity of petroleum products are stored on site?
- Do any underground tanks exist on site?
- Where are waste water discharge locations (point and non-point)?
- What permits are in place?
- Has there been any correspondence with the local sewer agency of the water board concerning compliance within the last two years?

24.19 ONSITE ACTIVITIES

Following the acceptance of the site agenda by the facility, the team members visit the site. The primary purpose of the visit is to review documents maintained at the facility, to interview facility employees, and to inspect thoroughly all-relevant operations conducted at the facility. This is an intensive and thorough period of collecting data.

The activities at site include

1. deriving material balance
2. identifying waste flow lines
3. monitoring of characteristics
4. evaluation of the performance of process and pollution control equipment/system
5. assessing environmental quality
6. holding discussions with the management and finally preparing the draft report.

348

24.20 TYPICAL AUDIT SCHEDULE

Day 1-Monday 9.00-10.00 AM
Meet with host Facility Manager and Environmental Manager to make introductions and review purpose of the audit. Refine schedules for interviews, tours, and special requirements. Receive overview presentation of facility operations.

10.00-11.30 AM
General tour of the facility to provide Audit Team orientation of overall facility. This is a brief "ride through" of key areas of the facility.

11.30-12.00 AM 1.00-5.00 PM
Conducting briefing for Facility Manager

Day 2-Tuesday 8.00-5.00 PM
Conduct audit of fuel/oil storage and soil response compliance. Physically inspect oil storage area, tank farms, and drum storage areas. Review oil spill contingency plans, spill reports, and spill prevention and staff control plans.

Day 3-Wednesday ---- Conduct interviews with key staff.

Day 4-Thursday 8.00-2.00 PM
Conduct audit of fuel! oil storage and spill response compliance. Physically inspect oil storage areas, tank farms, and drum storage areas. Review oil spill contingency plans, spill reports, and spill prevention and Control plans. Conduct interviews with key staff

Day 5-Friday 8.00 -10.00 AM
Audit Team makes an informal briefing to key staff on preliminary findings of audit.

11.00-12.00 PM
Conduct formal exit interview with Facility Manager.

The audit may take 3-10 days depending on the industry. Effectiveness of audit is a direct result of the qualification, confidence, training and proficiency of the personnel who conduct audits.

The team should understand regulatory requirements, relevant waste control technologies and their operations and process.

They should have capability to examine, question, sample and analyze waste and interpret data.

The management should be provided with a realistic assessment of environment performance, R & D environment management etc., so as to understand different operational mechanisms.

Having a fair idea on the manufacturing process, reconnaissance surveys should be made to be familiar with layout of the plant and process operation, and to understand possible impact on the surrounding environment.

Various activities to be carried out at site are discussed in detail in the following sections.

24.21 MATERIAL BALANCE

The entire manufacturing process of each product should be drawn into a process flow sheet representing various unit operations as blocks.

The quantities of inputs and outputs of each unit operation should be worked out for the entire process and data incorporated in the process flow sheet.

A comparison of these requirements with the actually used raw materials in the industry gives an indication whether they are used in excess of the requirements or used as per the requirements.

If the raw materials are used in excess, it may be presumed to be finding their way to air, water and soil thus causing pollution. Hence, it is important to reduce these excesses.

The unit operation should be checked up, to find out the cause of excess usage of the materials and accordingly modifications made.

Norms should then be fixed for performance of each of the unit operations, for wastes generated from each of these unit operations, the production and environment staff is simply to adhere to the norms.

The environment manager thus can have a control over production as well as wastes generation too.

24.22 EVALUATION OF WASTE TREATMENT FACILITIES

Performance of various pretreatment and final treatment facilities should be evaluated based on the analysis reports.

If the treated waste water, gaseous emissions and solid waste do not conform to the standards prescribed by the Pollution Control Board, reasons for the same should be diagnosed.

From the individual streams of waste water, recyclable and recoverable materials should be identified and provisions made for the same.

All the 'avoidable' wastes should be completely controlled and only the 'unavoidable' allowed for treatment.

Discussions with the staff, peruse of the records of the company and the reconnaissance survey will help in arriving at these flow sheets.

From these flow sheets, data sheets incorporating the raw material requirement, water consumption, waste and solid waste generation, and gaseous emissions should be worked out for each product manufactured.

24.23 WASTE FLOW

From water balance, the sources and quantities of generation of waste water, emissions and solid waste should be identified.

The waste pretreatment, final treatment and disposal path should be identified. The production staff should be consulted as these people are likely to know about waste discharge points and about unplanned waste generations such as spills, leaks, washings, other discharge points due to overflows, spills and other materials handling practices and sources should be accordingly finalised and waste flow sheet is prepared.

24.24 FIELD OBSERVATIONS

The entire plant should be inspected thoroughly.

The aspects of site layout, material handling and storage drainage system, safety aspects, lapses/negligence in operations and attitude of operators in process and waste treatment facilities handling of scrap and wastes, usage of sign boards, instruction colour codes etc., should be observed.

The attitude and technical capability of various staff including senior management should be observed as it is very critical in achieving the goal of safer environment. The training requirements can be assessed based on these observations.

24.25 MONITORING

The characteristics of the wastes as generated from the sources are important to understand its use for recycle, recovery or treatment.

The performance of the treatment facilities is to be monitored so as to check their efficiencies and to modify or install additional equipment facility, if necessary.

24.26 PREPARATION OF DATA SAMPLING

Proper environmental management results in too many records for an auditor to check all of them. The auditor, therefore, must sample from the whole in a representative manner.

The sample size and the sample itself can have a statistical basis or be based on the auditor's judgement. The method depends upon the objective of the protocol.

If the objective is to define the percentage of a population, which is not in compliance, the sample selection must be made in a statistically valid manner reflecting the potential for bias to enter the sampling process.

For example, if the population is distributed randomly, a random number table can be used. However, if the population is stratified (e.g., a facility has three separate operations each of, which files manifests, the sampling should be stratified accordingly).

If the objective is to discover any violation, the sampling may be concentrated away from subgroups expected to be in compliance. In any event, the auditor should collect the sample and not depend upon the facility to do it. The sampling process and the sample must of course be thoroughly documented.

The surrounding environment-groundwater, stream, soil surrounding and uses (residential agricultural etc.,) and ambient air quality should be monitored to determine the impact due to the industry.

With the above objectives, sampling points should identified and monitoring network established.

Parameters to be analyzed should be determined from material balances of the waste generated.

The frequency of sampling should be fixed so as to cover hourly and daily variation and characteristics.

Generally, two types of samples can be collected.
1. the grab sample and
2. the composite sample

The grab sample shows only the prevailing conditions at the time of sampling and cannot represent the average conditions.

However, grab samples are useful in determining the effects of extreme conditions of the waste during the time composite samples are being collected or when the waste water flow is intermittent.

24.27 EVALUATION OF AUDIT DATA AND PREPARATION OF AUDIT REPORT

After departing the facility, it is important to develop the audit report quickly. In developing the report and report format, style is important. The following guidelines should apply to any audit report.

- Be clear and concise
- Be sure to distinguish between isolated incidents and chronic problems
- State the facts as discovered
- Do not draw-unsubstantiated conclusion, limit findings to the facts
- State the nature of the problem clearly and exactly without using generalities.
- Give regulatory or management practice citations, very clearly citing where requirements have not been met. However be very careful not to draw legal conclusions.

In particular, do not state that there is a violation.

Two examples of an acceptable writing style are as follows:
1. A written waste analysis plan is required for all hazardous

wastes treated on site or stored on site. The facilities plan does not include the following two requirements listed in the regulation.

(a) Rationale for selection of parameters to be analyzed-parameters are listed.

(b) Frequency of review and repeat analysis-none is specified.

2. In reviewing a random sample of pH monitoring records for 15 days of the previous year for discharge # 1, three excursions were identified, which exceeded permit limits. We were not able to confirm that these incidents were properly reported.

An audit is an improvement tool. Therefore, the audit team should present recommendations in addition to a set of findings. It may be appropriate to present recommendations is a separate report so that they could be protected under the attorney work product privilege.

It is helpful to present any recommendations in a 'value-added format'–one that will provide the foundation for implementing corrective actions and ultimately designing an efficient, manageable facility compliance system. For each negative finding the report should provide.

- A description of the specific requirements
- Potential liabilities associated with the negative finding and
- A recommended corrective action

Although the recommendations should clearly outline the type of action needed to resolve any negative findings, they should not represent complete work plans.

24.28 DRAFT REPORT

After completing the above mentioned activities including determining material balance, identifying waste flow monitoring and analysis of various samples and field observations, a draft report should be prepared with findings and possible recommendations.

The draft report should be presented before the senior management and all concerned aspects should thoroughly be discussed.

The Management should put forward their views. The participation of the Management and their acceptance of various observations and recommendations make the task of implementation meaningful.

24.29 POST AUDIT ACTIVITIES

(a) Synthesis of Data

The requirement of various raw materials according to the mass balance of chemical equation involved in the manufacture of a product is called stoichiometric requirement.

(b) Final Report

Various aspects discussed above should be compiled and a final report prepared along with recommendations.

Action Plan

It can be seen that an audit program helps to troubleshoot potential violations of regulations and excursions from good management practices. However, without an action plan to address the problems and its cause the next audit may find the same shortcoming.

Therefore, after an audit report is completed the facility (not the audit team) must develop an action plan–a written response to the audit that specifies a clear set of actions and tasks with deadlines. It also clearly identifies the person responsible for implementing each action. A system of tracking deadlines to see that they are met should be established. Follow-through on audit findings is very important.

If a regulatory agency discovers that a facility has conducted an audit and did not attempt to resolve the findings, the agencies response will probably be much harsher than if the audit had never taken place.

The final report may, if necessary, be sent to the top management for comments so as to make further modifications.

24.30 FOLLOW-UP ACTIONS

Follow-up actions should be taken to check the progress of implementation of recommendations.

The Environment Division of the Industry should meet the other divisional heads periodically to review the progress.

24.31 CONCLUSION

The EA is considered for broad purpose i.e., to provide an indication to company management of how well environmental organization systems and equipments are performing and apply the best practicable means to preserve air, water, soil, plant and animal life from adverse effects of this operations and to minimize any nuisance that may arise.

EA should not be undertaken simply to facilitate compliance with law. It should be seen as a means to accomplish much more in as much as they provide an in-depth review of process in realizing long-term strategic goals.

Industry can be benefited from a critical self-examination of the purposes and technologies. It employs to see in, which area problems might arise, particularly with regards to human health. EA is clearly of that self-examination.

QUESTIONS

1. What are the types of environmental audits practised?
2. Explain in detail the steps involved in an environmental audit.

25 Waste Minimization and Clean Technologies

25.0 INTRODUCTION

Before making high investment in an external treatment, one has always to evaluate the possibility to reduce the emissions or discharge from the production or manufacturing unit. With less emissions or discharges from the production unit, the required external treatment would be less. In the extreme case, the internal measures can be alternative to external treatment. From an economic point of view, the optimum is often to have a combination of internal and external measures. When making such an evaluation one has to take into account that the internal measures often lead to reduced operating costs in the mill like raw material savings, water savings, etc.

The internal process evaluation should start with specifying the emission or discharge. The steps adopted are sources and the discharge points within the factory system (a) The discharge from storm water and low contaminated water can normally be sewered directly to the receiving waters (b) recycling of chemicals and process water leading to less effluent flow and less discharge of pollutants (c) process alteration by change of raw materials and chemicals which can reduce the amount of pollutants achieving increased yield which corresponds to less discharges (d) adopting process water saving measures change of cleaning procedures of process equipment, internal treatment of some separated effluent stream, which can result in process water and raw material recycling (e) improved operational routines like cleaning routines and prevention of accidental discharges and use of environmental control systems approach like use of simple control parameters (pH, conductivity, turbidity, temperature), alarm at overflow positions, spill collection systems and rapid feed back to operators and (f) by imparting environmental education of effluent treatment operators and other mill operators who have to be aware of environment and of the importance of spills. Industry has to realize that in any process, any material being discharged in a waste stream constitutes a raw material of negative cost. It is a sound business practice to recover these materials at economic costs, a situation that can be brought about by the input of right technologies. To start with it is necessary to conduct an in-plant survey.

25.1 IN-PLANT SURVEY

No two industries are alike as the processes and raw materials used are different. In the same type of industry, two different factories at two different places can discharge waste water of different composition. There will be diurnal, daily and seasonal variations in the quality and quantity of waste water discharged from an industry. It is therefore necessary to make flow measurements to determine peak, minimum and average flows and to find out the composition of the waste water from individual section and the combined final discharge.

Before proceeding to carry out flow measurement, information on (a) water using operations, water and material balance (b) sources and quantities of waste water generated and nature of pollutants in them (c) water and waste water quality (d) efficiency of existing treatment plant, if any (e) ultimate disposal of waste water on land, into sewer, into a stream or into a marine environment (f) raw materials/products and (g) site plan, plant layout, drainage map, process flow sheet, and building plan, have to be obtained.

25.2 FLOW MEASUREMENT

25.2.1. Flow Rates Measurement Based on Water Level Changes

There are two modes of waste water transport and two corresponding ways to measure the quantity of flow. These are (a)

Open channels, with the liquid surface exposed to the atmosphere. Here the quantity of flow (Q) is a function of the water depth Q = f(n) (b) Closed system, where the liquid is in a pipe and the pipe is full. Here the quantity of flow depends on the flow velocity Q = f(v).

The most commonly used equipment are wiers or parshall flumes which are placed in a channel as a constriction or dam causing changed level and velocity of the water.

Wiers are made of steel plates, planed wood or plastic. A wier is a constriction placed in a channel over which the water has a free fall. The flow is calculated from the geometry of the wier and the water level upstream. There are wiers without end-contraction, wier with end-contraction and Thompson Wier (V-notched). The edge of the wier plates in contact with the flowing medium is sharp and cut as to form an angle of 45° in the direction of flow. Wiers are mounted accurately at right angles to direction of flow and with the upper edge horizontal (spirit level).

The choice of suitable wier is made on the basis of channel width, the accessible dam height and the range within which the flow is expected to vary. Designs resultings in submerged flow situations should be avoided. A constriction in a channel cause an increase in the water level. For a suitably shaped constriction, the flow rate can be obtained with sufficient accuracy by measuring the level upstream the constriction.

Parshall flume is a convenient device for measuring flow in sewers. It is a modified venturi flume of standard dimensions. It consists of three parts, a converging section, a throat section and a diverging section. Future increase in flow should be given attention while designing a flume. The level of the floor of the converging section is higher than the floor in the throat and diverging sections. The head of the water surface in the converging section measures the flow through the flumes.

The elevation of the water surface should be measured back from the crest of the flume at a distance equal to 2/3 or the length of the converging section. The crest is located at the junction of the throat and converging section. The head should be measured in a stilling well instead in the flume itself as sudden changes in flow are dampened in stilling well. The size of the parshall flume should be determined during the preliminary survey. The general formula for computing the free discharge from a parshall flume is

$$Q = 4WH^N$$

Where Q = discharge, CfS
          W = width of throat in ft.
          H = head of water above the level floor in ft in the covering section
          N = $1.522W^{0.026}$

The flume may be built of wood, fiberglass, concrete, plastic or metal and can be installed at convenient locations such as a manhole. The parshall flume is used for sewer lines where continuous-flow measurements are desirable. The main advantage of the parshall flume over a wier is the self cleaning property of the flume. It is better to avoid submergence situations as far as possible by proper design.

Formula for parshall flume flow in m^3/s

$$Q = 0.1132 \times b \times (3.28 \times h^1)m \ m^3/s$$

Where m = $1.522 \times b \ 0.026$
          b = throat width (m)
          h' = water level at H before throat (m).

Several types of gauges can be used to register water level based on pressure, distance, time and capacitance. Generally gauges based on pressure are recommended for permanent installation. Venturi meter installed in a pipe line, consists of a throat of known diameter, converging from throat to pipe.

25.2.2. Flow Rate Measured Based on Velocity

If the flow area is known, the flow rate can be calculated from the velocity of flow, as measured with a velocity indicating instrument. Since the velocity varies over a cross-section of the flow, it is necessary to know the approximate velocity distribution across the section. From the velocity distribution the mean velocity is calculated. Velocity can be measured with a Pitot tube which measures the difference between the total and the static pressure in the system. The pitot tube is connected to a differential manometer.

25.2.3. Flow Rate Measurement Based on Dilution

When a salt solution or tracer dye is injected, the time for it to reach a given point or to pass between two given points is measured, the flow can be calculated from the result.

A concentrated solution of a substance (inorganic salt or dye tracer) that can easily be determined in low ion concentrations is injected into the medium at a constant known rate. The concentration of tracer is then measured and the flow is calculated from the formula,

$$Q = q \ (C_1)/(C_2 - C_3)$$

where Q = flow in check point, l/mm

q = injected flow, l/min

C_2 = concentration of tracer element in check point, mg/l

C_1 = concentation of tracer element in injected flow, mg/l

C_3 = concentration of tracer element in zero sample in check point in mg/l.

Good tracer elements are Lithium and K-salts. LiCI is commonly used for this type of investigations with good results. In this case the resulting concentration in check point should preferably be above 1 mg/l during a sampling time of about 10 min.

25.2.4. Flow Rate Measurement in Closed Systems

Several methods have been developed for flow rate measurements with a good degree of accuracy in closed systems. The methods may be divided into 3 main categories (a) measurement of a pressure difference (b) determination of the velocity and area of flow of the liquid and (c) methods based on dilution. The most common types of measuring devices used for continuous flow measurement in closed system are orifice plates, flow nozzles and entire tubes.

Installation of an orifice plate results in an increase in the speed of the flowing medium and a corresponding drop in the static pressure. The pressure difference is measured by manometers. A standardized orifice plate consists of thin plate with a central, sharp-edged hole which is mounted perpendicular to the direction of flow. The flow rate is highest and the concentration greatest just below the orifice. Orifice plates give a high level of accuracy but at the expense of a large pressure drop which increases the pumping costs. They do not function satisfactorily in the presence of suspended particulate and are therefore unsuitable for permanent installation in systems with fibres.

The liquid flow in a pipe can also be measured by a magnetic flow meter which consists of a straight pipe section fitted with standard flanges. An electric coil around the tube imposes a magnetic field which changes with the velocity of the liquid. Waste water flowing through a magnetic field produces a field voltage in proportion to velocity which is converted by electrical and mechanical means to indicate and record the flow. The supersonic meter is based on the measurement of velocity according to Doppler principle i.e. the frequency of sound wave is changed by reflection on air bubbles or particles in the fluid.

The dilution methods (salt solutions, chemical and dye tracer) can be applied to flow rate measurements in closed systems as described in earlier section.

25.2.5 *Indirect Measuring Methods for Flow Rate*

When, as is often the case, the flow cannot be measured directly, usually because the system is inaccessible for the installation of measuring equipment, other methods are resorted to.

If the system includes vats or containers, where the level can be measured, the flow can be calculated from the rate at which the level rises on closing the outlet value. Where feasibility exists the effluent may be diverted to a suitable container and the time required to fill the container can be measured. If there is a pump for pumping waste water, the pumping capacity and running time can give an idea of the quantity of waste water.

Most of the industries have water meters. If water consumption is known, waste water generated can be estimated by multiplying water consumption by 0.9.

By flow measurement it is necessary to measure peak flow, minimum flow and average flow.

25.3 COMPOSITION OF WASTE WATER GENERATED

The composition of the waste water generated in an industry varies hourly, daily and seasonally, Samples collected for analysis should represent the discharge of the waste water in the best possible way.

The following criteria should be satisfied while setting sampling points (a) It should be possible at that sampling point to determine the flow using either direct or indirect means (b) The effluent should have a homogeneous composition to ensure that the sample is representative and (c) the final discharge sampling point should be carefully selected and used for cross-checking the in-plant monitoring results.

At location having a relatively constant flow rate and composition, flow measurements and sampling can be done. Point of sampling should have good mixing conditions and represent total flow.

The sample collected from sampling point must reflect the variations that occur during the sampling period. At the final discharge point, the sample should preferably be collected automatically proportional to flow. At other sampling points, grab sampling methods are employed to get a composite sample.

Grab or composite or isokinetic samples form channels should be collected at some distance from the bottom in order to prevent sediment practicals from entering the sample in unrepresented quantities. Special care has to be exercised in sampling if the fluent contains dense suspended particles.

In the case of substantial variation in the flow rate, it is necessary that composite samples correspond to flow. Composite 8 hourly, 12 hourly or daily sample can be collected at regular time intervals if the flow rate and compositions are relatively constant. The volume of each sample should be large enough to ensure that a representative sample was obtained. The volume of the composite sample collected daily should be in the range 2-4 liters.

Samples should be handled in such a way that the sample characteristics are not distorted before analysis. Precautions to be taken include (a) the sampling bottles should be clean (b) samples should be analyzed as quickly as possible (c) the temperature and pH of the sample should be recorded immediately after sampling (d) samples might require refrigeration during the sampling period in order to minimize bacterial growth (e) samples should be mixed carefully before a portion is withdrawn for analysis and (f) samples that cannot be analyzed within an appropriate time should be refrigerated or frozen or preserved carefully.

Automatic samplers of different types are available in the market. Important points to be taken into consideration are representative sampling, frequency of sampling, accessibility to inspection and maintenance demands. Samples should be preserved properly and transported carefully to the laboratory for analysis.

25.4 ANALYTICAL METHODS RECOMMENDED FOR CHARACTERIZATION

All the parameters should be evaluated in accordance with well established standard analytical methods. In order to harmonize the results, procedure for each parameter suggested in Standard Methods for examination of water and waste water, American Public Health Association, Water Pollution Control Federation, has to be followed.

Portable test equipments are commercially available for spot analysis of many constituents. It is important that all field instruments be checked and recalibrated in the laboratory at frequent intervals.

The presentation of all experimental and evaluated results should be on a common from for easy understanding and comparision. It is necessary to check results by cross checking values. This will establish the reliability of the processes. Cross checks can be made on the basis of expected values, mass balances or technical data from own or others experience.

Types of discharges from an industry include (a) process related discharges-constant in time and directly related to the process and raw materials used (b) design related discharges-variable in time and used by errors in process and/or equipment design (c) break-down discharges-variable in time and caused by machinery malfunction and (d) operation-related discharges-variable in time and caused by human errors in coordination and operation of the industry.

The objectives of environmental monitoring program include (a) compliance with permits and regulations (b) control the efficiently of pollution abatement installations and (c) give information about environmental impacts, about variation of discharges throughout the operating time and about the relation between emissions and production. From in-plant survey data, pollution load calculations can be made. Efficiently of existing treatment plant, if any has to be studied. Based on all the data collected, treatment flow sheet has to be developed. Next steps include (a) functional design or unit sizes (b) detailed engineering drawings (c) construction (d) commissioning and (e) operation and maintenance manuals.

25.5 WASTE VOLUME AND STRENGTH REDUCTION

Volume of waste generated by an industry can be reduced by (a) segregation (b) conservation (c) reuse, recycle (d) in-plant control measures and (e) house-keeping.

25.6 SEGRGATION

It is cheaper to treat low volume of concentrated waste water than large volume of dilute waste waters. Cooling water is generally free from pollution. Waste water from process, cooling boiler blow down, sanitary waste, waste water from canteen and storm water should be segregated and treated separately. In the process, waste waters from some sections are stronger than those from the other sections. All these waste waters have to be sewered separately, if necessary.

25.7 CONSERVATION OF WATER

There is lot of scope for conservation of water in many industries like tanneries, textiles, paper mills, etc. In many industries, in summer the consumption of water is less. In water scarce areas the amount of water used is less per unit of product than in industries in other area. It is therefore possible to reduce water consumption in many industries without affecting the quality of the product. Important steps include (a) prevention of running taps, leaks, spills (b) alarm at overflow positions (c) spill collection systems (d) preventive maintenance and (e) no overloading. Modification of equipment and process automation has in many cases minimized operational errors, reduced spills and reduced waste generation.

25.8 REUSE AND RECYCLE OF WATER

First preference is reuse of waste water without treatment like reuse of textile mill wash waters. Second preference is reuse of waste water after partial treatment like reuse of paper machine waste water. Third preference is reuse of waste water after complete treatment. Improved operation routines like cleaning routines, analysis and prevention of spills (accidental discharges) and internal treatment of some separated effluent streams which can result in process water closure (recycling) and raw material recycling are important. Reclaiming water from sewage is being practiced using tertiary treatment methods in many countries. In many countries water reclaimed from sewage is being used in industries for various purposes like cooling, washing, etc.

25.9 IN-PLANT CONTROL MEASURES

In a sugar mill, for example, cooling oil used in roller mills for tandem cooling can be collected in a sump filled with bagasse, solid waste from the same industry. Bagasse absorbs large quantities of cooling oil which can later be used in boilers. The overflow from evaporators can be collected in a sump and recycled to clarification section. Press mud can be used as a soil conditioner and should not be allowed into drains. Proper storage of molasses is very important. Molasses spills have high BOD and COD.

In a pulp and paper mill, leakages and spillage of black liquor can be collected in a sump and pumped to soda recovery section. Chemical and process water recycling will result in less effluent flow and less discharge of pollutants.

25.10 HOUSE-KEEPING

Preventive maintenance, prevention of leaks and spills, cleaning schedule and clean environment are important house-keeping measures required to reduce the volume of waste water generated. Good house-keeping practices are very important and involve alteration of existing system to limit unnecessary generation of wastes attributable to human intervention.

25.11 WASTE STRENGTH REDUCTION

Strength of waste water generated from an industry can be reduced by (a) equipment change (b) process change (c) equalization and proportioning and (d) by-product recovery.

25.12 EQUIPMENT CHANGE

In tanneries, e.g. if drums and paddles are used instead of pits, the load of pollutants generated is decreased because of less chemical and water use. Save-alls recover fibers from waste water of paper machine which can be refused in the industry. Smoothnecked containers for collection of milk reduces milk lost during transfer into milk storage tanks in dairies. In textile industries, when Jiggers are used instead of intermittent Kiering, pollution load is reduced.

25.13 PROCESS CHANGES

Changes in raw materials and process may result in less pollutants. Increased yield will reduce pollutant discharged. More efficient is the process, less will be pollutants discharged. In a tannery, e.g., enzyme dehairing methods instead of use of lime and sulfide for dehairing, improvement of uptake of chromium in chrome tanning, refrigeration instead of using salt for drying of skins and hides will reduce the quantity of pollutants discharged. In pulp and paper mill, oxygen delignification process instead of Kraft process reduces the concentration of pollutants in the final effluent discharge. Sodium anhydride if used for pickling of metals instead of acids like sulfuric acid or HCI, pollutant discharge is reduced. Mercury discharge is eliminated if membrane technology is used in chlor-alkali industry.

A product can be manufactured by 2 or more distinct processes. For example, chloride process for production of titanium oxide generates considerably less waste than alternate sulfate process. Improving the efficiency of the process through modification of catalysts, reactor design and operating parameters reduces the quantity of waste water generated. Low waste or no-waste or clean technologies are now available.

25.14 EQUALIZATION AND PROPORTIONING

Equalization of waste water from different section may result in mutual neutralization and precipitation of pollutants. In a tannery, e.g. 40-60% of the pollutants are precipitated in the equalization tank itself.

Proportioning involves mixing industrial waste water with sewage in different proportions and then treating the mixed waste water in a combined treatment unit. Combined treatment of waste waters from a group of same or different industries is being practiced in many countries with success.

25.15 BY-PRODUCT RECOVERY

By-product recovery from the waste water reduces the strength of waste water to be treated. Fiber recovery, soda recovery and utilization of lime-mud are possible from pulp and paper industry waste waters and are being practiced in many countries. Form viscose rayon industry waste water it is possible to recover zinc, lignin sulfonate (sulfite process), furfural (from prehydorlysate waste). Food yeast can be grown in prehydorlysate waste water from viscose rayon pulp industry.

In a tannery, $33\frac{1}{3}\%$ by weight of skin or hide is solid waste. This solid waste can be used for glue, gelatin, carpets and leather board manufacture. In a textile industry, alkali can be recovered from pickling wastes. Phenol, ammonia, benzene and xylene can be recovered from ammonical liquor discharged from coke ovens. From molasses distillery waste waters, potash can be recovered. Compost can be made using spent wash and biogas can be obtained. Cattle feed can be made by multiple effect evaporation.

Mineral oil can be recovered from petroleum refinery wastes. From soap stock, fatty acids and glycerol can be recovered. By electrowinning process and ion-exchange, metals can be recovered from plating shops.

Recycling technologies include (a) distillation of solvent wastes, (b) de-chlorination of halogenated, non-solvent wastes and (c) metal concentration techniques such as ion exchange, evaporation, solvent extraction.

Some of the factors encouraging waste water minimization by the generator are shown in Table 1.

25.16 WATER CONSERVATION

Earth is a watery-planet, 71% of earth is water. Out of this water, 95% is in seas, 4% in ice-caps and snow. Only 1% of the total water (1,500 million km^3) is available as surface and ground water. Out of this only 0.2% is available for use. Where water is needed, water is not available. Therefore water resource management is becoming important.

Demand for water is on the increase. Limited capacity of underground water tables and ever growing pollution of surface water is causing the problem of scarcity of water. It is practically impossible to find a substitute for water.

By 2025, with the global population growing by 3.5 billion, some experts predict that over 90 countries in the world will be experiencing water shortage. As Sandra Postel, author of "The Last Oasis: Facing Water Scarcity" aptly puts it "Like oil, water is a strategic resource for which nations will compete fiercely as it becomes more scarce". The fear is that population growth and escalating per capita demand may severely squeeze the water resources. This will result in more water scarce areas if nothing is done to extend water supply which engineers feel, has limits, both economically and economically.

TABLE 25.1. Some Factors Encouraging Waste Minimization by the Generator.

Technical	Legislation/Policy
New processes available	Bans on specified wastes or raw materials
New chemicals available	Limits on waste production
New plant installed	Compulsory waste audits
Improved product design	Waste minimization is creiterion in plan
New raw materials	permits
Operational	**Management**
Regular maintenance	Waste policy adopted
Trained operators	Staff incentives
Printed company directives	Operational directives
Area set aside for collection,	Waste audit procedures
recovery	Positive publicity
Avoid over-ordering	Public scrutiny
Storage areas kept safe	Regular Manitoring
Disposal	**Economic**
lack of disposal sites	High disposal fees
Pre-treatment by generator	High dumping fines
required by authorities	High chemical costs
supplier obliged to accept	Incentives for new plant
return of surplus stock	

Estimates show some 300 million people in the world are currently living in regions of water scarcity.

Water has been cheap too long and it has never figured in the cost of production in any manufacturing industry. There is a need to establish optimum water requirements for each stage of processing in an industry and to set target figures at which to arrive. Minimum water quality requirements for each process have to be assessed and water balance calculations have to be made.

PREVENTION IS BETTER THAN CURE
↓   ↓

Cleaner Production   Remediation
Cleaner Technology   Clean-up Technology
Life Cycle Assessment

25.16.1 *Cleaner Production: (Pollution Prevention)*

A conceptual or procedural approach to production that demands that all phases of the life cycle of a product or of a process should be addressed with the objective of prevention or minimisation of short and long term risks to human health and environment.

25.16.2 *Cleaner Technology : (Pollution Prevention Technology or Service)*

Refers to avoiding environmental damage at source through use of materials, processes or practices to eliminate or reduce the creation of pollutants or wastes.

25.16.3 *Life Cycle Assessment (LCA)*

Environmental Life Cycle Assessment is a formal approach to defining and evaluating the total environmental load (Environmental impact and resource depletion as an aggregate) associated with a product or process or providing a benefit or service, by following the associated material and energy flows from their 'Cradle' (Primary Resources) to their 'Grave' (Ultimate Resting Place).

LCA Serves as a powerful decision support tool in determining whether a product or service genuinely reduces environmental load or whether the load is merely transferred from one source to other.

25.17. REMEDIATION

Repairing damage caused by past human activity or natural disasters.

CLEAN-UP TECHNOLOGY: (END OF PIPE) refers to reducing environmental damage by retrofitting, modifying or adding "end of pipe" pollution abatement measures to an, established plant or process.

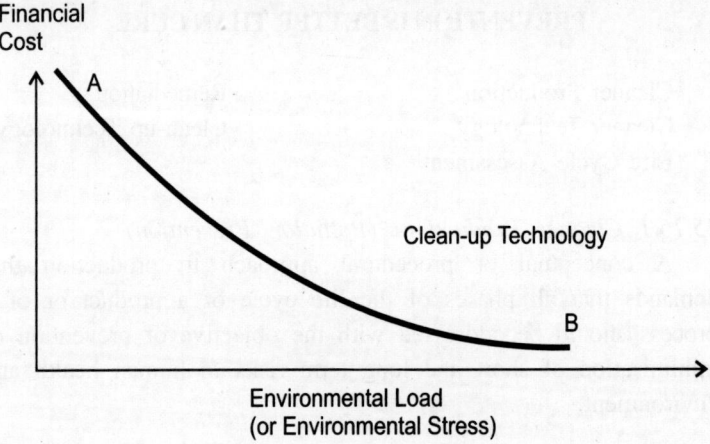

Figure 25.1: Clean-up Technology and the Efficiency of Environmental Improvements.

The asymptotic behaviour of the curve at low environmental load indicates that any human activity involves some environmental impact or resource utilization, so that environmental load cannot be completely eliminated.

Similarly, the asymptotic behaviour at low environmental cost indicates that even the most environmentally inefficient activity has a financial cost.

Clean-up technology involves increasing the financial cost to reduce environmental load. Point 'A' corresponds advanced countries.

Point 'B' corresponds to developing I underdeveloped countries where the environmental load is higher as well as the efficiency of expenditure on environmental improvements is also higher as evident from the gradient of the curve.

25.18 DISTINCTION BETWEEN CLEAN-UP TECHNOLOGY AND CLEANER TECHNOLOGY

Curve 1 represents an established technology to which clean-up can be applied.

Curve 2 represents a different technology which is cleaner.

As a typical example, an organisation operating at point 'C' has three options.

(i) It can adopt cleaner technology of curve 2 and can reduce environmental load without increase in cost (D).

(ii) It can adopt Curve 2 and reduce cost while retaining the environmental load (E)

(iii) It can adopt curve 2 and reduce both environmental load and cost. (F)

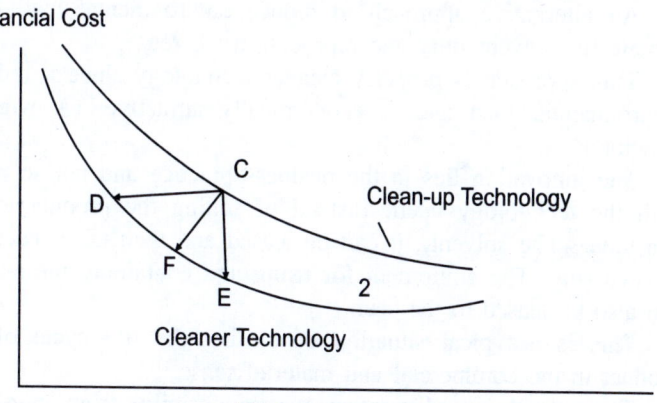

Figure 25.2: Distinction between Clean-up and Cleaner Technologies.

25.19 THE ADOPTABILITY OF CURVE 2 BY A PARTICULAR ORGANISATION DEPENDS ON

(i) Whether the cleaner technology is available

(ii) The organisational constraints in terms of the existing investment in technology 1 and the investment needed to exploit the technology 2.

A number of existing organisations can achieve improved

environmental performance in a short time by resorting to improvements to existing products and processes than adopting cleaner technology. (Cleaner Production and waste minimisation)

25.20 DISTINCTION BETWEEN PROVIDING A SERVICE OR BENEFIT RATHER THAN SUPPLYING A PRODUCT.

EXAMPLE

Traditionally, Organic solvents (Chlorinated) are used which result in substantial environmental load. New development to this is to use water-based solvents.

The environmental load arises not because of the usage of organic solvents, but from their release or escapee.

An alternative approach to reduce environmental load is to contain the solvent fully and reprocess it for reuse.

This approach is properly cleaner technology since it reduces environmental load and is economically attractive. (A win-win principle)

The innovation lies in the business practice and not so much with the technology itself. Instead of selling the product to the consumer (The solvent), it can be leased and then taken back for reprocessing. The equipment for using and containing the solvent can also be leased to the user.

This is a typical situation of closing the life cycle of the product in the commercial and material sense.

Typically, the supplier retains the responsibility from 'cradle' to 'grave'. A typical shift to a new service economy, since the service is traded and not the commodity.

Pollution Prevention Pays (3p)

Pollutants

+

Know How

↓

Potential resources

+

Profits

Hierarchy of Waste Management Options

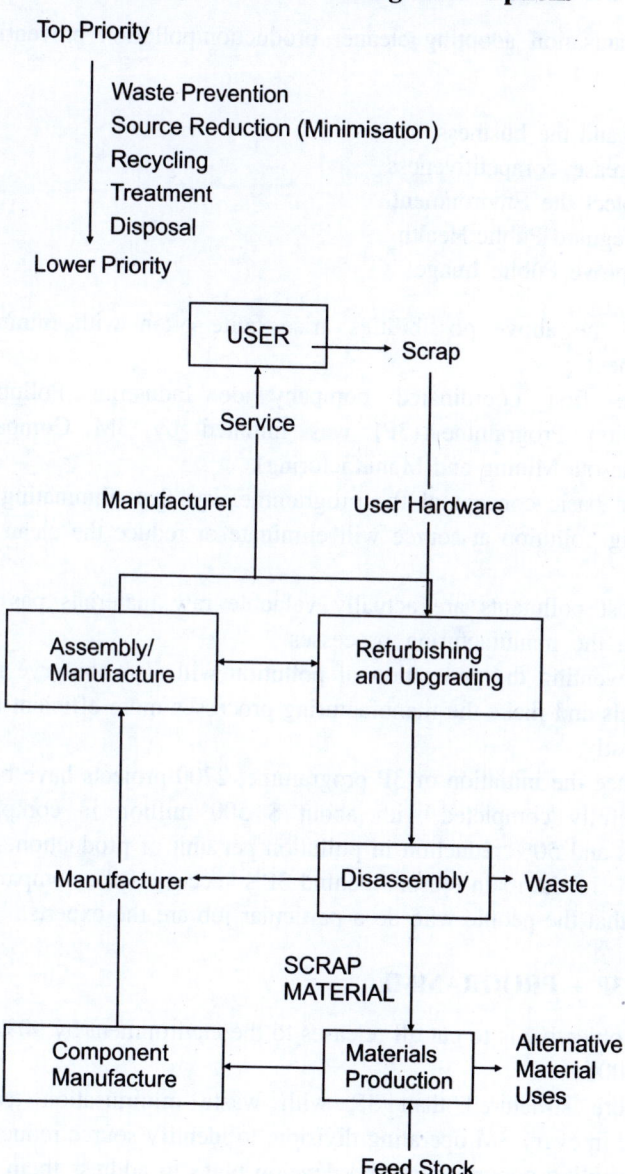

Top Priority

Waste Prevention
Source Reduction (Minimisation)
Recycling
Treatment
Disposal

Lower Priority

Figure 25.3: Materials Reuse : Dematerialisation of Economy.

25.21 POLLUTION PREVENTION/CLEANER PRODUCTION

An organisation adopting cleaner production/pollution prevention can

Expand the business
Increase competitiveness
Protect the Environment
Safeguard Public Health
Improve Public Image

All the above possibilities arise quite often with minimal investment.

The first coordinated companywide Industrial Pollution Prevention Programme (3P) was initiated by 3M Company (Minnnesota Mining and Manufacturing).

The basic concept of the programme was that eliminating or reducing pollution at source will eliminate or reduce the clean up costs.

Most pollutants are actually valuable raw materials passing through the manufacturing processes.

Preventing the generation of pollution will also conserve raw materials and make the manufacturing processes more efficient and less costly.

Since the initiation of 3P programme, 2700 projects have been successfully completed with about $ 500 million in company savings and 50% reduction in pollution per unit of production.

One of the main reasons behind 3P's success is the company's belief that the people who do a particular job are the experts.

25.22 3P + PROGRAMME

Major objective is to cut all releases to the environment by 90% by year 2000.

More structured than 3P, with waste minimisation teams formed in every 3M operating division, to identify source reduction and recycling opportunities and develop plans to address them.

Apart from meeting the return-on investment criteria as any

other project in 3P, 3p+ takes into consideration more intangible benefits, like the enthusiasm and morale of its employees.

25.23 "BEYOND 3P" PROGRAMME

Company believes that if its products compete on quality, performance and price for the sake of environmental improvements, then it will have competitive edge in future.

3M invested $ 170 m in effluent control equipment to-reduce hydrocarbon emissions by 70%.

EXAMPLE

A textile plant in India stopped using highly polluting sodium sulphide in the dyeing process and substituted hydro.

The substitute chemical (hydrol) was a waste stream from the maize starch industry. This process change required no capital expenditure.

The plant saved about 8 Lakhs in capital expenses and about 1.25 lakhs in annual running expenses since no additional treatment facilities were required.

The maize starch industry is also benefited economically and environmentally from this arrangement.

25.24 BARRIERS TO CLEANER PRODUCTION AND POSSIBLE SOLUTIONS

One may ask "if cleaner production is so good, why isn't every one doing it" ?

There are many factors that limit the diffusion and widespread use of cleaner production.

Current environmental legislative and regulatory systems quite often require or stimulate end of pipe pollution control or waste management technology. Hence regulations which have already proved successful in promoting sustainable production and consumption have to be more widely used and effectively enforced.

Cleaner production cannot be made mandatory and pricing is certainly the more powerful instrument to orient the consumption and behaviour of consumers and producers.

Another obstacle is the lack of awareness and expertise on the part of government and industry personnel. What we need is planners, managers and engineers who are educated and trained to integrate the environmental dimension in their strategies and choices, and in the design of processes and products.

Research has been focussed on end of pipe technologies. There is need to develop radically different technologies and the dematerialization of product and services.

Another barrier to cleaner production is the lack of capital for major cleaner production investments, even though they give good return on investments.

There is incredible human resistance to change. The unknown always seems risky and engineers often prefer to stick with well proven technologies rather than trying innovative ones. The solution lies in over coming these barriers.

QUESTIONS

1. What are the steps involved in waste minimization exercise?
2. What are the factors that encourage waste minimization by the generator?
3. Explain, by means of examples, how adoption of a clean technology is more economical than a clean up technology in industrial waste treatment.

Appendix

TABLE 1 : Useful Conversion Factors.

Multiply	by	To get	Multiply col. 3 by: to get Col. 1
1	2	3	4
Cm	0.033	ft	30.48
Cm^2	1.08×10^{-3}	ft^2	923.03
Cm^3	3.53×10^{-5}	ft^3	28317
Cm/sec	1.97	ft/min	0.508
Cm^3/sec	0.0021	ft^3/min	472
Cm^3/min	0.95	gpm (US)	1.051
Cm^3/sec	0.0159	gpm (US)	63.088
Centipoise	0.001	N Sec/m^2	10^3
Centistoke	10^{-6}	m^2/sec	10^6
dyne	10^{-5}	Newton	10^5
dyne/cm	0.067	lb/ft	14.9
$dyne/cm^2$	1.45×10^{-5}	psi	68947
$dyne/cm^2$	10^{-1}	N/m^2	10
$dyne/cm^2$	0.2953×10^{-4}	inch of Hg	3.3864×10^4
$dyne/cm^2$	0.987×10^{-6}	atm	1.013×10^6
gm	0.0022	lb	453.6
gm/cm	0.067	lb/ft	14.88
gm/cm^2	9.676×10^{-5}	atm	10335
gm/cm^3	62.4	lb/ft^3	0.016018
gm/litre	0.0624	lb/ft^3	16.02
gm/litre	8.34×10^{-3}	lb/gal (US)	119.8
hectare	2.471	acre	0.4047
hectare	0.0108×10^7	ft^2	92.67×10^{-7}
Kg	2.2046	lb	0.4536
Kg	9.8425×10^{-4}	ton (long)	1016
Kg	1.1025×10^{-3}	ton (short)	907

(contd.)

TABLE 1—(contd.)

1	2	3	4
Kg/m	0.672	lb/ft	1.4882
Kg/m^2	1.42 × 10^{-3}	psi	703.1
Kg/m^2	0.205	lb/ft^2	4.8824
Kg/m^2	3.284 × 10^{-3}	ft. of water	304.79
Kg/m^2	0.968 × 10^{-4}	atm	1.0335 × 10^4
Kg/m^2	2.8985 × 10^{-3}	inch of Hg	345
Kg/m^2	0.0394	inch of water	25.399
Kg/m^3	0.06245	lb/ft^3	16.012
Kg/m^3	0.8465	lb/gal (US)	1.18
Kg/cm^2	0.9675	atm	1.0335
Kg/cm^2	14.22	psi	0.0703
Kg/cm^2	32.84	ft. of water	0.03045
Kg/cm^2	28.96	inch of Hg	0.0345
Kg/kwh	1.6436	lb/hp-hr	0.6084
Kg/ha/day	0.8922	lb/acre/day	1.1208
Kg-m/sec	9.804 × 10^{-3}	Kw	102
Kg-Cal/min	0.0936	hp	10.688
Kg-Cal/min	0.309 × 10^{-4}	ft-lb/min	3.2389 × 10^4
Kg-Cal/min	0.0698	Kw	14.3334
Km	0.6214	mile	1.609
Km/hr	0.911	fps	1.0973
Km/hr/sec	0.911	ft/sec^2	1.0973
Km^2	0.386	sq. mile	2.590
Kw	0.442	ft-lb/min	2.26 × 10^{-5}
Kw	737.6	ft-lb/sec	1.356 × 10^{-3}
Kw	102	Kg(f)–m/sec	9.804 × 10^{-3}
Kw	1000	Nm/Sec	10^{-3}
Kw	1.3413	hp	0.7457
KWH	2.655 × 10^6	ft/lb	3.766 × 10^{-7}
KWH	1.3413	hp-hr	0.7457
litres	0.035	ft^3	28.32
Kw/m^3	0.038	hp/ft^3	26.316
litres	0.264	gal (US)	3.785
litres	0.22	gal (Imp)	4.5460
litres/sec	15.85	gpm (US)	0.063
litres/mm	3.806 × 10^{-4}	mgd (US)	2628.3
litres/min	5.886 × 10^{-4}	ft^3/sec	1699
litres/min	1.168 × 10^{-3}	acre-ft/day	856.5
litres/day/m	0.0805	gpd/ft (US)	12.422
litres/day/m^2	0.0245	gpd/ft^2	40.82
metre	3.282	ft	0.3048

(contd.)

TABLE 1—(contd.)

1	2	3	4
metre/min	0.0547	ft/sec	18.288
metre of water	2.82	inch of Hg	0.3546
m^2	2.471×10^{-4}	acre	4047
m^2	10.76	ft^2	0.0929
m^2/sec	10^4	Stoke	10^{-4}
m^3	35.3	ft^3	0.02832
m^3	264.2	gal (US)	3.785×10^{-3}
m^3	8.11×10^{-4}	acre-ft	1233.5
m^3/sec	0.212×10^4	ft^3/min	4.719×10^{-4}
m^3/sec	69.93	acre-ft/day	0.0143
m^3/sec	15.85×10^3	gal/min (US)	6.309×10^{-5}
m^3/sec	22.83	mgd	0.044
m^3/m^3	13.37×10^{-4}	ft^3/gal	744.9
m^3/min/m	10.76	ft^3 min/ft	0.0929
m^3/day/m^2	24.51	gal/day/ft^2	0.0408
m^3/kg	16.012	ft^3/lb	0.06245
mm of Hg	1.316×10^{-3}	atm	760
mm of Hg	0.045	ft of water	22.42
mm of Hg	0.039	inch of Hg	25.4
Newton	0.102	Kg (force)	9.807
Newton	4.448	lb	0.2248
Newton	10^5	dyne	10^{-5}
N/m^2	0.021	lb/ft^2	47.88
N/m^2	0.75×10^{-2}	mm of Hg	1.333×10^2
N/m^2	0.987×10^{-5}	atm	1.013×10^5
N/m^2	0.145×10^{-3}	psi	6.895×10^3
N-Sec/m^2	10	poise	10^{-1}
N-Sec/m^2	2.089	lb–sec/ft^2	0.4788
N/m	0.0685	lb/ft	14.59
Watt	1.34×10^{-3}	hp	0.746×10^3

$$°F = 32 + \frac{9}{5}(°C) \; ; \; °C = \frac{5}{9}(°F - 32)$$

1 kg force = (1 kg mass) (accl^n due to gravity)

= (1 kg mass) 9.807 m/sec^2

= 9.807 kg (mass) m/sec^2

= 9.807 Newton.

TABLE 2: Saturation Values of Dissolved Oxygen in Fresh Water, Exposed to an Atmosphere Containing 20.9% Oxygen under a Pressure of 760 mm of Mercury.

Temperature °C	0	1	2	3	4	5	6	7	8	9	10
D.O. mg/1	14.7	14.3	13.9	13.5	13.1	12.8	12.5	12.1	11.8	11.6	11.3

Temperature °C	11	12	13	14	15	16	17	18	19	20	21
D.O. mg/1	11.0	10.8	10.5	10.3	10.0	9.8	9.6	9.4	9.2	9.0	8.8

Temperature °C	22	23	24	25	26	27	28	29	30
D.O. mg/1	8.7	8.5	8.3	8.2	80	7.9	7.7	7.6	7.4

TABLE 3: Physical Properties of Water.

Temperature °C	Density, gm/cm^3	Absolute viscosity, Centipoise	Kinematic viscosity, centistoke.
0	0.99987	1.7921	1.7923
2	0.99997	1.6740	1.6741
4	1.00000	1.5676	1.5676
6	0.99997	1.4726	1.4726
8	0.99988	1.3872	1.3874
10	0.99973	1.3097	1.3101
12	0.99952	1.2390	1.2396
14	0.99927	1.1748	1.1756
16	0.99897	1.1156	1.1168
18	0.99862	1.0603	1.0618
20	0.99823	1.0087	1.0105
22	0.99780	0.9608	0.9629
24	0.99733	0.9161	0.9186
26	0.99681	0.8746	0.8774
28	0.99626	0.8363	0.8394
30	0.99568	0.8004	0.8039

! Centipoise = mN Sec/m^2; 1 Centistoke = mm^2/sec.

References

1. Andrew. D. Eaton: Standard Methods for Examination of water and waste water (2005)
2. Arcadio. P. Sincero: Environmental Engineering: A design approach, Prentice Hall (1996)
3. Bruce. E. Rittman: Environmental biotechnology: principles and applications (2000)
4. Colin Biard, Michael Cann: Environmental Chemistry (2004)
5. Charles. S. Revelle: Civil and Environmental Systems Engineering (2003)
6. Clair. N. Sawyer: Chemistry for Environmental Engineering and Science, McGraw Hill (2003)
7. George Tchobanoglous, Franklin L. Burton, H. David Stensel: Waste water Engineering-treatment and reuse (2002)
8. Gilbert. M. Masters: Introduction to Environmental Engineering and Science, Prentice Hall (1998)
9. Joseph. A. Salvato, Nelson L. Numerow and Franklin J. Agardyr: Environmental Engineering, Wiley publications (2003)
10. Joseph P. Reynolds, Louis Theodore and John S. Jeris: Handbook of Chemical and Environmental Engineering calculations, Wiley Intersciences Pub (2002)
11. Danny D. Reible: Fundamentals of Environmental Engineering, Academic Press (1998)
12. Jay H. Lehr and Janet K. Lehr: Standard Handbook of Environmental Science, health and technology, McGraw Hill (2000)

382

13. Janick Artiola, Ian L. Pepper and Mark L. Brusseau: Environmental monitoring and characterization, Academic Press (2004)
14. David H.F. Lin: Waste water treatment, CRC Press (1999)
15. Paul Bishop: Design of anaerobic processes for treatment of industrial and municipal wastes, CRC Press (1992)
16. Ronald W. Crites, Sherwood C. Reed and Robert Bastian: Land treatment systems for municipal and industrial wastes, McGraw Hill (2000)
17. Frank R. Spellman: Handbook of water and waste water treatment plant operations, CDC Press (2003)
18. Duncan Mara and Nigel J. Horan: Handbook of water and waste water microbiology, Academic Press (2003)
19. Sybil P. Parker and Robert A. Corbitt: McGraw Hill Encyclopedia of Environmental Science and Engineering (1993)
20. Gretchen Smith Biotechnology in Industrial waste treatment and bio-remediation, CRC Press (1995)
21. The Fu Yenr: Environmental Chemistry, Prentice Hall (1998)
22. Neal K. Ostler: Industrial waste sream generation, Prentice Hall (1998)
23. Larry D. Benefield: Biological process design for waste water treatment, Prentice Hall (1980)
24. Paul N. Cheremisinoff: Biomanagement of waste water and wastes, Prentice Hall (1993)
25. Metcalf and Eddy: Waste water treatment and reuse, McGraw Hill (2002)
26. Shun Dar Lin: Water and Waste water Calculations Manual
27. Ronald W. Crites: Small and decentralized waste water management systems.
28. C. Leslie Grady: Biological waste water treatment
29. Tom D. Reynolds, Paul Richards: Unit operations and processes in Environmental engineering (1995)
30. Raina M. Maier: Environmental Microbiology (2000)
31. Mackenzie L. Davis et. al. Principles of Environmental Engineering and Science, McGraw Hill (2003)
32. Neil K. Ostler, John T. Nielsen: Waste Management Concepts, Prentice Hall (1997)

Subject Index

384